FOOD IRRADIATION

ELSEVIER APPLIED FOOD SCIENCE SERIES

Microstructural Principles of Food Processing and Engineering
J.M. AGUILERA and D.W. STANLEY

Biotechnology Applications in Beverage Production
C. CANTARELLI and G. LANZARINI (Editors)

Food Refrigeration Processes
ANDREW C. CLELAND

Progress in Sweeteners
T.H. GRENBY (Editor)

Food Gels
P. HARRIS (Editor)

Thermal Analysis of Foods
V.R. HARWALKER and C.-Y. MA (Editors)

Food Antioxidants
B.J.F. HUDSON (Editor)

Development and Application of Immunoassay for Food Analysis
J.H.R. RITTENBURG (Editor)

FOOD IRRADIATION

Edited by

STUART THORNE

Department of Food Science,
King's College London, London, UK

ELSEVIER APPLIED SCIENCE
LONDON and NEW YORK

ELSEVIER SCIENCE PUBLISHERS LTD
Crown House, Linton Road, Barking, Essex IG11 8JU, England

Sole Distributor in the USA and Canada
ELSEVIER SCIENCE PUBLISHING CO., INC.
655 Avenue of the Americas, New York, NY 10010, USA

WITH 45 TABLES AND 53 ILLUSTRATIONS

© 1991 ELSEVIER SCIENCE PUBLISHERS LTD

British Library Cataloguing in Publication Data

Food irradiation.
1. Food. Irradiation
I. Thorne, Stuart
664.0288
ISBN 1-85166-651-6

Library of Congress Cataloging-in-Publication Data

Food, irradiation/edited by Stuart Thorne
p. cm. — (Elsevier applied food science series)
Includes bibliographical references and index.
ISBN 1-85166-651-6
1. Radiation preservation of food. I. Thorne, Stuart.
II. Series.
TP371.8.F63 1991
664'.0288   dc20   91-13147   CIP

No responsibility is assumed by the publisher for any injury and/or damage to persons or property as a matter of products liability, negligence or otherwise, or from any use or operation of any methods, products, instructions or ideas contained in the material herein.

Special regulations for readers in the USA

This publication has been registered with the Copyright Clearance Center Inc. (CCC), Salem, Massachusetts. Information can be obtained from the CCC about conditions under which photocopies of parts of this publication may be made in the USA. All other copyright questions, including photocopying outside the USA, should be referred to the publisher.

Printed and bound in Great Britain by Hartnolls Ltd.

PREFACE

There is no doubt that irradiation can extend the useful storage life of many foods. This is effected by destruction of micro-organisms and inhibition of biological processes such as sprouting. Although it has been attempted, irradiation sterilisation of foods has not been widely accepted, mainly because the large irradiation doses involved produce unacceptable off-flavours and other changes in the product. Combinations of irradiation with other techniques to endow foods with long storage lives have, however, been shown to be effective. A synergistic effect has been reported for the combined use of irradiation and heat to sterilise foods; the combination being reported to be more effective than either treatment alone. But perhaps the most important application of food irradiation is at lower doses to effect specific changes in foods, reduction in surface flora in strawberries, mushrooms and seafood or inhibition of sprouting in potatoes. It is probably to such operations that we must look for the main long-term future of food irradiation. Extending the storage life of mushrooms, strawberries and similar short-life produce from 1 or 2 days to 1 week or so has huge potential benefits both for industry and for the consumer.

The technology of food irradiation is simple, effective and well-established. The restriction on its large scale application is neither technological nor economic but consumer resistance. Radiation has had a bad press for almost half a century and all processes, whether atomic power generation or food irradiation, are suffused in the common mind with fear and suspicion. The acceptance of food irradiation is hampered by confusion with atomic weapons and the nuclear power station-accidents at Three Mile Island and Chernobyl. Food irradiation, of course, is on a different scale and does not involve large masses of fissionable material. But, nevertheless, it is irrationally tainted by comparison.

v

This confusion manifests itself not as might have been expected with the possibility of accidents at food irradiation plants but with scare-mongering about the effects of irradiation on food safety. The possibility of accidents is, of course, negligibly small. Unwarranted concerns about the safety of irradiated food falls into four categories. First, that the food might become radioactive. Second, that toxic or carcinogenic products might be produced. Third, that some sort of horrible mutant micro-organisms might result from irradiating foods. And, fourth, the rather curious claim that irradiation should not be permitted because it is difficult to detect! Irradiated food does not become radioactive. Extensive toxicological investigations have revealed little evidence of toxic products. Neither is their any evidence for dangerously mutant organisms. The worst that appears to happen is the loss of some vitamins; a characteristic shared with all other preservation processes.

Irradiation could reasonably be compared with thermal processing. This, too, induces changes in food such as vitamin loss. But, because it has been used since time immemorial, it is accepted. Indeed, some of the changes induced by heating — browning, textural changes, flavour changes — are considered essential to acceptability, but we know comparatively little about their toxicology. And where thermally-induced changes such as caramelisation have been investigated there is often far less certainty about their safety than there is with irradiation-induced changes. There has been far less investigation of the effects of heat on food safety — with the exception of thermal sterilisation — than into the safety of food irradiation. We probably know far less about the effects of heating on foods than about the effects of irradiation. But tradition is a powerful if irrational factor in acceptance. Would, for example, nutmeg or cochineal be accepted if they were proposed as new ingredients, today?

It is, of course, quite proper that we should both continue to use traditional processes and ingredients and insist that the new should be adequately tested. But the problem is convincing the consumer that testing of new processes has been adequate and fair and that the prospect of commercial gain has not blinkered assessment of results. The problem lies with science. No hypothesis can be proved true; it can only be refuted. If our initial hypothesis is that an irradiated food is safe, we can only prove it unsafe. That we have singularly failed to do so means that the probability that the process is safe is very high, but we cannot prove it.

In the mind of the man in the street, science provides proof — 'Scientific proof' is the stuff of newspaper headlines. The problem with the acceptability of irradiated foods is one of communication. The consumer does not understand the scientific method and the scientist is so imbued with it that it is never adequately explained to the layman. Television and newspaper 'science' programmes and articles do not help either; they are usually concerned with inventions and confuse science with technology. Neither does primary or secondary education help, in Britain at least. 'Science' is a mainstay of the National Curriculum, but it is concerned with observation of natural phenomena and with established 'facts' in chemistry, physics and biology; it is not really concerned with the scientific method at all. I do not know what the answer to the communication problem is. Perhaps the media — particularly television — could be persuaded to explain what science is really about. And many protesters want to protest and remain blinkered to reality.

What can be done is to ensure that scientists put their case on the media fairly and properly and explain how they work and what are the limitations of science. Professor Smith appears on television and announces that he is 99.9% certain that irradiated mangosteens are harmless. 'Ah', says the representative of the Consumer Protection Commission, who is paid to object, 'What if I am one of the other 0.1%?'. Professor Smith rarely gets the opportunity to explain that he has not implied that 0.1% of consumers will succumb to the evil effects of irradiated food, but that science never allows him to be absolutely confident. If absolute certainty were a pre-condition of introducing any development then no progress would ever be made.

The essential consideration about any new food process or ingredient is confidence that its use will ensure far more benefit than hazard. It is not adequate to consider the possibility of adverse effects of a preservation process or food ingredient without simultaneously considering the potential benefits in terms of food safety and quality. If, for example, nitrites in some meat products were not permitted, we might reduce the small probability of nitrosamine-induced carcinomas, but we would undoubtedly greatly increase the probability of botulism.

The most curious objection to food irradiation is that it is very difficult to detect. Surely this is grounds not for objection but for commendation. If much research over many years has found only limited methods for detecting irradiation in individual foods using the most esoteric analyses, then surely this is *prima facie* evidence for the safety of the process and its acceptability. The objection really is that it is difficult to detect whether irradiation has been used to make unacceptable food pass safety tests. There have, for example, been reports of seafood imported to Britain being rejected on bacteriological grounds and subsequently being accepted on import after irradiation. This, if true, is unethical and reprehensible. But it does illustrate well the abilities of irradiation and, presumably, the re-imported irradiated food was safe. It could equally be said of thermal processing that it could be used to make unacceptable food acceptable. I am sure that I could devise a process to heat the surface of reject seafood to reduce its surface flora to an acceptable level. No one has ever objected to heating on these grounds.

In spite of the problems of convincing consumers of the safety of food irradiation, it is finding increasing application. The great benefits that it offers will undoubtedly ensure its eventual acceptance.

This volume reflects the current state of food irradiation. Three chapters are concerned with applications of irradiation to food processing; one general and two about important specific applications. The larger part of the book discusses the major problems of irradiation; acceptance and detection. Consumer acceptance varies from country to country, dependent upon attitude to new processes in general and 'atomic' processes in particular. Chapters have, therefore, been included to discuss acceptance and status in the United States, in Europe and in developing countries.

STUART THORNE

CONTENTS

ix

LIST OF CONTRIBUTORS

C. BANDITSING
Biological Science Division, Office of Atomic Energy for Peace, Ministry of Science and Technology, Bangkok, Thailand

K.W. BÖGL
Institut für Strahlenhygiene des Bundesgesundheitsantes, Postfach 1108, D-8042 Heuherberg, Germany

R.J. BORD
Department of Sociology, 206 Oswald Tower, The Pennsylvania State University, University Park, PA 16802, USA

D.A.E. EHLERMANN
Institut of Food Engineering, Federal Research Centre for Nutrition, Engesserstrasse 20, D-7500 Karlsruhe, Germany

M.H. FEENSTRA
SWOKA, Institute for Consumer Research, Kon. Emmakade 192-195, 2518 JP The Hague, The Netherlands

J.B. FOX, JR.
Eastern Regional Research Center, United States Department of Agriculture, Agricultural Research Service, 600 E. Mermaid Lane, Philadelphia, PA 19118, USA

S. HACKWOOD
Department of Food and Nutritional Sciences, King's College London, Campden Hill Road, London W8 7AH, UK

C. HASSELMANN

Analytical Chemistry Laboratory, Faculty of Pharmacy, 74 Route du Rhin, 67400, Illkirch-Graffenstaden, France

T. HAYASHI

National Food Research Institute, Ministry of Agriculture, Forestry and Fisheries, Kannondai, Tsukuba, Ibaraki, Japan

L. HEIDE

Institut für Strahlenhygiene des Bundesgesundheitsantes, Postfach 1108, D-8042 Heuherberg, Germany

B. HOZOVA

Department of Chemistry and Technology of Saccharides and Foods, Slovak Technical University, Radliskeho 9, 812 37 Bratislava, Czechoslovakia

P. KIATSURAYANONT

Food and Drug Administration, Ministry of Public Health, Devaves Palace, Bangkok 10200, Thailand

L. LAKRITZ

Eastern Regional Research Center, United States Department of Agriculture, Agricultural Research Service, 600 E. Mermaid Lane, Philadelphia, PA 19118, USA

E. MARCHIONI

Association d'Etudes et de Recherches pour l'Ionisation en Alsace, CRN, 23 rue du Loess, 67037 Strasbourg Cedex, France

G.H. PAULI

Division of Food and Color Additives, Food and Drug Adminstration, 200 Cst. S.W., Washington, DC 20204, USA

P. POTHISIRI

Food and Drug Administration, Ministry of Public Health, Devaves Palace, Bangkok 10200, Thailand

A.H. SCHOLTEN

SWOKA, Institute for Consumer Research, Kon. Emmakade 192—195, 2518 JP The Hague, The Netherlands

L. SORMAN (deceased)
Department of Chemistry and Technology of Saccharides and Foods, Slovak Technical University, Radliskeho 9, 812 37 Bratislava, Czechoslovakia

D.W. THAYER
Eastern Regional Research Center, United States Department of Agriculture, Agricultural Research Service, 600 E. Mermaid Lane, Philadelphia, PA 19118, USA

Chapter 1

AN INTRODUCTION TO THE IRRADIATION PROCESSING OF FOODS

S. HACKWOOD

Department of Food Science, King's College London, Campden Hill Road, London W8 7AH, UK

1 INTRODUCTION

The food industry has used a variety of methods over the years to preserve or extend the shelf life of food. These have included cooking, packaging, smoking, chilling, freezing, dehydrating and using chemical additives. More recently, ionising radiation has been used to extend the storage life of foods.[1]

More research has been focussed on the effects of irradiation on foods than has been directed at any other form of food processing. This research has spanned 40 years and has been carried out in many countries.

The World Health Organisation (WHO) has declared irradiation to be a 'powerful tool against preventable food loss and food borne illness' and the British Government Advisory Committee on Irradiation (ACINF) reported in 1986 that with 'up to an average dose of 10 kGy the technique was effective in food preservation and would not lead to significant changes in the natural radioactivity of foods or prejudice the safety or wholesomeness of foods'. It was unfortunate that the committee reported just days before the explosion at a nuclear power plant in Chernobyl. This event fuelled the consumer's confusion between the undoubtedly dangerous 'radioactivity' and the food preservation technique 'irradiation'.[2]

Food irradiation can be used to: (a) inhibit the sprouting of vegetables; (b) delay the ripening of fruits; (c) kill insect pests in fruit, grains or spices; (d) reduce or eliminate food spoilage organisms; (e) reduce food poisoning bacteria on some meats and sea food products.

1

Irradiation has been hailed as a safer alternative to other methods of preservation such as the use of chemical additives. Other claims are that the process is completely safe and that consumers will benefit from reduced wastage, greater convenience and better quality food. Conversely, there is a body of opinion that claims a variety of adverse effects of irradiation, such as: (a) unique chemical changes; (b) loss of vitamins; (c) off-flavours and aromas; (d) a limited range of applicability; (e) the necessity for use of additives to offset undesirable effects; (f) adverse health effects in animals and humans fed on irradiated foods.[1]

The failure of food irradiation to gain wider acceptance is not difficult to fathom. Negative public attitudes towards virtually everything associated with radiation are found all over the world. In millions of people's minds radiation is associated with war on a scale the earth has never previously seen, with accidents that pose health threats lasting for generations and with nuclear wastes that will still be dangerous 10 000 years from now. Even recognising that radiation is an invaluable aid in diagnosing and treating disease, sterilising medical devices and pharmaceutical products and producing many kinds of manufactured goods, vast numbers of people are genuinely afraid of anything that would appear to increase the risk of exposure to irradiation.[3]

Supporters of irradiation say it will revolutionise the way we eat; food would stay edible for longer, prepared meals could sit on the larder shelf for months before being heated by microwaves for an instant hot dinner. Its detractors say it could result in unidentified toxins remaining in seemingly 'fresh' food and that abuse of irradiation to disguise decay could result in food poisoning and even death.[1] Before food irradiation can be universally accepted, it must be ensured that it brings real benefits to consumers, that there are few risks to the public or to those who work in the food industry, and that an effective system enforcing these regulations is established to prevent abuse.[1]

2 HISTORICAL BACKGROUND

Roentgen discovered X-rays in 1895 and Becquerel discovered radioactivity the following year. In 1896, it was suggested that ionising radiation could be used to kill micro-organisms in food, but it was not until 1921 that a practical use for food irradiation was established and Schwartz obtained a US Patent on the use of X-rays to kill the parasite *Trichinella spiralis* in meat — a cause of worm infection in humans. In 1930, Wust obtained a French Patent for the preservation of foods by irradiation.

In 1948, work on food irradiation started at the Low Temperature Research Station at Cambridge, England. Considerable scientific research was performed during the 1950s and early 1960s elsewhere in the United Kingdom. In the mid-1950s the US Army Quartermasters Corps became interested in food irradiation and sponsored research as part of President Eisenhower's 'Atoms for Peace' policy. Throughout the 1960s and 1970s research into the wholesomeness of irradiated foods and into the technological aspects of food irradiation became even more widespread.[6]

In 1970, 19 countries launched the International Food Irradiation Project (IFIP) financing research into combined processing using irradiation in conjunction with other preservatives. The project was completed in 1982, and replaced in 1984 by formation of the International Consultative Group for Food Irradiation (ICGFI), established by the Food and Agriculture Organisation (FAO), the International Atomic Energy Authority (IAEA) and the World Health Organisation (WHO) and comprising 26 member states of these three United Nations organisations. The ICGFI's brief was to undertake and sponsor research. Joint FAO/IAEA/WHO Expert Committees on the Wholesomeness of Irradiated Foods (JECFI) concluded in 1980 that irradiation of any commodity up to an average dose of 10 kGy neither presented any toxicological hazard, nor introduced any special nutritional or microbiological problems.[7] In 1983, the Codex Alimentarius Commission adopted a revised 'General Standard for Irradiated Foods' and a revised 'Recommended International Code of Practice for the Operation of Radiation Facilities Used for the Treatment of Foods' which incorporated the main conclusions of the 1980 JECFI.[6]

3 GENERAL ASPECTS OF IRRADIATION

3.1 Technique

In deciding to allow irradiation, governments have undoubtedly been influenced by the need to combat the growing incidence of food poisoning. In this context it should be regarded as an additional method of food sterilisation. There are two methods of irradiation used for food sterilisation:

(i) *Gamma Rays.* Electromagnetic radiations produced during the decay of certain radio-isotopes, e.g. Cobalt-60 (a cobalt isotope produced by irradiating cobalt metal in a nuclear reactor) or Caesium-137 (present as a fission product in the fuel elements used in nuclear reactors).

Cobalt-60 is usually used. This is a nuclear process and must comply with the regulations governing Radioactive Substances, Ionising Radiations (Sealed Sources). Gamma rays are very penetrating and can be used for irradiating products in their final shipping cartons.

(ii) *Electron Accelerators* produce high-energy electron beams and accelerate them to very high speeds, producing millions of kGys in fractions of a second. This process is a non-nuclear method but the degree of irradiation penetration is no more than a few centimetres.[4]

Food is irradiated in an irradiation unit — a high energy source situated in a room possessing thick concrete walls. The energy source is either a machine source of electrons or a radioactive element e.g. Cobalt-60.[5] The Cobalt-60 is contained in a series of stainless steel tubes which are mounted in a frame to produce a uniform field of emitted radiation. When not in use, the source is lowered into a concrete enclosed pit or a deep pool of water within the cell.[4]

Food must be fresh and of good quality before irradiation. It can be packaged and frozen or chilled as appropriate. Once food has been transported to the irradiation unit, it is loaded on to a conveyor belt which passes into the irradiation unit, carrying the food. At no time will there be contact between the food and the energy source.[5] The food remains within the unit for a predetermined time; the speed of the conveyor determines the exposure time of the product within the cell, and hence the dose delivered to the product. Control of this, combined with accurate methods of measuring the absorbed dose in the product, forms the basis of quality assurance procedures.[4] It then leaves the unit, can be handled immediately, and distributed to the wholesaler and retailer in the normal way. (The food is not radioactive.)[5] There are 11 gamma plants in the United Kingdom handling this type of work. Isotron PLC operates five of these and is the only independent processor.[4]

3.2 Cost

Both the gamma ray method and electron accelerators are costly processes to install. For example, to run a gamma ray installation the cost of the Cobalt-60, which has a half-life of 5·3 years, would be approximately £400 000. The plant to house the operation would involve a capital cost of approximately £1·25 million and the associated machinery involved would require a further investment of £0·75 million. The total capital cost would therefore be nearly £2·5 million. The capital cost involved in the installation of an electron accelerator would be in the region of £2 million, so there is little to choose between the two processes in capital costs.[4]

3.3 Dose

Different doses of radiation can be applied to foods to achieve different aims (Table 1). Low doses can be used to inhibit sprouting of vegetables, to delay ripening of fruit and vegetables and (as an alternative to chemicals such as ethylene oxide or ethylene dibromide) to achieve insect disinfestation.

TABLE 1
IRRADIATION DOSES RECOMMENDED FOR VARIOUS PURPOSES

Process	Approximate dose range (kGy)
Inhibition of sprouting	0·05–0·15
Delay ripening of various fruits	0·20–0·50
Insect disinfestation	0·20–1·00
Elimination of various parasites	0·03–6·00
Shelf life extension by reduction of microbial load	0·50–5·00
Elimination of non-sporing pathogens	3·00–10·0
Bacterial sterilisation	up to 50·0

Higher doses can be used to reduce the food's microbial load in a way analogous to thermal pasteurisation (for shelf life extension and elimination of some pathogens) and to kill parasites in meat. Doses higher than 10 kGy are usually needed to achieve bacteriological sterilisation. The food industry in the United Kingdom has expressed an interest in using irradiation up to a maximum overall average dose of 10 kGy for the treatment of food intended for general sale.[6]

3.4 Applications

Radiation produces chemical changes in the food and these in turn induce biological changes. The following effects can be produced:

A. *Radurisation* low doses, usually below 1 kGy: (a) Sprouting of vegetables such as potatoes and onions can be inhibited to extend their storage life. (b) Ripening of fruits can be delayed to increase storage life. (c) Insect pests in grains such as wheat and rice, or in spices and some fruits can be killed. This might replace current methods involving gas storage or fumigation treatments that are hazardous to workers, and could reduce losses of foodstuffs.

B. *Radicidation* medium doses, between 1 kGy and 10 kGy: The number of food spoilage micro-organisms such as yeasts, moulds and bacteria could be reduced to extend the life of foods or reduce the risk of food poisoning. This might be important in the case of *Salmonella* in chicken or fish, etc.

C. *Radappertisation* high doses above 10 kGy: These extremely large doses — higher than the proposed 10 kGy limit — can sterilise food completely by killing all bacteria and viruses. This would be used mainly for meat products allowing them to be kept almost indefinitely if adequately packaged.[1]

Food irradiation processing causes virtually no increase in temperature in the treated products and is therefore often termed a 'cold' process. It has been used for some 20 years for the sterilisation of disposable medical supplies, as well as pharmaceuticals that are ingested or injected and pro ducts such as sutures or implants which remain in the body. Of such products, 30% are presently sterilised by irradiation in about 150 plants and used commercially throughout the world. Another application for the benefit of humans is the radiation sterilisation of meals for inclusion in diets of patients lacking the usual immunological defences.

Post-harvest storage losses may in certain regions of the world be as high as 40%. Reduction of such losses by irradiation could contribute significantly to food availability, thereby meeting the world's growing need for food. It is sounder policy to conserve what is produced rather than to produce more to compensate for subsequent losses. Prolongation of storage life can facilitate a wider distribution of foodstuffs by enabling exports to those countries where transportation time previously made such distribution impossible. Wider distribution not only has an impact on export potential, but also provides a more varied and nutritionally superior diet to populations.[8]

The current awareness of food poisoning might convince the public that there is a place for irradiation in the armoury of techniques available to make food safer. It has been suggested that food irradiation might be used to 'cover up' poor hygiene during food production and manufacture. This argument could equally be applied to thermal processing. It is anticipated that all manufacturers wishing to use irradiation will be licensed and the licence fee will fund an inspectorate. The British Medical Association has stated that irradiation is probably the only way to eliminate *Salmonella* from chickens. Also of interest is the extension of the shelf life of chilled meals, particularly where temperature control prior to serving is inadequate. Such products, because of the extensive handling requirements during preparation are particularly prone to accidental contamination.

In almost all cases, foods must be irradiated in their final packaging in order to prevent recontamination. The form of packaging used is of importance as some plastics irradiated in contact with food materials lead to an increased risk of taint transfer. For this reason, it is expected that an approved list of packaging materials suitable for use during the irradiation of foodstuffs will be set up.

One food poisoning issue that will not be cured by irradiation is the presence of *Salmonella* in eggs. At very low doses, the egg white mucoproteins are broken down and the white loses viscosity. Off-flavours also develop in the egg yolk due to the high fat content. For the same reason, other fatty foods such as dairy products, are less suitable for treatment by irradiation. The treatment of exotic fruits to delay ripening is expected to increase the range of produce that could be imported; in addition, the produce that is imported should maintain a better quality in transit due to reduced chances of rotting, insect damage and physical damage (as the fruit would be firmer). Ripening will eventually occur as enzymes are not usually affected by irradiation. Also, factors such as moisture loss are not influenced by irradiation.[2]

The range of products for which irradiation is permitted varies very considerably from country to country, reflecting both variation in the general acceptability of irradiation and individual states' food production and policy.[7] Inhibition of sprouting in potatoes, onions and shallots is permitted in almost all countries that permit irradiation at all, reflecting the considerable quality improvement that can be effected with small doses. Sprout inhibition in potatoes has been permitted by almost 90% of all countries that have a permitted list of applications. The use of irradiation to reduce the microbial flora of spices is also widely permitted; the microbial quality of spices is often suspect and there is no other satisfactory way of improving this. Other permitted applications are many and varied. The Benelux countries, South Africa and Thailand have the longest lists of permitted applications, though the situation is changing rapidly.[7]

4 SCIENTIFIC AND TECHNOLOGICAL ASPECTS OF FOOD IRRADIATION

4.1 Chemical Effects

As with heat processing, radiation treatment induces chemical changes in the individual components of food: lipids, proteins, carbohydrates, water, vitamins, etc. Such changes often involve decomposition reactions, which give rise to new chemical species, including various volatile compounds of smaller molecular weight than the substrates and non-volatile compounds of similar or higher molecular weights than the substrates. It is mostly the volatile components in food which are responsible for its flavour and off-flavour. Ions and excited molecules are the first species formed when ionising radiation is absorbed by matter. A high-energy charged particle loses energy, in a series of small steps, by electric interaction with electrons in the absorbing material. This interaction involves a large number of atoms or molecules, distributed along the particle's track, raising them to excited levels. If these levels are above the ionisation potential of the atom or molecule, ions are formed. Excited molecules could also result indirectly by neutralisation of the ions formed.

The free radicals formed by the dissociation of excited molecules and by ion reactions are largely responsible for the observed chemical changes and generally dominate the mechanisms for the formation, upon irradiation, of stable radiolytic products. They may combine with each other in regions of high radical concentrations or may diffuse into the bulk of the medium and react with other molecules.[9]

4.2 Detection and Monitoring of Irradiation in Food

As the harmonisation of European legislation becomes a reality, the United Kingdom may be forced to come into line with its continental counterparts on the subject of food irradiation. The key issue is the development of practical and economic methods for detecting irradiated food products acceptable to all member states.[10] To enforce legislation and permit consumer choice we need to be able to test food for irradiation. Properly done, irradiation produces little change in appearance or taste. So how is it possible to identify food that has been irradiated? Under the sponsorship of the British Ministry of Agriculture Fisheries and Food (MAFF), active Research and Development into the detection of irradiated foods is underway in Britain. MAFF will be spending £150 000 per year (for 2–3 years) on research into detection methods using a variety of chemical and physical methods. Similar projects are underway throughout the European Community and in the United States.

There is no single method of detection to cover the broad range of foodstuffs, and it is also difficult to determine an irradiated product if it is used as an ingredient such as spices. The Paterson Institute of Cancer Research and Medical Oncology applies the technique of Electron Spin Resonance Spectroscopy (ESR) in the analysis of poultry products — a category of food most likely to undergo irradiation treatment.[10] The irradiation-induced changes are long lived, survive cooking and provide unambiguous evidence of irradiation to doses as low as 200 or 300 Gy. Electron spin resonance is now being used as evidence of whether bone-containing foods have been irradiated or not, and also shows promise of giving a semi-quantitative indication of the dose.[11] However, the spectrometer costs £50 000—100 000 making the technique impractical for routine testing. Commercial laboratories charge £15—20 for the analysis of a single sample. Successful introduction of irradiation depends upon the development of an acceptable, successful and viable system of monitoring and control — so far, one is not available.[10]

In common with all premises which contain radioactive isotopes or in which there is the possibility of exposure to ionising radiation, food irradiation plants are subject to inspection and control by regulatory authorities in order to make certain that the relevant national and international radiological safety standards are being achieved and maintained. These controls ensure the health and safety of workers in irradiation plants, and also protect the general public, for example in preventing contamination of the environment by material from radioactive sources.[6]

The basis of controls operated in countries allowing irradiation are those of the Codex Alimentarius Commission, a joint WHO/FAO body set up to prepare international food standards. In 1983, the Commission adopted both a General Standard for Irradiated Food as well as a Code of Practice for the operation of food irradiation facilities. At the heart of this standard is the requirement that facilities are licensed and registered by a competent national authority. The basis of the controls now proposed for the United Kingdom are:

(a) Applications to apply the process should be subject to detailed examination and approval by a competent national authority;
(b) The food to be irradiated should be of sound microbiological quality;

(c) Information about the nutrient content of the food both before and after treatment should be provided so that any nutritional changes can be monitored;

(d) The premises must be open to inspection to verify that the appropriate irradiation dose is being correctly applied, that hygiene conditions are satisfactory and that good manufacturing practices are being followed;

(e) Comprehensive records should be kept covering all aspects of the food, its packaging, the treatment and full details both of the consignor and consignee;

(f) Comprehensive documentation should accompany the irradiated produce for identifying in particular the plant and the treatment;

(g) There should be full and clear labelling;

(h) Imports should be dealt with by ensuring that controls applied in countries wishing to export to the United Kingdom are on a par with those proposed for this country.[12]

5 MICROBIOLOGICAL ASPECTS OF FOOD IRRADIATION

The major benefit of irradiation is its effect on micro-organisms — inactivated predominantly because the charged particles generated by it, although they may be very short-lived, cause breaks in deoxyribonucleic acid (DNA). Every living bacterial, yeast or mould cell contains at least one very large molecule of DNA which contains the genetic information necessary to create progeny, and this must remain intact if the micro-organism is to remain capable of growth and multiplication. Whether or not a particular DNA molecule is broken, and therefore whether or not the micro-organism containing it is inactivated, is a chance event depending on whether or not that DNA molecule was unlucky enough to have experienced an ionising 'hit'.

Minor by-products of the growth of micro-organisms in foods damage flavour, texture and colour. Major by-products include acids and gases, while slimes on meat, hazes in drinks and moulds on bread and preserves all cause spoilage and loss. Some micro-organisms, e.g. the *Salmonella* group are infective, while others such as *Staphylococcus aureus* produce toxins with subsequent illness to those who consume the contaminated food.

With the proposed treatment of up to 10 kGy, a high rate of microbial destruction can be obtained producing products effectively free from such pathogens. Irradiation has little effect on bacterial toxins, so cannot correct food already spoiled or previously made dangerous by the proliferation of a toxin-producing micro-organism.[13]

Since irradiation can and does cause mutations, it is pertinent to consider whether the use of the process could produce microbial mutants of increased virulence, or antibiotic resistance or increased radiation tolerance or greater pathogenicity. Radiation resistant mutants can be produced, but only under very special laboratory conditions which could hardly arise in commercial situations.

Irradiation may also be used in combination with other techniques. For example, to inactivate enzymes, heat plus irradiation may be required, or, where delicacy of flavour is at stake, irradiation of vacuum packs or in a nitrogen atmosphere may be indicated — so irradiation is not applied in a wholesale way, but only to a fraction of the total food supply. Where a conventional alternative is as effective it will continue to be used in most cases, as irradiation is unlikely to be a cheap process.[14]

6 TOXICOLOGICAL ASPECTS OF FOOD IRRADIATION

It is essential to determine whether the consumption of irradiated food products could lead to either short or long-term toxic effects, due to the chemical changes that may have occurred. The Committee on Toxicity of Chemicals in Food, Consumer Products and the Environment (COT) decided that the toxicological safety of food irradiation should be evaluated in three ways:

(a) First, by evaluating the toxicity of individual radiolytic products which had been detected in foods.
(b) Second, by considering the effects in animals of experiments where the whole animal diet had been irradiated.
(c) Third, by looking at experiments in which individual irradiated food items had been fed to animals.

Mutagenic hazards arising from the consumption of irradiated food have also been extensively investigated. This was considered to be the most appropriate way of determining the overall toxicological consequences of food irradiation. The extensive data on irradiated food was adequate to allow a proper evaluation of its toxicological safety and there was no evidence to suggest that any toxicological hazard to human health would arise from the consumption of food irradiated up to an average dose of 10 kGy. At higher doses there is no evidence for toxicological hazard, but the data are more limited and not adequate to allow a general clearance for food irradiated at these doses. COT has pointed out that new toxicological data on food irradiation will continue to become available and that the use of irradiation may be extended to a wider range of foods in the future.[6]

7 NUTRITIONAL ASPECTS OF FOOD IRRADIATION

Irradiation is not unique in having the capacity to produce nutritional changes in food. Cooking and other heat treatments cause some degree of destruction of nutrients.[6]

7.1 Macronutrients

Proteins serve primarily as a source of amino acids in the diet and these are little changed at doses of up to 10 kGy. Some purified carbohydrates show small changes but these are of no significance in actual foodstuffs. Polyunsaturated fats may undergo a degree of oxidation sufficient to distinctly reduce their nutritional advantages, but saturated fats are mainly resistant to damage.[14]

7.2 Micronutrients

These are minor components of foods which have little or no energy value but which are essential for health. They include vitamins, minerals and certain polyunsaturated fatty acids. Minerals are unaffected, but there are sometimes losses of vitamins and polyunsaturated fatty acids.[6] Experimental irradiation of vitamins in isolation is not always useful as results are different when the vitamin is present as a food constituent, where they are to some extent protected at the matrix of the food and the presence of other constituents. Also, the same vitamin may show differing susceptibilities in different foodstuffs.

Data is available on the effects of irradiation on many fruits and vegetables: potatoes are perhaps the most important as contributors of vitamin C in European and American diets. They are irradiated solely to inhibit sprouting and only small doses of around 0.1 kGy are required. Vitamin C levels decrease during normal storage, and immediately following irradiation some workers have reported falls in the level of vitamin C (about 28% at a 0·1-kGy dose) while others failed to detect any change. Other evidence suggests that at the end of 4–9 months of winter storage there was little difference in vitamin C content between the treated and untreated batches. There is also evidence that suggests some of the ascorbic is converted into dehydroascorbic acid on irradiation and while this substance has equal vitamin activity to ascorbic acid itself, it may have escaped detection.

Data is also available on onions, carrots, bananas, mangoes, papaya, dates, tomatoes, strawberries and mushrooms, after irradiation either to extend storage life or to eliminate insect infestation or to inhibit sprouting. In most cases the doses required were 1 kGy or less. The vitamin C losses were studied most and turned out to be less than with more conventional treatments. Investigations with strawberries demonstrated that vitamin C losses in stored irradiated strawberries were less than in the stored unirradiated controls.

The extension of storage life and the reduction or elimination of pathogens such as *Salmonella* in meat and poultry has been investigated. Higher doses up to about 7 kGy were required. Thiamin loss after treatment was reported, but it was found that this depended upon the method of analysis used. The heat processing of pork, beef and chicken caused similar or greater losses, but in the case of meat and poultry an irradiation loss is likely to be followed by a further cooking loss.

With fish, 2–3 kGy is sufficient to reduce spoilage and pathogenic contamination. The vitamins are virtually unchanged except there may be some loss of thiamin. However, if the irradiation is carried out in the absence of air at low temperatures, such losses are eliminated. With fatty fish such as herring and mackerel, irradiation may induce peroxide formation and some losses of fat soluble vitamins, depending on conditions.

Cereals and grains may be irradiated at levels of 1 kGy or less to eradicate insect infestation. At this level, there is no significant effect on the nutritional value or functional properties of the macronutrients and no effect on the B-group vitamins in the bread baked from irradiated wheat.[14]

In some countries, including the United Kingdom, irradiation at doses of 25 kGy or more has effectively been used to sterilise the diets of patients suffering from diseases or undergoing treatments making them particularly susceptible to infection. Irradiated food has also been used in the preparation of specialised diets for, e.g. astronauts from the USA and USSR. The consumption of irradiated food did not cause any detectable adverse nutritional effects in this closely monitored group.[6]

8 CONSUMER ATTITUDES

The London Food Commission has carried out surveys suggesting that the following groups want and do not want food irradiation:

Who does want it?

(a) Some sections of the food industry who want it to extend shelf life of some foods.
(b) Some sections of the food industry research.
(c) The irradiation industry that already has spare capacity and is building even more plants in anticipation of an early decision to legalise it in Britain.
(d) The nuclear industry that stands to gain from an extension of the 'atoms for peace' concept into food processing, and perhaps more directly, from growth in isotope production or uses for currently problematic nuclear wastes.
(e) Some consumers' organisations whose main concern is labelling.
(f) The British Government.

Who doesn't want it?

(a) Some food importers who see it as undermining food hygiene standards.
(b) The Food and Health Lobby which is concerned about losses in food's nutritional value and other possible injury to health.
(c) Some retailers and manufacturers who are concerned both about being able to give assurances about safety and about public opinion. Some British retailers oppose irradiation in principle or argue that it has no place in their policy of selling quality fresh food to consumers on a fast turnover. They claim to have no need to extend shelf life.
(d) Other organisations who, while not opposed to irradiation in principle, are concerned over supposed scientific uncertainties, the lack

of systems for monitoring and controlling the technology, and the way the public debate and consultation process is being handled.[1]

9 CURRENT STATUS AND LEGISLATION

Irradiation is currently permitted in over 30 countries. The USSR was the first country to legislate — potatoes in 1958. In Western Europe, Holland is the major user, an extensive PR campaign having led to general acceptance of the process by the public in the late 1960s.

Within the European Community, France, Denmark, Belgium, Luxembourg, Spain and Italy permit irradiation while Ireland, Greece and Portugal have no rules either permitting or forbidding use.[7] In December 1989 in the United Kingdom, the Food Safety Bill was being put through the House of Lords, and proposals were put forward to adopt food irradiation in the United Kingdom.

Outside the European Community, the United States and Japan authorise the process as do Argentina, Bangladesh, Brazil, Canada, Chile, China, India, Israel, Mexico, Norway, New Zealand, the Philippines, South Africa, Thailand, Uruguay and various of the Eastern bloc countries. But clearances given vary tremendously from permission for irradiation for experiments, for test marketing, for export only, for provisional periods, to unconditional authorisation.[7]

The disharmony in existing legislation or regulations related to irradiated foods is a major constraint in the commercialisation of food irradiation. Lack of harmonisation constitutes a barrier to international trade. Food exporting countries find it difficult to permit food irradiation if their main trading partners do not accept irradiated foods. Harmonisation of national legislations and regulatory procedures will enhance confidence among trading nations that foods irradiated in one country and offered for sale in another country have been subjected to commonly acceptable standards of irradiation and manufacturing practices and irradiation treatment control.[8]

It was recommended that food be irradiated at levels of up to 10 kGy and at energies not exceeding 10 MeV for accelerated electrons or 5 MeV for gamma radiations, and this will not lead to any significant change in the natural radioactivity or prejudice the safety and wholesomeness of the food.[14]

10 LABELLING

Irradiated foods cannot be recognised by sight, smell, taste or feel. The only sure way for consumers to know if a food has been processed by irradiation is for the product to carry a label that clearly announces the treatment in words, a symbol, or both. The symbol shown in Fig. 1 is gaining increasing acceptance as a means of informing the public that a food product has been treated with ionising radiation.[3] Irradiated food need not be labelled on health grounds. However, there is strong consensus that the consumer is entitled to know if the food has been processed in order that consumers can exercise their freedom to make informed choices of food products.[15]

FIG. 1. The internationally recognised symbol for irradiated foods.

Opponents of food irradiation in the United States, have proposed a bill to the House of Representatives Subcommittee on Health. This would force food processors to label all irradiated foods as well as foods containing irradiated ingredients. In addition, it would force restaurant operators to mark all menu items containing irradiated foods and include a complete explanation on their menu.[16]

To be of genuine value to consumers, labelling of irradiated food must be supported by public information and education campaigns designed initially to help consumers decide whether they want to be able to buy radiation-processed foods and subsequently to help them make wise decisions in the selection and use of irradiated food products. National steering committees composed of representatives of all interests could prove most useful in coordinating educational activities, by ensuring that the information materials developed and distributed to the public are accurate, comprehensive and consistent.[3]

11 CONCLUSION

The relatively new process of preserving food by irradiation complements, rather than competes with, the presently available traditional methods. It offers an alternative for foods for which heat treatment is not appropriate or where there may be concern about residues from chemical additives. However there is room for improvement. In the area of radiation chemistry we need to know more about the interactions between the different food components on their radiolytic products, the nature and quantity of the non-volatile products, the unusual sensitivity of dairy products, or the effects of possible combinations of irradiation with other food-processing treatments.[9]

In the long term, it will be the willingness of the food retailers to stock irradiated food that will determine the success of the technique in this country. Previous trials elsewhere in the world have shown that consumers soon recognise that irradiated produce is of a very high quality and demand for the products in all the trials outstripped supply.[2]

Food which is not eaten is of no nutritional value. If irradiation prevents wastage from insect infestation, from mould growth and from bacterial action, then this is nutritional gain. If irradiation extends choice by opening up possibilities for new trade in exotics this is further nutritional gain.[14]

Finally, even the best regulations on labelling will be of limited value unless they can be enforced. Until such time as a test for detecting previously irradiated food is developed, and the monitoring agencies are all trained to use it, there is a strong argument for being cautious in the use of food irradiation.

REFERENCES

1. Webb, T. and Lang, T. In: *Food Irradiation — The Facts,* Thorsons Publishing Group, London, 1987.
2. Eves, A. In: *Irradiation: Your Questions Answered, Food Manufacture,* August 1989, pp. 44–45.
3. Anon. In: *Food Irradiation,* World Health Organisation, New York, 1988.
4. Guise, B. Processing Practicalities, *Food Processing,* October 1989, pp. 53–54.
5. Wilkinson, V. In: *Food Irradiation in Practice,* Institute of Food Science and Technology Joint Technical Consultative Panel, London, 1986, pp. 169–170.
6. Anon. In: *Report on the Safety and Wholesomeness of Irradiated Foods,* Advisory Committee on Irradiated and Novel Foods, HMSO London, 1986.
7. eale, S., Marking time on irradiation, *Food Manufacture,* January 1988, pp. 36–39.
8. Van Kooij, J.G., National trends in and uses of food irradiation, *Food Reviews International,* 1986, pp. 1–17.
9. Nawar, W.W., Volatiles from food irradiation, *Food Reviews International,* 1986, pp. 45–77.
10. Uley, E., Irradiation: What's the Holdup? *Food Manufacture,* June 1988, pp. 39–40.
11. Swallow, A.J., Can we tell if our food has been irradiated? *Chemistry in Britain,* February 1988, pp. 102–103.
12. Anon. In: *Report on Food Irradiation,* Department of Health and Social Security, HMSO, London 1988.
13. Gould, G.W. In: *Food Irradiation: Microbiological Aspects,* Institute of Food Science and Technology, London, 1986, pp. 175–180.
14. Hawthorn, J., Safety and wholesomeness of irradiated foods, *BNF Nutr. Bull.,* 1989, **14** (3), 150–161.
15. Coles, L. In: *Food Irradiation: Labelling and Enforcement Aspects,* Institute of Food Science and Technology, London, 1986, pp. 185–187.
16. Unger, H., Industry scruples stall irradiation uptake, *Food Processing,* November 1986, pp. 41–42.

Chapter 2

DETECTION OF IRRADIATION BY THERMO- AND CHEMILUMINESCENCE

KLAUS WERNER BÖGL & LYDIA HEIDE
Institut für Strahlenhygiene des Bundesgesundheitsantes, Postfach 1108, D-8042 Heuherberg, Germany

1 INTRODUCTION

In a number of countries it is not currently allowed to bring irradiated food into the market. On the other hand, increasing numbers of permits have been issued in many countries during the past years for the use of ionizing radiation in certain areas of food technology, because national and international bodies recommended food irradiation to be accepted[1]. It is quite clear that this situation will not change completely in the near future, which means that the regulations concerning irradiated food will remain very different between individual countries during the coming years. This is the background for the increasing interest in developing methods that allow identification of irradiated food. For control purposes, the Codex Alimentarius Standard and Code of Practice for Irradiated Food rely on a tight system of licences and registrations of radiation processing facilities and on identification of radiation treatment in shipping documents. To enforce compliance with such regulations through the authorities, an identification method is needed by which the product itself may be examined. By the use of such methods, national food inspection authorities would be able to control imports. Labelling is another reason for the necessity of identification methods. During the past few years it has become increasingly clear that detailed labelling of irradiated food (at least for first generation products, but in some countries also for second generation food) will need to be included in food regulations in many parts of the world. Identification methods will be helpful in proving compliance with these labelling regulations. A further problem in international trade with irradiated foodstuffs will be the differing labelling regulations in different countries. Clearances of irradiated foodstuffs will also be accompanied by specific limitations. One example is the prohibition of re-irradiation of irradiated food. An additional aspect is that identification methods may prevent irradiation from becoming a substitute measure for low level hygienic procedures. In summary, there are different major reasons for the development of methods identifying irradiated food: (i) to ascertain that legally imposed food laws are not violated; (ii) to control international trade of irradiated food; (iii) to prevent irradiation from becoming a substitute treatment for low level hygienic procedures; (iv) to supervise correct labelling; (v) to avoid multiple irradiation; (vi) to control the absorbed dose and homogeneity of dose distribution.

In order to demonstrate whether a food has been treated with ionizing radiation or not, it is necessary that a radiation-specific change takes place in the food and that this change is measurable. However, those changes which in many cases take place after irradiation very closely resemble changes that also occur in consequence of other treatments. Measurement methods suitable for use in the course of inspection are, therefore, often not based on actual radiation-specific changes. Many investigations have been carried out in an effort to design reliable methods for detecting whether or not a food has been irradiated. Attempts have been made to apply physical, chemical and biological forms of measurement.

The results show that no general method exists that is applicable for all foods. For individual foodstuffs, several methods are possible, depending on the type of food, but none has as yet been internationally tested and accepted as reliable for all types of irradiated food. In fact, the development of identification methods is still scientifically at a basic level (with some exceptions), and only after further intensive development and standardization, may the methods provide a basis for laboratory inspection. So far it has likewise not been possible to measure the actual radiation dose with which foods were exposed; it is currently only possible to identify radiation treatment or to give a very rough dose estimate. Furthermore, it may be relatively complicated to detect irradiated food in a second generation, e.g. irradiated spices in meat and milk products or irradiated poultry in chicken salad.

So far, identification has been discussed in detail in not more than three meetings. Two meetings were organized by the Commission of the European Communities: the first Colloquium was held in Luxembourg in 1970[2], and the second in Karlsruhe in 1973[3]. A third meeting was organized by the WHO (Regional Office for Europe, Copenhagen,

Denmark) in cooperation with the Bundesgesundheitsamt (German Federal Health Office, BGA) and the Gesellschaft für Strahlen- und Umweltforschung mbH (GSF) in Neuherberg/Munich in November 1986[4]. Although no test has so far been accepted, considerable progress is expected to be made in the near future. Various ingenious methods have been reported in the literature and a number of new concepts and ideas are suggested.

Two techniques are currently very close to practical application. On one side, the luminescence techniques have been tested in a European collaborative study with good results, so that these luminescence techniques have been included into the official collection of food control methods in the Federal Republic of Germany in 1989. On the other side, electron spin resonance spectroscopy has been demonstrated as well suited for detection purposes[4] for different types of food containing bones, shells or seed, like chicken and strawberries. A European collaborative study is also in preparation. Other methods, such as the detection of o-tyrosine and the measurement of volatile compounds from the lipid fraction may likewise be developed for practical control purposes during the coming years.[4]

2 FUNDAMENTAL LUMINESCENCE MEASUREMENTS OF IRRADIATED FOOD

2.1 The Physical and Chemical Background of Thermo- and Chemiluminescence

2.1.1 Thermoluminescence (TL)

When a dielectric is exposed to ionizing radiation, a portion of the electrons or electron-holes subsequently released in solid matter may be trapped by certain imperfections. Ultraviolet light and, for some types of material, chemical reaction or mechanical stress may release a similar effect. At low temperature or for imperfections energetically close to the ground state, the binding of trapped electrons may remain unchanged for many years. The release of trapped electrons may occur spontaneously by thermic stimulation or intensive light irradiation. The return of electrons to their ground state is by exothermic reaction. The energy being released is largely turning into heat; a small portion of the energy may, however, be emitted as light, depending on the material between $< 10^{-8}$ and $> 1\%$ of the absorbed ionizing energy.[5,6] When measuring the emitted light as a function of temperature, the results are curves (TL-curves or glow-curves) that are more or less characteristic of the respective material. These curves often contain several maxima and their shape depends on several external criteria, i.e. type of radiation, radiation dose and heating rate. An increase in activation energy causes a TL maximum at higher temperatures. In any case, an increase in the heating rate is followed by an increase in TL maximum intensity and shifting of maxima towards higher temperatures. At constant experimental conditions (this also includes good substance homogeneity, as is, for instance, not always the case with spices), the TL curve of a specific substance is well reproducible. The light emission maximum as well as the total intensity of light (integral value) is in many cases proportional to the radiation dose.

TL-effects are measurable in many different types of substances. The examination of 3000 different substances by Becker and Scharmann[6] showed TL-effects in 75%. A multitude of organic substances have likewise been examined as to their TL-characteristics[7]. The majority of investigations were done on crystals containing metal-ions (e.g. LiF/Mg, Na, Ti; CaF_2/Dy; $CaSO_4$/Mn). Materials of this type are known in radiation protection as solid state dosemeters. In accordance with the claims Becker and Scharmann[6] compiled on the assessment of thermoluminescence dosimetry (TLD) materials, irradiated spices or other dry foodstuffs will be particularly approachable for TL measurements if the following conditions prevail:

- high concentration of imperfections and high probability of light emission during the recombination process, i.e. a high light yield per dose unit (sensitivity);
- good stability of the radiation induced defects in the material not only at room temperature but also at storage temperature of up to about 40°C, i.e. the principal maximum of TL-emission should not be lower than 150°C;
- the spectrum of the emitted TL-light must be in good agreement with the maximum sensitivity of the combined photomultiplier and filter device in the measuring equipment and must permit discrimination between the infrared heat radiation of the sample material and the heating equipment. Emission maxima between ~300 and 500 nm are usually the most suitable;
- the distribution of imperfections should be as uncomplicated as possible, since additional TL-maxima at low temperatures are

rapidly fading (return of electrons to stable condition without heating), while high temperature maxima are more difficult to heat out;

• the sample material should be relatively insensitive to light, humidity and other environmental influences and should release as few interfering side signals due to chemiluminescence, phosphorescence, etc., as possible.

2.1.2 Chemiluminescence (CL)

When treating solid samples of various substances with ionizing radiation, light is emitted by these samples as soon as they are being dissolved in water. The light is emitted in short pulses whereby the pulse length is determined mainly by the time interval during which the irradiated sample is dissolved in the solution. The integral light yield depends on the administered radiation dose and, in many cases, increases with the dose until a saturation value is reached.[8,9]

The light yield of the irradiated substance can be essentially increased by using a photosensitizer during the process of dissolution. One of the most frequently used photosensitizers is luminol, a cyclic hydrazide of 3-aminophthalic acid capable of dissociating two protons from the hydrazide structure in alkaline solution. The molecule disintegrates during oxidation and creates a CL that is particularly pronounced at pH-values > 10. The emission spectrum of the emitted light in aqueous solution reaches its maximum at a wavelength of 424 nm.[8,9]

Where the mechanism is concerned that causes the luminescence of the single molecule, it can be said that a triplet excited level of 3-aminophthalate shows the species to be responsible for luminescence and that 3-aminophthalate and molecular nitrogen are produced in the final oxidation phase.[8-13] The reaction process leading to light emission is demonstrated in Fig. 1. Luminol reacts in solution containing oxidation media, such as H_2O_2 or peroxide radicals, by forming molecular nitrogen. 3-Aminophthalate produced at the excited level is reversed to its ground level during light emission. When solid organic substances are irradiated or subsequently come in contact with water, other oxidizing agents such as hydrogen peroxide and organic peroxide radicals may develop as follows: carbon radicals will initially develop from irradiation. They can be converted by oxygen into peroxide radicals or may react in the luminol solution with water by forming OH-radicals. H_2O_2 will then be produced via the OH-radicals. If the irradiated substances are insoluble solids, e.g. pulverized spices, then upon contact with water the

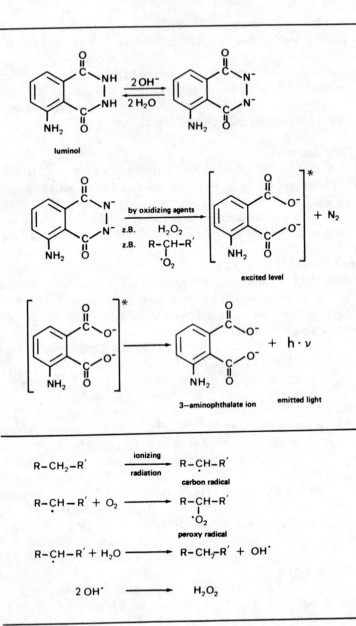

FIG. 1. The physical and chemical background of CL.

radicals will at first react only on the surface. Therefore, at a specific radiation dose, the amount of emitted light is very much larger when dealing with soluble substances in luminol solution, since all relevant reaction products are thereby able to react with luminol in a short period of time.

2.2 Light Emission Curves of Different Irradiated Foods

Figure 2 shows TL and CL glow curves for unirradiated and irradiated celery powder.

For CL measurements, a freshly prepared solution was used, which in general contained 0.7 mM luminol (Sigma), 3·8 μM hemin (Sigma, Typ I) as catalyst and 11·8 mM sodium carbonate (Merck) at a pH of 10, 10·5 or 11 according to the type of food. The sample dosage in most experiments was 5–30 mg per analysis. All tests were done at room temperature. To measure the emitted light, the LKB 1251 luminometer was used. The reaction of the samples with the luminol solution was initiated by automatic dispensation of 0·2 ml reagent solution to the examined sample. At the start of the reaction, the light was integrated for 5 s and recorded. Integral values (mV · s) and maximal values (mV) of the light yield were registered.

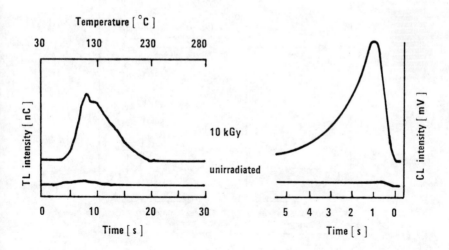

FIG. 2. TL and CL glow curves of unirradiated and irradiated celery powder.

For TL measurements, the intensity of the emitted light was measured
as a function of temperature by linear sample heating to 280°C within 30
s (for spices). The amount of the sample (5–25 mg) was determined by
a series of optimization experiments for each food. The TL analyzer
2000A/2080 (Harshaw) was used for light emission measurements. The
measuring chamber was flushed with nitrogen as inert gas by a constant
flow of 4 l/min. To avoid evaporation of components during heating, the
sample was covered with a glass plate. After the heating cycle had been
initiated, the light intensity (nC) was recorded and integrated.

The CL glow curve of the irradiated celery powder shows very clearly
the fast reaction of the irradiated product with the luminol solution
resulting in a short light pulse. The intensity of the emitted light is among
other factors a function of solubility of the irradiated product in the
luminol solution, because only radiation induced products from the

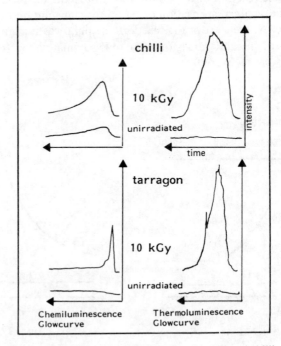

FIG. 3. CL and TL glow curves of unirradiated and irradiated chilli and tarragon
powder.

surface of the material can contribute to the light emission, if the irradiated substance is not soluble in the luminol solution. Another factor which can reduce the CL intensity is a quenching of the emitted light due to coloured substances which can be extracted from the irradiated food through the luminol solution parallel to the reaction between the irradiated food and the solution. The TL glow curves can be controlled with the heating rate; heating of the sample to 280°C can be done in times shorter and longer than 30 s. Using constant experimental conditions the shape of TL glow curves are characteristic for the examined substances, but working with irradiated samples of the same type of food the shape of the TL glow curves can vary from measurement to measurement because of differences in the heat contact between sample and heating equipment.

The CL and TL glow curves are relatively similar for a lot of different spices as can be seen in Fig. 3; the experimental conditions were identical compared with the measurements in Fig. 2. Figures 4 and 5 demonstrate that strong TL effects can also be measured in irradiated fresh products. It can be seen that even after a storage period of some hours (room temperature) the effects are still available; the difference of the TL glow curves between the two mushroom measurements may be based on a variable heating contact. The irradiation of spices, fresh mushrooms and fresh apple peels were done at room temperature.

In Figs. 6 and 7 CL glow curves of deep frozen (−20°C) chicken, shrimps and fish are summarized. The time scales of both figures are identical. The reaction of deep frozen samples with the luminol solution was initiated again by automatic dispensation of the reagent solution to the sample at room temperature. This may be an explanation for the relatively long light emission process of the deep frozen products, due to a slow melting of the samples in the luminol solution. The reproducibility of the CL glow curves is relatively good for shrimps and fish, as can be seen in Fig. 7.

2.3 The Effect of Radiation Dose

Figures 8–10 summarize some of the results concerning measurements of luminescence intensities as function of radiation dose. The figures demonstrate very clearly dose response curves of tarragon, basil, milk powder and brown onion-skin. The effect of radiation dose was very similar for a lot of other types of foodstuffs.

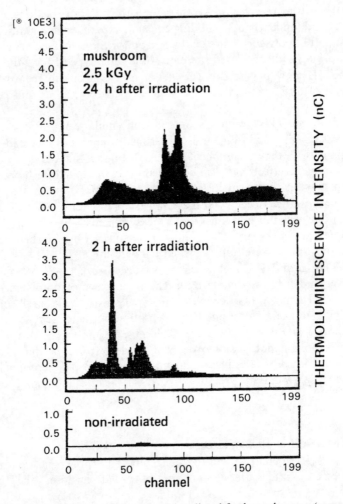

FIG. 4. TL intensity of unirradiated and irradiated fresh mushrooms (measuring time 80 s = channel 0–199, maximum temperature 250°C, heating rate 100°C/35 s).

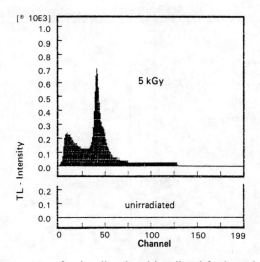

FIG. 5. TL glow curves of unirradiated and irradiated fresh apple peel (measuring time 320 s = channel 0–130, maximum temperature 150°C, heating rate 10°C/s).

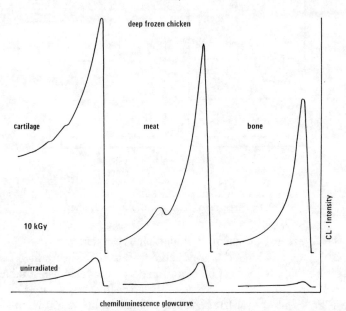

FIG. 6. CL glow curves of unirradiated and irradiated deep frozen chicken (cartilage, meat, bone).

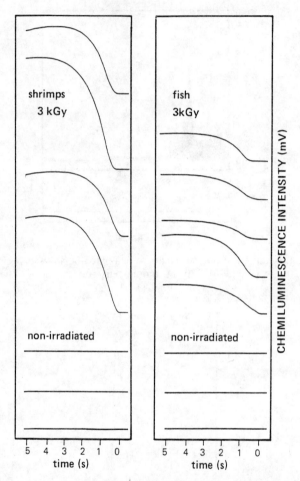

FIG. 7. CL intensity of unirradiated and irradiated deep frozen shrimps and fish (fillet); three resp. four parallel measurements.

2.4 The Effect of Storage after Radiation Treatment

By developing any identification method, it must be ascertained that the effect is not only subject to the radiation dose but is also measurable even after relevant storage times. Some of the respective results are summarized in Figs. 4 and 11–13. Irradiated dried mushrooms, e.g., are still detectable with the two luminescence techniques 1 year after the radiation treatment (Fig. 11). This effect is clear, because the radiation induced

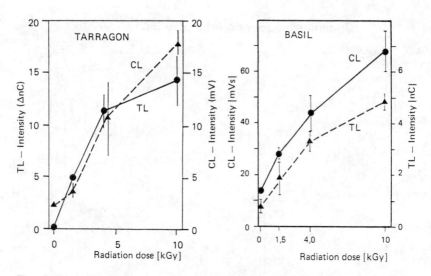

FIG. 8. Dose response curves of tarragon and basil powder determined by CL and TL measurements.

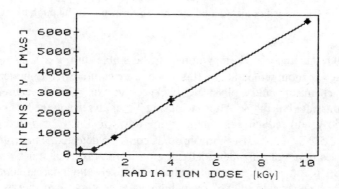

FIG. 9. CL intensity of irradiated milk powder as a function of radiation dose.

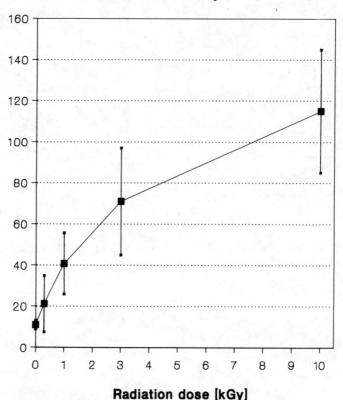

FIG. 10. CL intensity of irradiated onion-skin (brown) as a function of radiation dose.

products in the dry champignons can be stable during a long time of storage at room temperature. It is much more complicated to understand the chemical and physical background of the measurements demonstrated in Fig. 4. After increasing TL analyzer voltage from 690 V to 850 V and reducing the heating rate down to 100°C/35 s irradiation with 2·5 kGy of even fresh mushrooms could be detected. The irradiated samples were not only identified 2 h after irradiation, but still 1 week after the treatment. These results indicate relatively stable imperfections in the fresh material in spite of a very high water content. A further example to identify irradiated samples after a long storage time is demonstrated

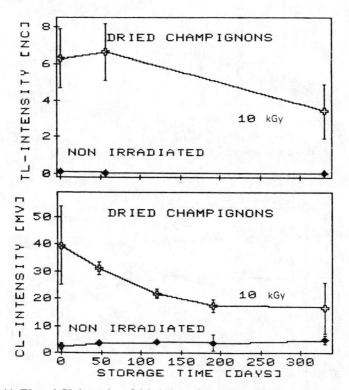

FIG. 11. TL and CL intensity of dried champignons as a function of storage time after irradiation.

in Fig. 12. After a storage period of 2 h the TL identification of irradiated ginger is still very simple, whereas CL is not suitable for this type of spice. The last figure of this group (Fig. 13) shows that also different parts of irradiated deep frozen chicken show a relatively strong CL after more than 2 weeks storage time at −20°C. Other authors published as well increased CL- and TL-intensities of irradiated fresh mushrooms, fresh potatoes, fresh onions, and deep frozen chicken and fish.[14,15] The storage stability of radiation induced effects in deep frozen products can

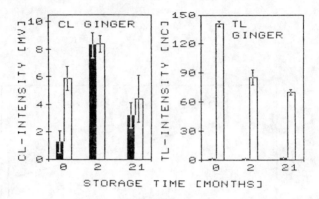

FIG. 12. CL and TL intensity of unirradiated and irradiated (10 kGy) ginger powder as a function of storage time after irradiation.

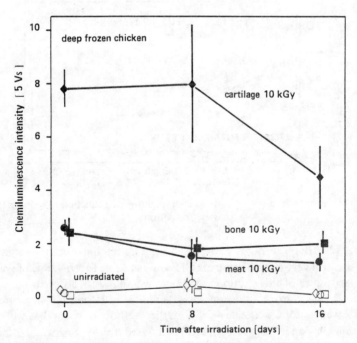

FIG. 13. CL intensity of irradiated deep frozen chicken (cartilage, bone, meat) as a function of storage time after irradiation.

be explained with the decreased velocity of chemical and physical processes in the frozen state.

2.5 The Effect of Ethylene Oxide

When developing any identification method for irradiated foods it is also important to ensure and prove that changes which occur in usual food processing cannot induce similar effects. Parameters that may enhance the degree of light emission are: heat treatment, air exposure, UV light, ethylene oxide, grinding or particle size. On the other hand, it is possible that further processing after irradiation (moisture, heat treatment) reduces the specific effect and renders an identification impossible. Deliberate manipulation of detection parameters should therefore be impossible with these methods of detection. Some experimental results concerning these influences are given in the following.

Spices might be fumigated with ethylene oxide (ETOX). ETOX has been used for some decades as a technique to reduce the microbial contamination of spices. Today, this treatment is forbidden at least in most of the West European countries. Therefore, the extent to which ETOX affects luminescence was examined. A quantity (15) of spices were exposed to ETOX for 6 h. The gas concentration was 750 g (90% ETOX, 10% CO_2) per m^3 and the temperature was 22°C. Measurements were carried out 2 weeks after treatment. No effect was observed in marjoram and juniper berries (Table 1). Onions, paprika and pepper showed very little increased CL and TL intensities, but the values were still within the range of unirradiated samples. Therefore, no unknown albeit ETOX-fumigated sample of the spices mentioned above would be falsely declared as treated by ionizing radiation.

2.6 The Effect of Molecular Oxygen

In a few products containing components sensitive to oxidation, it was observed that CL increases without heating while a thin layer of the sample is exposed to air; this increase in CL intensity may be caused, for example, by air oxidation of the lipid fraction. Figure 14 shows the CL intensity as a function of air exposure of milk powder and sesame. It must be stated that generally for products sensitive to oxidation, the CL technique is limited for detection purposes.

2.7 The Effect of Heat Treatment

It is known that heat treatment can repair radiation damage in many solid substances. It was tested whether heat treatment may decrease the

TABLE 1

CL AND TL INTENSITIES OF UNTREATED AND ETOX-TREATED SPICES (THREE DIFFERENT BATCHES I, II, III)

Spice	CL intensity (mV)		TL intensity (nC)	
	Untreated	ETOX treated	Untreated	ETOX treated
Onion				
I	3·5 ± 0·4	16·9 ± 3·6	0·31 ± 0·04	0·59 ± 0·14
II	9·2 ± 0·3	22·4 ± 4·1	0·45 ± 0·01	0·76 ± 0·07
III	7·5 ± 0·2	17·6 ± 1·5	1·00 ± 0·20	0·59 ± 0·04
Paprika				
I	3·3 ± 0·3	10·4 ± 0·5	1·76 ± 0·05	2·03 ± 0·11
II	2·6 ± 0·4	8·8 ± 0·3	0·90 ± 0·08	1·51 ± 0·39
III	2·5 ± 0·5	10·7 ± 1·0	0·62 ± 0·04	0·85 ± 0·10
Pepper				
I	4·5 ± 0·7	7·1 ± 0·3	0·29 ± 0·01	0·71 ± 0·17
II	3·8 ± 0·2	5·9 ± 0·1	0·41 ± 0·11	0·82 ± 0·26
III	4·3 ± 0·8	5·7 ± 0·3	0·59 ± 0·02	0·98 ± 0·11
Marjoram				
I	1·5 ± 0·2	1·5 ± 0·0	1·58 ± 0·10	1·43 ± 0·20
II	1·5 ± 0·1	1·7 ± 0·1	1·42 ± 0·40	1·87 ± 0·41
III	1·5 ± 0·0	1·5 ± 0·1	1·86 ± 0·20	1·29 ± 0·16
Juniper berries				
I	1·2 ± 0·1	1·3 ± 0·1	—	—
II	1·5 ± 0·4	1·5 ± 0·4	—	—
III	1·0 ± 0·1	1·3 ± 0·4	—	—

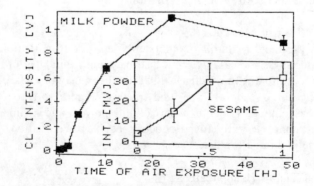

FIG. 14. CL intensity of milk powder and sesame as a function of air exposure.

luminescence intensity using 11 species for TL and six spices for CL. No changes in TL intensity were usually found in unirradiated samples which had been heated, but the intensity of irradiated samples was decreased. The amount of decrease depends on the temperature and heating time (Tables 2 and 3). In most of the investigated samples, TL enabled the identification of irradiated spices even after a 1-h-treatment at 100°C. The situation is different with the CL technique; CL is more strongly influenced by exogenic effects (e.g. oxidizing effects). Especially the CL intensity of unirradiated milk powder shows a maximum increase after a 16-h treatment at 50°C (Fig. 15). This is one feature by which the signal may be associated with the formation of oxidizing products of the lipid fraction. Spices are not so sensitive; 60°C heating has no effect on unirradiated tarragon, paprika, cinnamon and pepper, but treatment at 100°C for 16 h increased the CL intensity of tarragon and paprika. Heating of irradiated samples caused in most cases a decrease of CL intensity. The results of these experiments are summarized in Table 4. As it is proved that the quality of the spices is injured by heat (if it is sufficiently high to quench the radiation effect) due to loss of colour, flavour and consistency, such treatment, which could eliminate radiation effects, is unsuitable for practical application.

2.8 The Effect of Water Vapour Treatment

It is known that free radicals remain stable in dry solids, but are removed on contact with water. Therefore, the effect of a post-irradiation water-vapour treatment (17 h at 25°C) was examined in paprika, tarragon, cinnamon and milk powder by measuring CL and in paprika and tarragon by the TL technique. The wet irradiated samples showed no further CL-increase, whereas TL was only slightly reduced. For example, results from investigations with paprika powder are given in Fig. 16. Because the practical application of such a treatment is rather complicated and expensive (the products must be dried afterwards) and identification by TL is still possible, it is not necessary to take this effect into consideration.

2.9 The Effect of UV-irradiation

Spices may also be exposed to sun light or UV radiation.[16,17] Therefore the extent to which such treatment would affect luminescence was examined. UV-C radiation (sterisol lamp, 20 W, 50 Hz, 10 cm distance) was applied to a thin layer of unpacked spices for 1 h and measurements were carried out 24 h later. The exposure results showed either no change or

TABLE 2

TL INTENSITY OF UNIRRADIATED AND IRRADIATED SPICES AFTER HEAT TREATMENT AT 100°C AT DIFFERENT PERIODS OF TIME

Spice	Radiation dose	TL intensity (nC)			
		0 min	10 min	30 min	60 min
Aniseed	0 kGy	1·5 ± 0·3	1·2 ± 0·1	1·1 ± 0·1	—
	10 kGy	9·1 ± 0·7	4·8 ± 0·4	4·2 ± 0·1	—
Basil	0 kGy	0·3 ± 0·2	0·8 ± 0·0	0·7 ± 0·0	—
	10 kGy	6·9 ± 1·1	6·2 ± 2·1	7·4 ± 1·3	—
Chive	0 kGy	0·2 ± 0·0	0·6 ± 0·0	0·7 ± 0·0	—
	10 kGy	2·9 ± 2·5	1·2 ± 0·2	1·3 ± 0·3	—
Curcuma	0 kGy	0·4 ± 0·1	0·7 ± 0·3	0·4 ± 0·0	0·6 ± 0·1
	10 kGy	196·7 ± 32·4	122·0 ± 26·5	52·1 ± 0·7	33·0 ± 1·6
Garlic	0 kGy	0·6 ± 0·1	0·9 ± 0·2	1·0 ± 0·1	—
	10 kGy	8·3 ± 0·9	3·3 ± 1·1	2·5 ± 0·5	—
Onion	0 kGy	0·2 ± 0·0	0·5 ± 0·1	0·6 ± 0·0	0·6 ± 0·0
	10 kGy	22·3 ± 1·1	6·8 ± 2·8	4·1 ± 0·8	3·5 ± 0·6
Paprika	0 kGy	0·4 ± 0·2	0·5 ± 0·0	0·9 ± 0·0	0·5 ± 0·2
	10 kGy	70·5 ± 9·2	32·9 ± 3·8	30·4 ± 7·0	13·1 ± 3·3
Parsley	0 kGy	0·3 ± 0·1	0·3 ± 0·0	0·2 ± 0·0	0·2 ± 0·0
	10 kGy	12·8 ± 0·8	4·1 ± 1·0	4·6 ± 0·1	2·6 ± 0·3
Pepper	0 kGy	0·5 ± 0·2	0·4 ± 0·1	0·4 ± 0·0	0·5 ± 0·1
	10 kGy	29·8 ± 10·9	31·7 ± 7·6	28·4 ± 11·3	18·4 ± 2·8
Sage	0 kGy	0·6 ± 0·1	0·6 ± 0·1	1·3 ± 0·1	1·3 ± 0·3
	10 kGy	21·2 ± 2·7	6·0 ± 0·8	8·5 ± 0·7	6·2 ± 1·1

TABLE 3

TL INTENSITY OF PAPRIKA AS A FUNCTION OF HEAT
TREATMENT FOR 10 MIN AT DIFFERENT TEMPERATURES

Temp. (°C)	TL intensity (nC)	
	Unirradiated	Irradiated
25	0·6 ± 0·1	60·2 ± 7·7
100	0·9 ± 0·1	42·7 ± 7·0
125	0·7 ± 0·1	21·0 ± 4·8
150	1·1 ± 0·3	8·2 ± 1·3
175	1·0 ± 0·0	8·1 ± 1·4
200	0·3 ± 0·0	0·6 ± 0·0

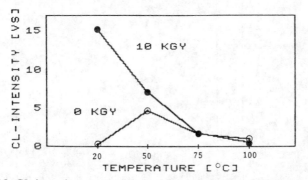

FIG. 15. CL intensity as a function of heat treatment of milk powder.

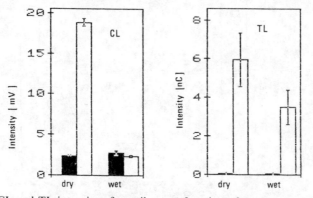

FIG. 16. CL and TL intensity of paprika as a function of a water vapour treatment (18 h at 25°C, ■ unirradiated, □ 10 kGy).

TABLE 4

CL INTENSITY AFTER HEAT TREATMENT AT 60 AND 100°C AT DIFFERENT PERIODS OF TIME

	No treatment	60°C/16 h	100°C/10 min	100°C/60 min	100°C/16 h
Paprika I					
0 kGy	3·3 ± 0·1	—	2·6 ± 0·2	—	—
10 kGy	9·0 ± 1·0	—	6·4 ± 0·4	—	—
Paprika II					
0 kGy	2·5 ± 0·2	2·5 ± 0·3	—	—	14·5 ± 0·3
10 kGy	18·0 ± 0·5	14·3 ± 1·0	—	—	15·7 ± 1·0
Pepper					
0 kGy	4·5 ± 0·7	—	—	5·5 ± 0·7	—
10 kGy	10·1 ± 1·8	—	—	8·4 ± 0·3	—
Tarragon					
0 kGy	10·1 ± 0·8	9·3 ± 0·5	—	—	17·4 ± 1·9
10 kGy	17·5 ± 1·3	13·3 ± 1·3	—	—	16·7 ± 0·7
Cinnamon					
0 kGy	2·9 ± 0·3	4·7 ± 0·1	—	—	3·9 ± 0·2
10 kGy	2000 ± 165	1380 ± 222	—	—	800 ± 210

TABLE 5
CL AND TL INTENSITIES OF UNTREATED AND UV-C IRRADIATED SPICES

Spice	CL intensity (mV)		TL intensity (nC)	
	Untreated	UV irradiated	Untreated	UV irradiated
Mushrooms	1·34 ± 0·09	3·06 ± 0·27	0·25 ± 0·04	0·21 ± 0·03
Carrots	1·08 ± 0·03	2·22 ± 0·08	1·53 ± 0·03	7·07 ± 0·32
Horse-radish	3·26 ± 0·07	3·10 ± 0·18	0·70 ± 0·03	2·35 ± 0·05
Asparagus	4·06 ± 0·35	3·97 ± 0·57	0·94 ± 0·02	1·42 ± 0·02
Paprika	2·31 ± 0·16	14·66 ± 1·57	0·41 ± 0·04	0·95 ± 0·02
Cardamom	0·78 ± 0·01	1·01 ± 0·06	3·51 ± 1·26	7·69 ± 1·01

only a slight increase compared to the untreated reference sample (Table 5). Further investigations with UV-irradiated (UV-AB and UV-C) cinnamon and milk powder showed similar results.[16] If, however a whole package is UV-irradiated and samples for food control are taken from the inner part, no UV effect will occur, due to the low range of UV radiation.

2.10 The Effect of Grinding

The dependence of particle size and the effect of grinding on the TL and CL intensity is of interest. Grinding the irradiated material (leaves, seeds or whole pieces) below 250 μm yielded an increase in TL efficiency, whereas the intensity of untreated samples was not at all or only little affected. Due to this positive effect, it is possible to identify irradiated seeds like fennel and caraway. In addition, the TL light yield of herb leaves was increased and the peak maximum of the glow curves became sharper. The results are summarized in Table 6. Some experiments with basil and tarragon showed lower CL intensities of irradiated ground samples compared with unground material;[18] an explanation may be a higher absorption of the emitted light in the luminol solution when using ground spices.

TABLE 6
TL INTENSITY OF UNIRRADIATED AND IRRADIATED
SPICES CORRELATED WITH THE PARTICLE SIZE

| Spice | TL intensity (nC) | |
	Unirradiated	10 kGy
Caraway		
Seed	0·9 ± 0·1	5·6 ± 2·2
< 250 μm	2·5 ± 0·1	29·8 ± 8·8
Chive		
> 1400 μm	0·3 ± 0·1	3·0 ± 0·2
< 250 μm	0·3 ± 0·0	7·4 ± 1·4
Cinnamon		
> 1400 μm	0·0 ± 0·0	4·0 ± 2·2
< 250 μm	0·2 ± 0·0	8·5 ± 2·0
Fennel		
Seed	0·9 ± 0·1	13·5 ± 1·4
< 250 μm	4·0 ± 0·7	218·4 ± 25·3
Leek		
> 1400 μm	1·3 ± 0·4	7·8 ± 1·7
< 250 μm	0·3 ± 0·0	4·0 ± 1·1
Mushroom		
> 1400 μm	0·2 ± 0·1	4·4 ± 1·0
< 250 μm	0·4 ± 0·0	25·7 ± 1·8
Parsley		
> 1400 μm	0·4 ± 0·1	9·9 ± 1·9
< 250 μm	0·3 ± 0·0	29·9 ± 2·8
Savory		
> 1400 μm	0·6 ± 0·1	40·0 ± 14·9
< 250 μm	0·5 ± 0·1	110·3 ± 11·6
Thyme I		
> 1400 μm	1·2 ± 0·2	38·6 ± 10·2
< 250 μm	1·8 ± 0·2	95·1 ± 12·2
Thyme II		
> 1400 μm	1·5 ± 0·2	108·4 ± 25·8
< 250 μm	1·2 ± 0·1	236.0 ± 31

3 APPLICATIONS FOR DETECTION OF IRRADIATION

3.1 Estimation of Threshold Values for Differentiation between Irradiated and Non-irradiated Samples

Since the methods for detecting irradiated samples had been proved reliable,[16,18-31] it was necessary to ascertain how broad the scale of luminescence intensity is if different varieties of one spice are tested.[32,33]

This is very important because thresholds have to be fixed above which the samples must be declared as irradiated. Since luminescence fades in irradiated spices during storage, the maximum light emission of the un-treated sample is the determining value. As expected and demonstrated by three examples in Fig. 17, the CL and TL intensities of unirradiated and irradiated samples from batches of each of the different spices are widely scattered. No value was found to be characteristic of a particular spice. In nearly all cases the luminescence intensity of the irradiated sample is increased, compared to the corresponding unirradiated sample. For thyme, it is not possible to identify irradiated samples by CL measurements, while TL measurements are successful. In some exceptional cases, irradiated samples of laurel and mace may be identified by both CL and TL measurements.

Furthermore, the luminescence fading effect is to be considered. This means that prior to practical application of such methods, investigations have to be carried out which consider both aspects, the fading and the intensity range of products from different producers. The following two examples show results from investigating different batches of hot paprika and mushrooms. The intensity ranges of the unirradiated and irradiated samples are given in Figs. 18 and 19. The first example, hot paprika (Fig. 18), shows that in most cases it is not possible to identify irradiated samples by CL measurements. But there is a good chance for a successful identification by TL over long storage periods. Irradiated mushrooms (different species: champignon, chanterelle, edible boletus and morel) show increased CL intensity (Fig. 19). More than 50% of the irradiated samples will still be identifiable 1 year after treatment. The same applies to the TL method.

Now, an example is given as to how thresholds or reference values for the differentiation between unirradiated and irradiated spices can be defined. The highest of the detected intensity values for unirradiated samples of one type of spice is taken, and this value is multiplied by a factor 2–4, depending on the tolerance range the food inspection is willing to set. The resulting luminescence intensity determines the value above which a sample is to be identified as irradiated.

The practical procedure would be as follows: assuming the highest TL intensity measured for unirradiated hot paprika (Fig. 18) was 2 nC; 2 nC multiplied by, e.g. factor 2 equals 4 nC. This means, the TL intensity of 4 nC is the threshold for hot paprika. If any unknown examined hot paprika sample shows a TL intensity above this value, this proves that it was irradiated. By fixing the threshold in this manner, an unirradiated

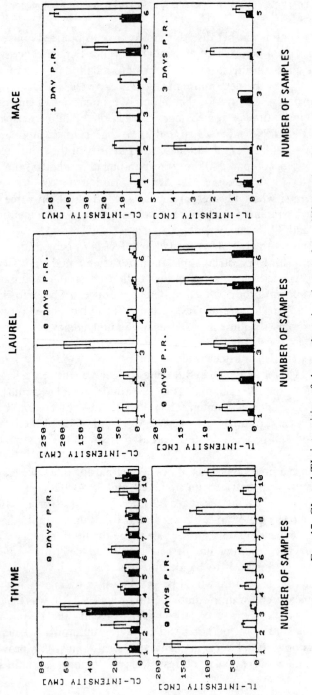

FIG. 17. CL and TL intensities of thyme, laurel and mace from different batches.

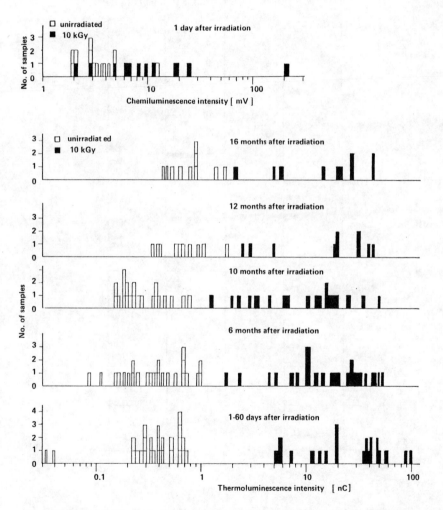

FIG. 18. CL and TL intensities of different batches of hot paprika at various periods after irradiation.

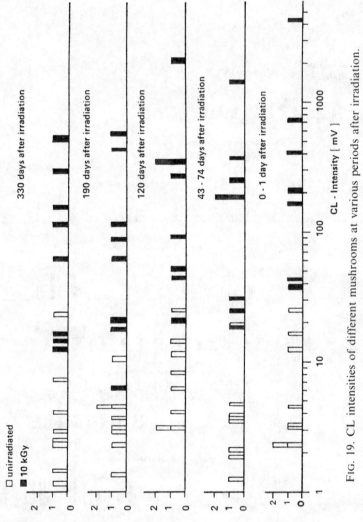

FIG. 19. CL intensities of different mushrooms at various periods after irradiation.

sample will not be identified as irradiated, but it may occur that some irradiated samples are not detected. For hot paprika, all irradiated samples could be identified, if the investigations are carried out soon after irradiation: 6 months later, still more than 90% are detectable. In this way, threshold values and identification rates can be set for every spice. In many cases it is possible to differentiate all irradiated samples easily, provided the analyses were carried out within 1 or 2 months after irradiation. Later, the fading effect occurs, i.e. the intensities of the irradiated samples decrease and the two intensity ranges are partially overlapping with the consequence that some of the irradiated samples cannot be identified. All investigated spices and corresponding CL and TL intensity ranges for unirradiated and irradiated samples during the storage period after irradiation are listed in Tables 7 and 8. The number of tested spice batches and investigations of these batches are included. A summary of the proximity of usability of the luminescence methods for identifying irradiated spices is shown in Table 9. As a consequence of the positive results from these investigations, the luminescence techniques will be included into the official collection of food control methods in the Federal Republic of Germany. For this reason, the Federal Health Office will keep a reference sample set of unirradiated spices and other dried foodstuffs on hand for use at the food control laboratories. This might be helpful in setting individual thresholds for the differentiation between irradiated und unirradiated samples in reference to individual measuring equipment.

3.2 Reference Sample Measurement Trials

Since TL and CL measurements had been proven as methods for rapid identification of irradiated dried food during long-term experiments[32,33] and in a first interlaboratory study,[34,35] which was not a real blind trial, a second and more extensive blind study was organized. Twelve institutions, including the German Federal Health Office (BGA) took part: seven institutions responsible for food control, five from the Federal Republic of Germany, one from Switzerland and one from Denmark; four research groups, two from the Federal Republic of Germany, one from Hungary and one from Scotland; finally, the BGA. Seven laboratories wanted to carry out TL and CL measurements, three CL and two TL measurements, only. For each method 10 different spices were chosen to be analysed; five samples of each spice type were used. These batches of samples had been prepared and packed at BGA by the end of 1987. The only information provided at the outset was sample type and a statement

TABLE 7

CL INTENSITIES OF UNIRRADIATED AND IRRADIATED SPICES AND THE CORRESPONDING NUMBER OF INVESTIGATED BATCHES (DIFFERENT VARIETIES OF THE SAME TYPE OF SPICE), SAMPLES AND PERIODS OF EXAMINATION

Spice[a]	CL intensity range unirradiated sample (mV)	CL intensity range irradiated sample (mV)	No. of types	No. of investigations	Examined period after irradiation (days)
Allspice, berry	1·3–4·7	3·9–19·5	1	11	201
Allspice	2·3–4·7	4·3–9·3	1	4	47
Aniseed, seed	0·8–6·8	21·4–60·7	1	18	617
Aniseed	1·5–3·6	7·9–22·7	1	2	60
Asparagus	2·8–9·7	3·4–70·6	6	14	288
Basil	1·1–12·0	3·2–24·7	3	18	225
Caraway	1·1–4·6	3·4–10·8	1	9	281
Cardamom	0·7–7·8	2·8–203·4	8	29	360
Carrots	0·8–7·8	52·1–3 382	3	11	361
Celery bulb, ground	1·5–43·5	5·0–1 657	7	14	110
Celery leaves	2·3–5·2	2·9–9·3	3	6	110
Celery seed	2·8–6·9	4·3–30·0	2	4	110
Chilli	4·9–19·3	11·1–35·5	1	6	227
Chive	1·7–2·7	4·3–6·5	1	3	77
Cinnamon	9·7–84·9	15·5–1 432	10	20	578
Cloves, bud	0·6–1·3	2·7–43·8	1	12	238
Cloves	0·8	0·8–1·0	1	3	20
Coriander, seed	1·3–3·9	8·9–83·1	2	14	219
Coriander	0·9–10·2	6·0–20·9	1	8	223
Cumin	1·2–6·5	8·8–48·1	1	14	413
Curry	0·7–14·7	1·8–2 154	5	25	331
Dill	1·5–9·5	2·5–10·6	10	20	375
Fennel, seed	0·7–10·1	4·7–65·1	3	18	432
Fennel	1·1–4·4	4·1–10·1	1	9	251
Garlic	9·1–97·4[b]	21·3–1 918[b]	11	62	330
Ginger	1·3–8·3	5·1–12·8	4	12	254
Horse-radish	1·5–90·7	12·3–615·2	7	17	289
Juniper berry	1·0–127·2	3·0–1 845	8	50	1 144
Juniper berry, ground	1·2–6·8	26·1–94·4	1	8	251
Laurel	0·7–17·0	1·1–195·0	6	24	919
Lavender	3·7–15·8	8·8–11·6	1	3	360

Leek	1·7–63·5	12·6–102·3	6	12	370
Lemon peel	4·2–72·3	3·0–3 066	7	21	373
Lovage	1·0–3·6	1·9–1 181	7	14	60
Mace	1·4–7·9	5·0–25·5	4	8	60
Marjoram	1·2–57·5	1·3–46·1	10	30	402
Mushrooms					
Champignon	2·3–7·4	16·6–39·5	4	8	330
Chanterelle	2·5–24·7	509·0–4 267	1	5	330
Edible boletus	1·2–25·4	6·3–737·8	5	21	330
Morel	1·2–19·2	14·4–440·0	3	11	330
Mustard seed	1·4–14·0	7·7–32·7	7	7	1
Nutmeg	1·2–6·2	1·9–7·6	10	30	362
Onion	2·3–14·7	6·2–46·5	3	15	218
Orange peel	2·1–34·6	3·2–41·8	7	21	300
Oregano	1·4–16·3	1·2–14·1	10	30	400
Paprika, hot	1·9–12·0	2·4–200·0	12	34	487
Paprika, sweet	1·5–17·0	2·0–32·0	12	41	503
Parsley	3·3–12·0	7·9–26·6	5	5	9
Pepper, black	3·7–21·4	7·2–23·0	10	20	550
Pepper, green	2·0–20·8	5·5–36·7	7	28	881
Pepper, white	1·4–11·9	2·3–17·7	5	20	212
Rosemary	1·2–12·8	1·3–16·6	10	10	1
Sage	1·4–8·6	1·7–8·1	10	20	360
Savory	1·7–19·7	2·8–21·3	10	20	379
Sesame seed	1·1–13·7	7·7–123·1	1	14	390
Shallot	3·9–19·5	5·5–105·5	3	8	190
Tarragon	1·8–5·6	4·6–29·6	3	21	196
Thyme	6·7–42·8	8·4–63·9	10	20	387
Turmeric (curcuma)	0·8–2·3	4·0–17·0	1	10	228

[a]If not otherwise decribed, the examined spices were either powdered or used as very small pieces.
[b]Measuring units = mVs.

TABLE 8

TL INTENSITIES OF UNIRRADIATED AND IRRADIATED SPICES AND THE CORRESPONDING NUMBER OF INVESTIGATED BATCHES (DIFFERENT VARIETIES OF THE SAME TYPE OF SPICE), SAMPLES AND PERIODS OF EXAMINATION

Spice	TL intensity range unirradiated sample (nC)	TL intensity range irradiated sample (nC)	No. of types	No. of investigations	Examined period after irradiation (days)
Allspice	0·01–0·5	0·03–0·9	1	8	316
Aniseed	0·36–1·5	2·8–25·2	2	4	328
Asparagus	0·01–0·9	0·5–110·0	4	12	283
Basil	0·14–1·6	0·7–6·9	1	15	715
Caraway	0·02–1·7	0·2–9·5	2	14	611
Cardamom	0·59–9·2	3·1–926·0	7	28	360
Carrots	0·49–3·8	5·0–18·5	2	8	365
Celery bulb	0·31–1·7	0·6–16·6	5	10	83
Celery leaves	0·14–0·2	0·7–7·9	3	6	83
Celery seed	0·53–1·5	20·6–130·9	4	6	83
Chilli	0·00–1·0	2·6–17·1	1	15	697
Chive	0·06–0·3	0·3–7·4	2	10	120
Cinnamon	0·02–0·8	0·3–48·0	10	50	1 212
Cloves	0·00–0·8	0·03–0·8	1	6	97
Coriander	0·26–1·2	1·6–9·0	1	8	318
Cumin	0·11–1·1	1·2–16·4	1	10	552
Curry	0·15–1·8	5·0–129·2	5	25	364
Dill	0·27–1·4	0·7–46·3	10	40	372
Fennel	0·12–1·8	1·2–6·4	1	10	528
Garlic	0·11–0·9	0·2–18·9	10	60	337
Ginger	0·69–2·3	17·6–168·1	6	10	78
Horse-radish	0·15–1·0	1·0–40·5	5	20	800
Juniper berry	0·65–1·9	0·9–2·8	1	4	78
Laurel	1·17–8·0	4·2–15·2	6	18	296
Lavender	10·8–12·0	12·6–14·8	1	3	363
Leek	0·15–5·3	2·1–347·0	6	24	523
Lemon peel	0·57–1·5	0·6–14·1	5	14	372

Lovage	0·13–1·0	1·9–105·5	7	14	56
Mace	0·12–1·6	0·7–5·1	4	8	56
Marjoram	1·00–6·0	6·7–114·2	10	60	792
Mushrooms					
Champignon	0·01–0·3	3·4–25·7	5	11	539
Chanterelle	0·53–0·8	6·1–26·0	2	5	530
Edible boletus	0·01–0·2	0·5–45·2	4	14	530
Morel	0·01–0·3	4·7–263·8	2	6	530
Nutmeg	0·01–0·2	0·1–1·0	10	60	887
Onion	0·04–1·2	0·2–22·3	4	17	202
Orange peel	0·10–1·2	0·23–12·4	7	21	299
Oregano	0·66–2·8	2·3–127·9	10	60	895
Paprika, hot	0·09–1·8	1·8–102·3	10	60	986
Paprika, sweet	0·03–1·8	1·1–33·5	10	53	1 030
Parsley	0·10–0·4	0·4–29·9	3	17	612
Pepper, black	0·18–0·8	0·7–41·2	10	28	219
Pepper, white	0·18–2·2	0·4–86·0	5	16	301
Rosemary	2·89–14·4	7·0–29·2	10	40	834
Sage	0·49–5·2	4·3–93·0	12	62	962
Savory	0·39–3·9	6·3–191·3	10	40	377
Shallot	0·06–0·7	0·1–3·6	3	9	750
Tarragon	0·02–0·7	2·4–22·4	1	17	617
Thyme	0·60–6·1	5·9–171·9	10	50	895
Turmeric (curcuma)	0·00–0·5	5·6–350	5	16	553

TABLE 9
PROXIMITY TO USABILITY OF TL AND CL MEASUREMENTS TO IDENTIFY
IRRADIATED SPICES

Identification[a]	TL	CL
Very good (factor > 50)	Turmeric Savory Mushrooms Celery seed Ginger Paprika	Juniper berries Carrots Mushrooms
Good (factor > 10–50)	Pepper Garlic Onion Cinnamon Leek Dill Chilli Cumin Marjoram Horse-radish Oregano Sage Thyme Fennel	Cardamon Garlic Laurel Celery
Limited (factor > 2–10)	Carraway Cardamom Carrots Garlic Parsley Rosemary Lovage Orange peel Basil Celery	Onion Horse-radish Asparagus Shallot Coriander Fennel
Bad (factor < 2)	Juniper berries Laurel Cloves Shallot Nutmeg Mustard seed Coriander Mace	Leek Paprika Cloves Pepper Cinnamon Mustard seed Ginger Mace Savory Basil Dill Sage Marjoram Oregano Thyme Rosemary Orange peel Nutmeg Lovage Parsley

[a]Factor in brackets: luminescence intensity of irradiated samples divided by luminescence intensity of unirradiated samples.

that each group of five contained at least one irradiated sample (10 kGy) and one unirradiated sample. As an aid to comparability, four standard samples were enclosed with their TL or CL intensities measured at BGA. The standard samples were supported to equalize our threshold values with other measuring equipment to judge whether blind samples had been irradiated or not. Measurements were taken using different types of TL or CL readers. The participants were asked to do 3-fold analyses and in cases of poor reproducibility due to sample heterogeneity, 5-fold analyses in which the highest and lowest value should be neglected so that the mean value and standard deviation was calculated by three values. The results of TL measurements are as follows: operating with factor 3 (the threshold value to distinguish irradiated from unirradiated samples has been defined by multiplying the highest TL intensity of a reference group of unirradiated samples by a factor 3), all samples except one irradiated sample of mushrooms had been identified (Fig. 20). Consequently, the recovering rate for the samples examined in this collaborative study by TL measurements is better than 99%. Generally, the CL results are not as good as the TL results, because the CL is limited besides other factors by a faster fading effect also. In order to prevent that unirradiated samples are estimated as irradiated samples, calculations with factor 2 were not sufficiently accurate. By evaluation with

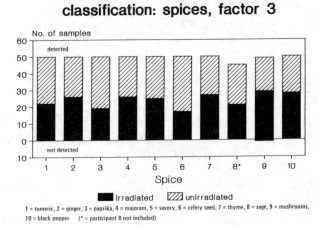

FIG. 20. Results of the European intercomparison concerning the identification of irradiated spices with TL measurements.

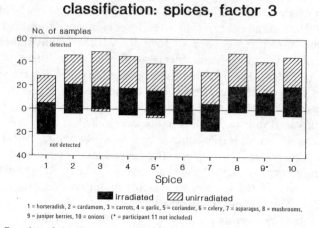

CL measurements
classification: spices, factor 3

Fig. 21. Results of the European intercomparison concerning the identification of irradiated spices with CL measurements.

factor 3 (Fig. 21), two unirradiated samples (carrots and coriander) were declared irradiated, but by considering the standard deviation, these samples would have been identified correctly. In total, 80 out of 228 irradiated samples were not found.

The results of this intercomparison demonstrate very clearly that mainly the TL technique is ready for a practical application. For the CL analyses it was shown that care must be taken not to declare unirradiated samples as irradiated, but additionally it became quite obvious that also the CL technique can contribute to the identification of irradiated samples of spices in spite of some doubts[36-40] and a higher or smaller number of irradiated samples (depending on the type of spice) which cannot be detected.

4 FURTHER POSSIBILITIES TO DEVELOP THE APPLICABILITY FOR IDENTIFICATION PURPOSES

4.1 Application for Fresh, Refrigerated and Deep-frozen Types of Irradiated Food

Some of the presented experimental results demonstrate that the potential of the luminescence technique might not be limited to the routine identification of irradiated spices and other types of dried foodstuffs. [14,15,21,29] A practical application for detecting irradiated fresh potatoes, fresh mushrooms, deep-frozen chicken, shrimps and fish and also for other types of foodstuffs seems to be possible. The main advantage of the luminescence techniques, compared with other methods of detection, [4] is the simplicity of the analyses resulting in a very short time for analyses (max. some minutes) for each food sample. In order to prove the practical applicability of the luminescence techniques for routine detection of irradiated fresh, refrigerated and deep-frozen types of foods, a lot of additional experimental investigations are necessary, e.g. dose response experiments, storage experiments with different storage temperatures, better knowledge of interfering effects, like influence of oxidation, the light emission variability of different samples and batches of the same type of foodstuffs and optimizing experiments to develop the best experimental conditions. Also a special luminescence analyser with the possibility of measuring large pieces of chicken, meat, mushrooms and fruits would be an advantage.

4.2 Thermoluminescence Glow Curve Deconvolution

The TL glow curve can contain different light emission peaks, as can be seen in Figs. 4 and 5, but the different peaks can also overlap so that only one peak maximum is observable. The physical background of different maxima in one glow curve are different excited states of the trapped electrons and electron-holes. The low temperature peaks in glow curves fade in a shorter time compared with high temperature peaks. One possibility to reduce errors in routine food control is the utilization of only the high temperature peaks because of their very flow fading with time. Some of the glow curves of irradiated and unirradiated spices contain different separated peaks, but in most cases they are overlapping so that the corresponding glow curves seem to be homogeneous. In such cases a computerized glow curve deconvolution (CGCD) may be helpful; suitable computer programs are available. First investigations concerning the number of peaks in TL glow curves showed the following results: ginger, 1 peak; savoury, 3 peaks; celery, 3 peaks; pepper, 2 peaks; paprika, 2 peaks; sage, 2 peaks; curcuma, 2 peaks (the heating rate was 3°C/s up to a maximum temperature of 280°C) (Fig. 22) (pers. comm.). Besides slower fading with time when using only the high temperature peaks, measurements are also necessary concerning the light emission difference between irradiated and unirradiated samples when not comparing the total light emission, but only the light emission of single high temperature peaks or the high temperature area of the glow curve. It might be that

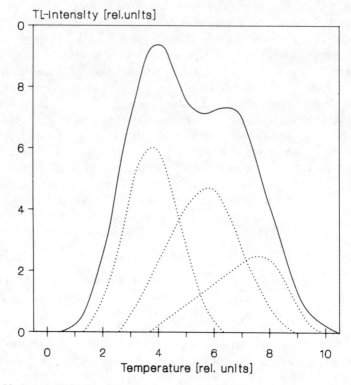

FIG. 22. Integral TL glow curve and computerized glow curve deconvolution of irradiated savory (10 kGy).

the factor 'intensity of the irradiated sample/intensity of the unirradiated sample' can be increased in this way; this would also contribute to a more precise identification of irradiated samples. Also an estimation of the difference in time between measurement and irradiation may be possible when comparing the low and high temperature peaks in a TL glow curve.[41] Because of the different fading of high and low temperature peaks, the ratio of the peak areas is a function of time; this function may allow an estimation of the time between irradiation and TL measurement.

4.3 Measurement of Thermoluminescence Wavelengths
Measurements of the TL light emission spectrum, scanned at individual temperatures, may provide more information and a better differentiation

between untreated and irradiated samples, also in regard to radiation dose and storage time, but the equipment for such investigations is complicated and expensive.

ACKNOWLEDGMENTS

The authors wish to thank Mrs. S. Albrich, Mrs. R. Guggenberger, Mrs. E. Mentele and Mrs. A. Spiegelberg for their very extensive and helpful technical assistance and Mr. H. Kirsch and Mr. A. Wolf, Gesellschaft für Strahlen- und Umweltforschung mbH, for irradiating the samples.

REFERENCES

1. Anon., *Wholesomeness of Irradiated Food.* Report of a Joint FAO/IAEA/WHO Expert Committee. World Health Organization, Technical Report Series 659, Geneva, 1981.
2. Smeets, G. (ed.), *The Identification of Irradiated Foodstuffs.* Commission of the European Communities, EUR 4695 d-f-e, Luxembourg, 1970.
3. Anon., *The Identification of Irradiated Foodstuffs.* Commission of the European Communities, EUR 5126 d/e/f/i/n, Luxembourg, 1974.
4. Bögl, K.W., Regulla, D.F. and Suess, M.J. (eds.), Health impact, identification, and dosimetry of irradiated foods. In: *Report of a WHO Working Group (Neuherberg/Munich, 17.-21.11.1986) on Health Impact and Control Methods of Irradiated Foods,* Published on behalf of WHO, Regional Office for Europe, Copenhagen. Report of the Institute for Radiation Hygiene of the Federal Health Office, ISH 125, Neuherberg/Munich, 1988.
5. Horowitz, Y.S., *Thermoluminescence and Thermoluminescent Dosimetry.* Vol. 1, CRC Press, Boca Raton, 1984.
6. Becker, K. and Scharmann, A., *Einführung in die Festkörperdosimetrie.* Verlag Karl Thiemig, München, 1975.
7. Cooke, D.W., Thermally stimulated luminescence in biochemical systems, *Radiat. Prot. Dosim.,* 1984, **8**, 117–138.
8. Ettinger, K.V. and Puite, K.J., Lyoluminescence dosimetry. Part I. Principles, *Int. J. Appl. Radiat. Isot.,* 1982, **33**, 1115–1138.
9. Puite, K.J. and Ettinger, K.V., Lyoluminescence dosimetry. Part II. State-of-the art, *Int. J. Appl. Radiat. Isot.,* 1982, **33**, 1139–1157.
10. White, E.H. and Bursey, M.M., Chemiluminescence of luminol and related hydrazides: The light emission step, *J. Am. Chem. Soc.,* 1964, **86**, 941–942.
11. White, E.H., Zafiriou, O.C., Kägi, H. and Hill, J.H.M., Chemiluminescence of luminol: The chemical reaction, *J. Am. Chem. Soc.,* 1964, **86**, 940–941.
12. White, E.H. and Bursey, M.M., Analogs of luminol. Synthesis and chemiluminescence of two methoxy-substituted aminophthalic hydrazides, *J. Am. Chem. Soc.,* 1966, **31**, 1912–1917.

13. Roswell, D.F. and White, E.H., The chemiluminescence of luminol and related hydrazides, *Methods in Enzymology*, 1978, **57**, 36.

14. Rippen, G. and Kaltenhauser, A., Rapid method to characterize irradiated food by means of chemiluminescence. In: *Report of a WHO Working Group (Neuherberg/Munich, 17.-21.11.1986) on Health Impact and Control Methods of Irradiated Foods*, K.W. Bögl, D.F. Regulla and M.J. Suess (eds.), Published on behalf of WHO, Regional Office for Europe, Copenhagen. Report of the Institute for Radiation Hygiene of the Federal Health Office, ISH 125, Neuherberg/Munich, 1988, pp. 245–247.

15. Anon., *Nachweismethoden zur Erkennung von mit ionisierenden Strahlen (Gamma-, Elektronen- und Röntgenstrahlen) behandelten Lebensmitteln*, Battelle-Institut e.V. Frankfurt, BF-R-66 651, Frankfurt, 1988.

16. Bögl, K.W. und Heide, L., Die Messung der Chemilumineszenz von Zimt-, Curry-, Paprika- und Milchpulver als Nachweis einer Behandlung mit ionisierenden Strahlen. In: *Report of the Institute for Radiation Hygiene of the Federal Health Office*, ISH 32, Neuherberg/Munich, 1983.

17. Kolbak, D., UV light affecting thermoluminescence signals in spices, *Radiat. Prot. Dosim.*, 1988, **22**, 201–201.

18. Heide, L. und Bögl, K.W., Chemilumineszenzmessungen an 20 Gewürzsorten - Methode zum Nachweis der Behandlung mit ionisierenden Strahlen, *Z. Lebensm. Unters. Forsch.*, 1985, **181**, 283–288.

19. Bögl, K.W. und Heide, L., Nachweis der Gewürzbestrahlung. Identifizierung gammabestrahlter Gewürze durch Messung der Chemilumineszenz, *Fleischwirtsch.*, 1984, **64**, 1120–1126.

20. Bögl, K.W. and Heide, L., The identification of gamma irradiated foodstuffs by chemiluminescence measurements. In: *Analytical Applications of Bioluminescence and Chemiluminescence*, L.J. Kricka, P.E. Stanly, G.H.G. Thorpe and T.P. Whitehead (eds.), Academic Press, London, 1984, pp. 573–576.

21. Bögl, K.W. and Heide, L., Chemiluminescence measurements as an identification method for gamma-irradiated foodstuffs, *Radiat. Phys. Chem.*, 1985, **25**, 173–185.

22. Heide, L. and Bögl, K.W., Die Messung der Chemilumineszenz von 16 Gewürzen als Nachweis einer Behandlung mit ionisierenden Strahlen. In: *Report of the Institute for Radiation Hygiene of the Federal Health Office*, ISH 53, Neuherberg/Munich, 1984.

23. Heide, L. and Bögl, K.W., Die Messung der Thermolumineszenz — Ein neues Verfahren zur Identifizierung strahlenbehandelter Gewürze. In: *Report of the Institute for Radiation Hygiene of the Federal Health Office*, ISH 58, Neuherberg/Munich, 1984.

24. Albrich, S., Stumpf, E., Heide, L. and Bögl, K.W., Chemiluminescent- and thermoluminescent-measurements to identify radiation-treated spices. A comparison of the two methods. In: *Report of the Institute for Radiation Hygiene of the Federal Health Office*, ISH 74, Neuherberg/Munich, 1985.

25. Heide, L. and Bögl, K.W., Die Messung der Chemilumineszenz von Feststoffen zur Dosisermittlung nach einer Strahlenexposition am Beispiel von Kaliumjodid und Zucker. In: *Report of the Institute for Radiation Hygiene of the Federal Health Office*, ISH 41, Neuherberg/Munich, 1984.

26. Heide, L. and Bögl, K.W., Identification of irradiated spices with thermo- and chemiluminescence measurements, *Int. J. Food Sci. Technol.*, 1987, **22**, 93–103.
27. Heide, L., Stumpf, E., Albrich, S. und Bögl, K.W., Die Identifizierung bestrahlter Lebensmittel mit Hilfe von Lumineszenz-messungen, *Bundesgesundhbl.*, 1986, **29**, 51–56.
28. Heide, L. and Bögl, K.W., Rapid identification of irradiated spices and condiments. In: *Proceedings of the Fourth European Conference on Food Chemistry*, Vol. 1, W. Baltes, P. Baardseth, R. Norang and K. Söyland (eds.), Norwegian Food Research Institute, As, 1987, pp. 255–259.
29. Heide, L. and Bögl, K.W., Thermoluminescence and chemiluminescence investigations of irradiated food — A general survey. In: *Report of a WHO Working Group (Neuherberg/Munich, 17–21.11.1986) on Health Impact and Control Methods of Irradiated Foods*, K.W. Bögl, D.F. Regulla and M.J. Suess (eds.), Published on behalf of WHO, Regional Office for Europe, Copenhagen. Report of the Institute for Radiation Hygiene of the Federal Health Office, ISH 125, Neuherberg/Munich, 1988, pp. 150–206.
30. Heide, L. and Bögl, K.W., Fortschritte bei der Identifizierung bestrahlter Gewürze durch Messung der Chemilumineszenz, Thermolumineszenz und Viskosität, *Fleischwirtsch.*, 1988, **68**, 1559–1564.
31. Moriarty, T.F., Oduko, J.M. and Spyrou, N.M., Thermoluminescence in irradiated foodstuffs, *Nature*, 1988, **332**, 22.
32. Heide, L., Albrich, S., Mentele, E. and Bögl, K.W., Thermolumineszenz- und Chemilumineszenzmessungen als Routine-Methoden zur Identifizierung strahlenbehandelter Gewürze. Untersuchungen zur Festlegung von Grenzwerten für die Unterscheidung bestrahlter von unbestrahlten Proben. In: *Report of the Institute for Radiation Hygiene of the Federal Health Office*, ISH 109, Neuherberg/Munich, 1987.
33. Heide, L. and Bögl, K.W., Thermoluminescence and chemiluminescence measurements as routine methods for the identification of irradiate spices, In: *Report of a WHO Working Group (Neuherberg/Munich, 17.-21.11.1986) on Health Impact and Control Methods of Irradiated Foods*, K.W. Bögl, D.F. Regulla and M.J. Suess (eds.), Published on behalf of WHO, Regional Office for Europe, Copenhagen. Report of the Institute for Radiation Hygiene of the Federal Health Office, ISH 125, Neuherberg/Munich, 1988, pp. 207–232.
34. Heide, L., Delincée, H., Demmer, D., Eichenauer, D., v.Grabowski, H.-U., Pfeilsticker, K., Redl, H., Schilllng, M. and Bögl, K.W., Ein erster Ringversuch zur Identifizierung strahlenbehandelter Gewurze mit Hilfe von Lumineszenzmessungen. In: *Report of the Institute for Radiation Hygiene of the Federal Health Office*, ISH 101, Neuherberg/Munich, 1986.
35. Heide, L. and Bögl, K.W., Routine application of luminescence techniques to identify irradiated spices — A first counter check trial with 7 different research and food control laboratories. In *Report of a WHO Working Group (Neuherberg/Munich, 17–21.11.1986) on Health Impact and Control Methods of Irradiated Foods*, K.W. Bögl, D.F. Regulla and M.J. Suess (eds.), Published on behalf of WHO, Regional Office for Europe, Copenhagen. Report of the Institute for Radiation Hygiene of the Federal Health Office, ISH 125, Neuherberg/Munich, 1988, pp. 233–244.

36. Delincée, H., Use of chemiluminescence for identifying irradiated spices. In:*Report of a WHO Working Group (Neuherberg/Munich, 17–21.11.1986) on Health Impact and Control Methods of Irradiated Foods*, K.W. Bögl, D.F. Regulla and M.J. Suess (eds.), Published on behalf of WHO, Regional Office for Europe, Copenhagen. Report of the Institute for Radiation Hygiene of the Federal Health Office, ISH 125, Neuherberg/Munich, 1988, pp. 248–265.

37. Meier, W. and Zimmerli, B., Experiments with chemiluminescence measurements. Preliminary results with imported spices. In: *Report of a WHO Working Group (Neuherberg/Munich, 17.-21.11.1986) on Health Impact and Control Methods of Irradiated Foods*, K.W. Bögl, D.F. Regulla and M.J. Suess (eds.), Published on behalf of WHO, Regional Office for Europe, Copenhagen. Report of the Institute for Radiation Hygiene of the Federal Health Office, ISH 125, Neuherberg/Munich, 1988, pp. 266–268.

38. Delincée, H., Ist die Bestrahlung von Gewürzen durch Chemilumineszenz nachweisbar? *Fleischwirtsch.*, 1987, **67**, 1410–1418.

39. Sattar, A., Delincée, H. and Diehl, J.F., Detection of gamma irradiated pepper and papain by chemiluminescence, *Radiat. Phys. Chem.*, 1987, **29**, 215–218.

40. Meier, W., Konrad-Glatt, V. and Zimmerli, B., Nachweis bestrahlter Lebensmittel: Chemilumineszenzmessungen an Gewürzen und Trockengemusen, *Mitt. Gebiete Lebensm. Hyg.*, 1988, **79**, 217–223.

41. Moscovitch, M., Automatic method for evaluating elapsed time between irradiation and readout in LiF-TLD, *Radiat. Prot. Dosim.*, 1986, **17**, 165–169.

Chapter 3

CONSUMER ACCEPTANCE OF IRRADIATED FOOD IN THE UNITED STATES

R.J. BORD

Department of Sociology, 206 Oswald Tower, The Pennsylvania State University, University Park, PA 16802, USA

1 INTRODUCTION

Assessing the degree of consumer acceptance of a novel, risky technology is, itself, a risky venture incorporating elements of both science and art. Available data suffers from a plethora of problems and, even when of good quality, requires a great deal of interpretation as to its meaning. Furthermore, the leap from survey data to assumptions about public acceptance must be made with considerable caution. The marketing of a somewhat novel technology is made in an arena that incorporates multiple actors: food producers, the media, anti-nuclear activists, and others play at least as central a role as does the 'public'. What follows is an analysis of the public acceptance of irradiated food issue in the United States as it stands in late Summer, 1990. This analysis will rely on survey and marketing data but will go beyond those findings to assess other impacts and the interaction of multiple factors.

This assessment will proceed along three tracks: first, public opinion surveys and marketing research will be overviewed in an effort to discern patterns and trends; second, recent popular literature on food irradiation will be discussed in terms of the images being presented to the reading public; finally, the potential for future shifts in consumer acceptance will be assessed in light of what is known about the impact of public interest groups, the media, and decision making concerning risky technologies. A summary and conclusion section will provide the synthesis.

2 IRRADIATED FOOD AND PUBLIC ATTITUDES

The assessment of consumer attitudes is important for many reasons: (1) opponents want to increase their political clout by fostering the belief that food irradiation is widely opposed; (2) the product irradiation industry and its investors want an accurate assessment of the potential market for irradiated food; (3) politicians want a barometer by which to guide their decision making; (4) government regulatory agencies need feedback to help inform their rule-making; and (5) the uncommitted and uninformed elements of the general public uses public opinion information to help shape their own attitudes.

Any nuclear technology bears the burden of being associated with characteristics that tend to stimulate fear: the technology is highly complex and both relatively unknown and poorly understood; exposure to radiation is involuntary and difficult to detect; its long-term health impacts are uncertain; and it is associated with cancer and other 'dreaded' outcomes.[1] Furthermore, events such as the Three Mile Island reactor accident, the subsequent Chernobyl disaster, the recent acknowledgement by the US Department of Energy of long-term, high-level radiation releases in the Hanford, Washington area, and the continuing unsuccessful struggle to find repositories for low and high level nuclear waste do little to inspire public confidence in the nuclear industry. Pessimism about anything nuclear has become the expected public response in the United States.[2]

In the book, *Food Irradiation: Who Wants It,* several antinuclear activists present the pros and cons of food irradiation and contend that 'studies of consumer attitudes have been done and show that consumers overwhelmingly do not want irradiated food.'[3] Most of the research cited by those authors to support that assertion comes from England. Studies done on the North American continent, however, do not support 'overwhelming' rejection of the irradiated food concept.

The most comprehensive overview of the American experience with the food irradiation concept up to 1986, The Brand Report,[4] reached the following conclusions: The period of the early 1960s was characterized as a time of optimism and a focus on the benefits of food irradiation; the end of this period, and the beginning of a period of disinterest, came in 1968 when the FDA quashed the Army's petition to irradiate ham; and finally, the modern period of renewed interest began in the 1980s and was signalled by the FDA's 1984 announcement of new, more relaxed, regulations. The subsequent overview will focus on the modern period: studies done in the 1980s.

2.1 Results of Public Opinion Polls
It is necessary to keep in mind that no public opinion studies are without flaws, some of them very serious. Poll results are subject to the vagaries of sampling error, respondents' answers are very much conditioned by the wording of the question, and the meaning of answers about poorly understood topics is open to conjecture. All of the national surveys on food irradiation incorporate samples that are predominantly white and disproportionately older and better educated. On the one hand the case could be made that this type of sampling error is positive because these types of people are most likely to be involved in an issue such as food irradiation. On the other hand, there is evidence that those with lesser education are more opposed and that blacks and other minorities are becoming increasingly involved in facility siting conflicts and other environmental issues.[5]

Furthermore, there are dubious assumptions made about the answers to survey questions. Perhaps the most serious of these assumptions is that there is some sort of strong correspondence between the respondents' answers and their future behavior regarding the attitude object. The theoretical and research challenges to that assumption are far too numerous to mention here. Suffice it to point out that overt behavior is more likely to parallel survey responses under the following conditions:

- the time between attitude measurement and overt behavior is short;
- the situational constraints favor attitude-behavior correspondence;
- the attitude is measured at the same level of specificity as the target overt behavior; and,
- the respondent has a set of well-formed beliefs about the attitude object.[6]

Many of these conditions are not met in the studies to be reviewed and these shortcomings will be discussed in the final section of this paper.

This overview will be structured according to the major, consistent findings of the opinion polls done between 1984 and 1990. Each finding will be presented and then discussed in terms of variations in the research and the implications for decisions about public acceptability.

In surveys asking intent to purchase or use irradiated food a majority or near-majority of respondents fall into the middle, 'Probably' or 'Uncertain', categories while ~ 10–25% make up the accepting and rejecting ends of the scale.

Bruhn and Schutz assert: 'Consumer attitude and marketing studies show that, given information about irradiation, the majority will choose irradiated food.'[7] That may be an overly optimistic prediction. Perhaps the most frequently quoted estimate of how the public is distributed with regard to attitudes toward food irradiation is the Brand Group typology: 'Rejectors' were estimated to be that 5–10% of the population who have strong ecological concerns and are anti-nuclear; 'Undecideds' make up that 55–65% of the population who are not comfortable with their level of knowledge about the technology; and, the 'Acceptors' comprise 25–30% of the population who, the authors suggest, may have a tenuous hold on their positive attitudes. This typology is based on a multi-item index which includes questions about emotional reactions and intent to try irradiated food and to serve irradiated food to the family.

Perhaps the most significant finding in the Brand Group study is the degree of uncertainty surrounding the irradiated food issue. This finding haunts all similar studies and warrants careful consideration. Table 1 presents the results of four studies which asked intent to purchase questions.

TABLE 1

RESULTS OF FOUR SURVEYS ASKING HOW LIKELY THE RESPONDENT IS TO TRY OR BUY IRRADIATED FOOD (IN PERCENTAGES)

Gidwani[8]		Brand Group[9]		Schutz et al.[10]		Bord and O'Connor[11]	
Would buy	25	Def. will try	22	Very likely	15	Def. will try	14
Not sure	44	Prob. will try	47	Likely	30	Prob. will try	63
Would not	28	Prob. not	20	Uncertain	34	Prob. won't	17
		Definitely not	11	Unlikely	13	Def. won't	5
				Very unlikely	9		

If the focus in Table 1 is the percentage of respondents who say they will, or probably will, try irradiated food then the outcome is indeed optimistic: 25, 45, 69, and 77% so indicate. The larger percentages characterize the more recent studies. Conversely, 22, 22, 28, and 31% indicate that they probably or definitely will not purchase irradiated food. If, however, the position is taken that the only respondents who have committed attitudes, those likely to truly influence behavior, are those at the poles of the attitude continuum, then the conclusion is far less optimistic. Only 14, 15, 22, and 25% exhibit solid positive attitudes. If the middle responses are viewed as uncertain responses then 44, 67, 77, and 80% of these samples can be so characterized. Note that both of the latter conclusions complement the Brand Group's typology, one based on a sophisticated multi-item index, quite well.

The clustering of responses in the middle categories indicates the extent of uncertainty about food irradiation in the general population. Most studies report that a substantial minority of respondents have never heard of the concept. Studies in 1984 demonstrated that ~70% were unaware of food irradiation,[12] by 1988, 64–55% reported being unaware, and in 1989 only 40% indicated being unaware.[13] It is reasonable to assume that substantial segments of the population still know relatively little or nothing about food irradiation.

This ambivalence is indexed in other ways in surveys on food irradiation. While a substantial majority indicate they 'definitely' or 'probably' will try irradiated foods, 69% in the Brand study and 61% in the Bord and O'Connor study, only about 52% express comfort in serving irradiated foods to their families. Individuals are often willing to take risks that they would rather not expose their family members to. That is apparently the case with irradiated food. However, the fact that from 52 to 53% of the surveyed public would feel totally or somewhat comfortable serving irradiated food still seems to indicate a reasonably substantial base of potential support. However, just as in the intent to 'personally try' questions, a far smaller percentage exhibit solid attitudes: only 13 and 17% of the respondents say they would feel totally comfortable serving, or definitely would serve, irradiated food to their families.

Ambivalence is further indexed by questions asking if respondents would support legislative initiatives to ban irradiated food. The Brand Group study found that when asked whether they would vote for the acceptance of irradiated food in the United States 21% said 'yes', 64% said 'yes, with some conditions', and 15% said 'no'. The 'conditions' included labeling and further testing to establish safety. Bord and O'Connor asked respondents if they would support a legislative ban on irradiated food. Only 4% said they would 'strongly support' such a ban, 35% said they would 'probably support it', 49% chose 'probably oppose', 7% picked 'strongly oppose', and 5% did not answer the question. The Brand and the Bord questions seem to indicate that a majority of people would support food irradiation with their votes. However, Bord and O'Connor also asked respondents how they felt that others would respond to an attempt to ban all irradiated food from the market: 60% said others would 'strongly' or 'probably' support such a ban. The finding that individuals expect others to be opposed to this technology presents the possibility that there may be reluctance to publicly support food irradiation, especially if 'anti' arguments are dominant. This perceived majority opposition may indicate that it is easier to publicly elicit negative responses because that is believed to be the norm. Bord and O'Connor found that to be true in the focus groups that made up their research. Oftentimes the respondent who just made very favorable written responses to food irradiation would make very negative verbal responses to indicate agreement to perceived group norms. These participants did not know how each other actually felt about the issue but assumed that most of the others must be opposed. The Brand Group also concluded that there is a greater potential to shift consumers toward an 'anti', rather than a 'pro', position.

This uncertainty and ambivalence is a key finding. It is impossible for people to have a solid set of beliefs about something they know little about. Without a solid set of beliefs there can be only shaky attitudes and shaky intentions to behave relative to irradiated food. Shaky attitudes are ripe for influence and can change dramatically given new and dramatic information.[14]

Questions dealing with perceived hazards or general concerns about irradiated food result in a majority or near majority expressing considerable concern and doubt.

For example, in the Brand Group study 89% of the respondents reported having some or many doubts about irradiated food while 52% agreed that irradiation 'sounds like a dangerous technique'. Longitudinal data provided by the Opinion Research Corporation, for the Food Marketing Institute, further illustrates the public trepidation about food irradiation. The Food Marketing Institute sponsors annual telephone surveys using random digit dialling to elicit approximately 1000 respondents who are described as a group that 'closely approximates the US population'. Since 1986 a question about the perceived health hazard of food irradiation has been included. Each year, including 1990, ~40% of the sample picks 'serious health hazard' for irradiated foods, the range is from 36% in 1988 to 43% in 1987. A nationwide survey done by Wiese Research Associates, Inc, also demonstrated 40% of the sample indicating that their initial reaction to irradiated food was one of 'major concern'.[15] In 1990, as in other years, ~70% of the Food Marketing Institute sample picked either 'serious health hazard' or 'somewhat of a hazard.'

This level of concern and doubt, coupled with the uncertainty noted above, would appear to bode ill for proponents of irradiated food. Concern, however, must be placed in context before generalizations can be made. It is socially and psychologically 'easy' to be concerned. In fact, in an environmentally conscious age it is normative to express concern. In 1990, for example, one national poll found that 80% of its sample picked 'protecting the environment' over 'keeping prices down'.[16] Table 2 presents the Food Marketing Institute's data for six issues measured over 7 years. Some of these issues were not measured in 1984, 1985, and 1986. Table 2 dramatically illustrates that what appears to be a high level of concern for irradiated food, when placed in the context of other food

TABLE 2

RESPONSES TO THE FOLLOWING QUESTION: 'I'M GOING TO READ A LIST OF FOOD ITEMS THAT MAY OR MAY NOT CONSTITUTE A HEALTH HAZARD. FOR EACH ONE, PLEASE TELL ME IF YOU BELIEVE IT IS A SERIOUS HEALTH HAZARD, SOMEWHAT OF A HAZARD, OR NOT A HAZARD AT ALL?'

	Serious hazard						
	Jan. 1984	Jan. 1985	Jan. 1986	Jan. 1987	Jan. 1988	Jan. 1989	Jan. 1990
Residues such as pesticides and herbicides	77%	73%	75%	76%	75%	82%	80%
Antibiotics and hormones in poultry and livestock	X	X	X	61%	61%	61%	56%
Irradiated foods	X	X	37%	43%	35%	42%	42%
Nitrites in foods	X	X	X	38%	44%	44%	37%
Additives and preservatives	32%	36%	33%	36%	29%	30%	26%
Artificial coloring	26%	28%	26%	24%	21%	28%	21%

(In 1990, if the serious health hazard and somewhat of a hazard responses are combined then 97% express concern for pesticides and herbicides, 89% for antibiotics and hormones, 71% for irradiated foods, 81% for nitrites in foods, 88% for additives and preservatives, and 74% for artificial coloring.)

(Adapted from: Trends: Consumer Attitudes and the Supermarket, 1990)[17]

concerns, turns out to be quite moderate. Pesticides and herbicides in food and antibiotics and hormones in poultry and livestock generate significantly higher levels of concern than irradiated food. Nitrites in foods is comparable to irradiated food in the concern it elicits, while additives, preservatives, and artificial coloring arouse somewhat less concern. The note at the bottom of Table 2 points out that if both the extreme and moderate hazard responses are combined for the year 1990 then irradiated food actually fares better than any of the other options in terms of public concern.

Another noteworthy result in Table 2 is that a larger percentage of people choose 'not a hazard' for irradiated food than any other issue except artificial coloring. As previously noted, a relatively large percentage (18%) is simply 'not sure'.

There is another way of putting the concern levels of Table 2 into context. The Food Marketing Institute also asks an open-ended question of all respondents: 'What, if anything do you feel are the greatest threats to the safety of the food you eat?' This type of open-ended question does not lead the respondent as does the structured question about levels of concern. Open-ended questions demand recall and, it can be argued, are more likely to tap issues that are truly salient for the respondent. In 1990, this question only stimulated 1% of the sample to mention food irradiation as a 'greatest threat'. On the other hand, 29% mentioned spoilage/germs, 19% mentioned pesticides/residues/insecticides/herbicides, 16% mentioned improper packaging/canning, and 16% mentioned chemicals. Food irradiation, at this time, does not appear to be a salient issue for US consumers.

At this juncture we have a snapshot of a public that, for the most part, is not adamantly opposed to food irradiation. In fact, ~25% of all samples express no reservations about personally trying irradiated food. There is, however, substantial levels of concern and uncertainty. While high, concern over food irradiation is not as great as for chemical contaminants and additives in food. That observation leads to the next generalization.

Food surveys, surveys on toxic substances, and general environmental concern surveys indicate that the US public views hazardous chemicals as the most serious environmental problem affecting health and safety.

There can be little doubt that, at this time, the major environmental concern in the United States is toxic chemicals. This concern includes hazardous chemical waste sites, chemical manufacturing plants, pesticides and herbicides applied to crops, and to some extent, food additives and preservatives. Almost every survey used in this paper finds high levels of concern with chemicals and preservatives. Every study that has forced a comparison between chemicals or irradiation to control pests or improve food quality has favored food irradiation.[12] Americans clearly perceive a link between chemicals and cancer.[13] That does not mean, however, that every respondent enthusiastically endorses irradiation as a replacement for the use of chemicals in food. A *Good Housekeeping* consumer's poll indicated that when asked to select between chemical preservatives or food irradiation 23% chose irradiation while only 3% chose chemicals but, 44% expressed uncertainty and 27% wanted neither.[18]

The argument that food irradiation is a replacement for the chemical treatment of food is clearly one of the most powerful tools the proponents of food irradiation have. Opponents obviously recognize this and have countered with arguments that not only would food irradiation not significantly minimize the use of pesticides and herbicides but would increase the use of some post-harvest chemical additives.[19] If these counter arguments are effective they will substantially reduce the perceived need and support for food irradiation.

The impact of more information about food irradiation on consumers' attitudes is contingent on the type of information and the respondents' prior attitudes on a number of dimensions.

Given that many people have not even heard of food irradiation, and that of those who have heard about it many know very little about the process or its outcomes, providing more information should build a knowledge base upon which consumers can make more informed decisions. Bruhn and Schutz reach this conclusion in their article on consumer awareness of irradiated food.[20]

At least four studies done in the United States have attempted to assess the impact of more information on consumers' attitudes toward, and receptivity to, food irradiation.[21] In general, those respondents who are at the extremes of the attitude continuum, either definitely pro or anti food irradiation, do not change their positions after being exposed to more information. The conclusions of Bruhn *et al.* that educational efforts will not reduce concerns among those already opposed, appears to be sound. Those in the more undecided categories tend to shift, some shift more pro and some more anti. Two studies, Bruhn *et al.* and Weise Research Associates, Inc. demonstrate a shift to the anti position by those who are initially undecided after discussions or exposure to information about product attributes.

Bord and O'Connor,[22] using a factorial design, presented respondents with three types of information: a technical versus nontechnical discussion of the food irradiation process; an extended discussion of the pros and cons of food irradiation (either present or absent); and, a discussion of the history of use of food irradiation (either present or absent). The only manipulation that had a statistically significant effect on post-message attitudes was the history of use. This information discussed the use of irradiated food by astronauts and those with extreme immune deficiency diseases. The benefits implied by this message appeared to assure some respondents. Again, however, those who were initially opposed remained opposed.

The impact of prior awareness on attitudes produces somewhat conflicting results. The Brand Group and Bord and O'Connor found that prior knowledge correlated with somewhat more positive attitudes. Bruhn *et al.* found just the opposite, previous knowledge correlated with greater concern.

The apparent differences in results in these studies may be a result of the researchers inability to hold the type of prior information constant. In focus group discussions, Bord and O'Connor found that prior information tended to be one of two types: information about food irradiation gleaned from a newspaper or popular magazine which is neutral to positive in tone; or, information obtained from television news broadcasts, environmental activist publications, or from discussions with acquaintances who are nuclear opponents, which is negative in tone. Those exposed to neutral or positive communications tend to be less concerned while those exposed primarily to anti-nuclear arguments tend to be more concerned. The implications of these findings will be discussed in greater detail in the next section.

All studies indicate that consumers want more information about irradiated food. Some self-report studies indicate that if consumers were provided more information they would be more favorable. This appears to be an oversimplified generalization. Reactions to further information will depend on the receiver's prior attitude and strength of prior attitude, on the content and tone of the communication, on the trustworthiness of the communicator, and the social-political climate existing at the time of the communication. There is no reason to expect a strong linear relationship between education/exposure to information and attitudes toward food irradiation.

The above discussion indicates that opposition to food irradiation may be more complex than often depicted in the extant research literature. That issue forms the next topic for discussion.

Multi-variate analysis of determinants of acceptance-opposition to food irradiation indicate a complex pattern having more to do with social attitudies than knowledge or level of education.

There are very few multivariate studies of public reactions to food irradiation. Bruhn *et al.* found that preexisiting attitudes, prior knowledge, and age accounted for much of the variance in change in attitude toward food irradiation as a result of exposure to information. Also, willingness to buy was significantly related to prior knowledge, higher income, and being male.

Bord and O'Connor (forthcoming) used multiple regression to ferret out the relative impact of a number of independent variables on each of three dependent variables. The dependent variables were attitudinal items designed to tap different dimensions of the acceptability of irradiated food. The first dependent variable consisted of responses to a question asking to what extent the respondent intended to try irradiated foods. The second dependent variable was comprised of responses to a question asking how comfortable the respondent would be serving irradiated food to her family. All respondents were female. The third dependent variable consisted of responses to a question of support for a legislative ban on all irradiated food from the US market. The independent variables included the following: the different types of information noted previously; attitude scales assessing trust in government and industry, the extent to which respondents feel powerless in the face of big government and big business, perceptions that modern technology is too complex and fraught with unknown danger, and fear of radiation in general; a four-item scale assessing knowledge of food irradiation; a two-item scale dealing with the degree to which the respondent accurately ranks smoking and car accidents as major public health threats; and, an item asking if the respondent thinks that irradiated food will be more expensive than food treated by conventional methods. These independent variables include measures of knowledge of both irradiated food and societal risk, economic concerns, and various social attitudes. Demographics such as sex, age, education, and income were also included.

Six variables explain 44% of the variation in the question assessing the likelihood of the personal use of irradiated food. Trust in government and industry, alone, explains 36% of the variation in that question. Fear of nuclear technology explains another 5% of the variation. Together, trust and fear of nuclear technology explain 93% of the explained variation in willingness to personally try irradiated food. Four other variables account for ~1% of the explained variance each: the history of use message, the respondent's level of knowledge about food irradiation, general anti-technology attitudes, and an attitude scale measuring alienation from a society preceived as dominated by the wealthy and the powerful. A lack of trust in government and industry appears to be a major determinant of opposition to a risky technology. Similar findings characterize the results of a study examining public attitudes toward Superfund hazardous waste sites.[23]

The determinants of reactions to the question of whether one would feel comfortable serving irradiated food to the family are very similar to those noted above. Five variables explain 52% of the variation in responses to this question. Trust accounts for 74% of the explained variance and fear of nuclear technology and alienation from powerful institutions are also important. However, knowledge of food irradiation is the second most important variable, accounting for 20% of the explained variance. Those who have more accurate knowledge of food irradiation are more likely to be willing to serve irradiated food to their families. This factor, however, places a distant second to trust in government and industry.

The determinants of support for a legislative ban on irradiated food are different and add further complexity to the issue of public attitudes toward irradiated food. First, all of the independent variables account for only 21% of the variation in the dependent variable. In the two dependent variables discussed above the amount of explained variance in each was ~ 50%. This discrepancy in the amount of explained variance is probably a result of the relatively abstract nature of the decision to support or not support a legislative ban. The decisions to personally try irradiated food or to serve it to one's family are very personal and salient to one's view of him or herself. Attitudes that closely tie to self-concept tend to be highly committed attitudes. The decision to support or not support a legislative ban is several steps removed from personal considerations, it is a more abstract decision that probably inspires a more cool, calculated approach. This interpretation is supported by the fact that most of the explained variance in this dependent variable is accounted for by prior knowledge and estimated cost. Fear of nuclear technology, alienation from institutions, and distrust play a statistically significant role, but they account for <20% of the explained variance combined.

The results of this multi-variate approach have several important implications. First, questions of acceptability or concern will have different outcomes depending on their focus. People are apparently more willing to personally take a risk than to subject it to others close to them. Second, the more important determinants of concern or decisions to use are social attitudes rather than knowledge or education. That does not mean that knowledge did not have an impact: those who had prior knowledge of the food irradiation process tended to be more accepting. However, trust is by far the most important determinant. Third, personal decisions to support something like a legislative ban on irradiated food, while also involving social attitudes, may be based disproportionately on rational criteria such as cost and knowledge of the process.

The research by Bord and O'Connor suggests that normative factors are more important than technical information in structuring attitudes toward food irradiation. The discussants in the focus groups comprising this research stressed the complexity of modern technology, its uncertainty, greed on the part of its sponsors, and too little effective government control. The point was made that even if the technical plan is flawless, those managing the technology would create problems. Examples such as Bhopal, Chernobyl, Three Mile Island, and even Nixon's Watergate were used to make the point.

Demographic factors do not appear to have a powerful impact on attitudes toward food irradiation. However, gender appears to have a consistent impact: women tend to express greater concern than men. Studies of gender differences in reactions to nuclear technology and environmental concerns produces mixed results. In general, however, technologies and environmental hazards that pose a personal health threat or health threats to future generations produces higher levels of concern in women.[24] Women, for example, tend to express greater concern for toxic chemicals and nuclear power plants but not for acid rain. The research by Gidwani and that by Weise Research Associates indicates that women tend to express higher levels of concern over irradiated food than men. Also, most of the studies reviewed above show ~20–25% of the sample being 'acceptors' of irradiated food but the 1989 Bord and O'Connor study, which included women only, produced only 14% that said they would 'definitely' try irradiated food and 13% that picked 'totally comfortable' serving irradiated food to their families.

This is an important finding. While changes in gender roles have somewhat redistributed the tasks allocated to males and females in the United States, women still tend to be the major gatekeepers for food.[25] If women cannot be convinced of the utility and safety of food irradiation the technology is not likely to come into widespread use.

Marketing studies do not indicate widespread or vociferous opposition to irradiated fruit.

Fruits clearly labeled as irradiated have been successfully sold twice in the United States. Irradiated papayas were sold in Irvine and Anaheim, California on March 28, 1987 and irradiated mangoes were sold in Miami Beach, Florida on September 11, 1986. In both cases the media was not forewarned of the impending sale and no protestors were involved. The irradiated papayas outsold hot-water treated papayas by a ratio of better than 10:1[26] and the two tons of mangoes were sold within a week.[27]

These two events have highly suggestive implications. First, many Americans apparently have few qualms about purchasing irradiated food if the quality appears high. Second, if left simply to market forces it is probable that some irradiated foods would find a market niche. It is highly unlikely, however, that market forces will be given free rein. Opposition groups have already had a substantial impact on marketing decisions about irradiated food.

The implications of these events for the arguments developed previously are as follows: the large degree of uncertainty characterizing the public's attitudes about food irradiation indicate that there is substantial room for social influence in either the 'pro' or 'anti' food irradiation direction. Various factors appear to favor the 'antis'. This topic will be thoroughly covered in section three.

Consumers demand the proper labeling of irradiated food, more research on its health and nutrition effects, and more information on the topic.

All studies demonstrate an overwhelming consensus on the need for unambiguously identifying irradiated food, for conducting more research on its health and nutrition impacts, and for providing more information to the public, especially in the marketplace. These findings complement general social policy tendencies in the United States. The Western value of individual autonomy has been increasingly translated into legal mandates for full information disclosure, as a prerequisite to adequate decision making, in product advertising, in risks involved in the manufacture of chemicals, in health care delivery, and in research with human subjects.

Demands for full information interface with the trust issue noted previously. A lack of trust in business and industry explains much of the opposition to irradiated food. Given the cultural value of full information disclosure, any perceived attempt by industry or government to be less than totally forthcoming will provide opposition groups with a potent weapon to marshall public opposition. This theme is already being played out by those dedicated to stopping food irradiation.

2.2 Summary of Part Two

Statements about widespread public opposition to irradiated food are clearly exaggerations. Depending on how the available survey data is interpreted there is either modest support/interest in irradiated food or widespread uncertainty. The position taken here is that multiple indicators favor the uncertainty interpretation.

Research examining multiple correlates of attitudes toward irradiated food indicate that decisions to use or serve to family hinge heavily on trust in industry and government. Technical information or explanations of the food irradiation process may play a role in public acceptability but, multiple studies indicate that such information tends to shift some 'antis' to a positive position while shifting some of those favorable to an unfavorable position. The net outcome is often little different from the pre-information position. This does not mean, however, that information will not play a serious role in the eventual outcome of attempts to market irradiated food. Uncertainty breeds a sensitivity to information that can reduce that uncertainty. The issue then becomes the kind of information that is likely to be available to the public and how that information is processed by the public. The next section examines messages about food irradiation being provided by the print media and the likely impact of that information.

3 THE PRINT MEDIA AND FOOD IRRADIATION

A review of the food irradiation books and articles available in the popular, nontechnical literature was conducted to provide some insight into the nature of the messages targeted at the general public. Articles in food chemistry and food technology journals are not included in this review.

The articles in these types of periodicals are almost uniformly favorable toward food irradiation but they reach a very limited and specialized audience.

In conducting this review, particular attention was directed toward the nature of the message, its source, the probable audience, and the likely effect. While limited, there is a substantial and growing literature on food irradiation available to the general public.

This literature review builds on that compiled by the Brand report. The Brand report included a 126-item bibliography resulting from a thorough literature search. However, that report includes items only up through 1985. We added over 30 items to the Brand list. Each article was read carefully and coded for the following: the direction of the article, clearly pro, clearly con, or in between; the specific arguments used and the recurring content; and, the conclusions reached by the author. What follows is a summary of that literature search.

The Brand Group's report, encompassing literature through 1985, is worth quoting at this juncture:

'There is a significant body of non-technical irradiation available to the public. In addition to periodic news items related to FDA action on irradiation regulation, there are quite a few magazine, trade and newspaper articles which can be found in most public libraries.'

That conclusion is still appropriate 5 years later. The problem, however, is that the general public tends to read local newspapers, relatively few magazines, and fewer trade publications. Only those people who are already quite interested in food irradiation will go to the library to seek these sources out. These are not the large number of 'uncertains' noted in the survey analysis.

The Brand report goes on to delineate several distinct waves of literature on food irradiation. The modern period begins in the decade of the 1980s after various national and international organizations and agencies announced renewed interest in food irradiation. In particular, the FDA's 1984 announcement of its new proposed regulations generated a rash of popular articles. The Brand Group characterizes the literature of this period as 'adversarial'.

Actually, the popular literature is a mixture of pro, con, and basic information articles. Information, and information providers, of this period can be characterized as follows:

- The interest aroused by the new FDA regulations spurred a spate of articles from business oriented publications which, usually, briefly describe the technology and then go on to speculate on its future for investment purposes. These articles tend to depict food irradiation in a favorable light.
- Semi-technical and popular science magazines, and technical/science sections of major newspapers such as *The New York Times*, also had articles describing the technology and speculating on its future. The tone of these articles tended to be guardedly optimistic, but doubts about the technology were also aired. In some cases, opposition arguments were given more space the those of proponents. Opposition arguments tended to be more emotionally laden and appeared to have more 'shock value'. Examples will be provided subsequently.
- Health and environmental publications also began to attend to food irradiation during this period. Their response was almost uniformly negative.

3.1 Recent Themes in the Opposition Literature

The most logical beginning for this review is with the 156 page paperback book, *Food Irradiation: Who Wants It*.[3] Two of the authors of this book are with the London Food Commission while the third is with the Health and Energy Institute. Both organizations are well known for their antinuclear stance. The Health and Energy Institute claims as advisory board members some of the leading anti-nuclear activists in the United States including Rosalie Bertell, Helen Caldicott, and John Gofman. The Institute has published fact sheets on Colbalt-60 and Cesium-137 and anti food irradiation pamphlets.

Food Irradiation: Who Wants It discusses all the anti-arguments found, usually in part, in other publications. Its arguments can be classified as follows: food health and safety issues; environmental-operational issues; and, science policy/ethical issues. The food health and safety issues raise seven possibilities:

- that unique radiolytic particles are created with possible links to cancer and birth defects;
- that irradiation results in nutritional losses, including the loss of vitamins and damage to poly-unsaturated fats and fatty acids;
- that consuming irradiated food has negative effects on the body's immune system;

- that toxins will be hidden by the false appearance of freshness;
- that bacterial contamination may be concealed by the improper use of irradiation;
- that spoilage rates will increase in some foods;
- that irradiation may promote the use of different chemical additives to control some of the undesirable effects of the irradiation process itself.

Four environmental-operational issues are raised:

- that operating irradiation plants have already had accidents resulting in environmental contamination;
- that irradiation plants are hazardous to workers and have already overexposed some workers at operating facilities;
- that the 'massive' trucking of dangerous radioactive material over long distances will be needed;
- that there are insurmountable difficulties in adequately monitoring and regulating plant operations.

Finally, five science policy/ethical issues are presented:

- that the high degree of scientific uncertainty involved in estimating new technologies, such as food irradiation, warrants letting the public, not the experts, decide;
- that the FDA's labeling regulations are seriously flawed;
- that food irradiation is directly linked to DOE's need to dispose of nuclear waste from weapons production;
- that food irradiation will further concentrate the food market to the economic detriment of the consumer;
- that the FDA decision process was flawed and perhaps influenced by other governmental agencies and agenda.

The arguments presented above, and their variations, constitute the armamentarium of those opposed to food irradiation and subsets appear in almost any article on the subject, even those that attempt to present both sides of the issue or that lean in the pro direction. The general tone of the anti articles, Johnsrud, for example, is measured and somewhat cautious. The terms 'may' and 'could' appear frequently.

Of course, not all opposition arguments are so mellow. Gibbs, the founder of 'Medical Students Against Food Irradiation', writing in the

Progressive, describes the food irradiation process and then prefaces the rest of the article with: 'This is not the beginning of a science-fiction horror story.'[28] One periodical tells us that: 'A person exposed to 100 kilorads of radiation would be dead in less time than it takes to eat an apple. Yet the next apple you eat may have been irradiated with the same 100 kilorads.'[29] In general, popular health, exercise, organic food magazines, and those periodicals characterized by a strong anti-establishment orientation, mirror the sentiments expressed in the *Progressive* and *Science Digest* articles. Even ordinarily reserved periodicals, such as *U.S. News and World Report*, present the issue in terms that are clearly not neutral. A 1986 article talks about food irradiation as 'shrouded in fear and scientific unknowns'. It goes on to point out that the FDA rule making on irradiation had drawn more mail from concerned consumers than any other modern issue.[30] However, the Brand report pointed out that the comments received by the FDA were the result of a highly organized letter writing campaign that cannot be taken as an accurate reflection of public opinion.

Even articles that attempt to present both sides of the issue, many of which appear in major newspapers, tend to give more space to opposition arguments and, often, present highly inflammatory quotes from opponents. For example, *The Atlanta Constitution* quotes Denis Mosgofian, director of the National Coalition to Stop Food Irradiation, as saying: 'Food irradiation is a disposal plan disguised as a food treatment plan.'[31] This statement was in reference to the Department of Energy's plan to build demonstrator food irradiation plants and to use byproducts of weapons production as the energy source. It would be inaccurate to depict these articles as anti-food irradiation. Opposition arguments, and opposition spokespersons, tend to be more attention-getting. Their arguments are more vivid, they target popular villains such as big government and big business, and they claim to represent the people. Even growth in the number of groups opposing food irradiation warrants extensive news coverage.[32] Proponents arguments, by contrast, are quite bland. Opposition makes news, support tends not to.

Only one clearly supportive article published for the general reading public was located for the last 2 years. Paul SerVaas, writing in the *Saturday Evening Post*, argues the need for food irradiation as a post-harvest disinfestation and preservation alternative to chemicals and raises the possibility that the United States will eventually be at a competitive disadvantage abroad with regard to irradiated food. He ends his article with the following statement:

'The reality is we will never have, and have never had, absolute guarantees about the food we eat. Food, or the lack of it, can already kill us or make us sick. All of us thus must decide if the devil we know is better than the devil we don't.'[33]

It is difficult to assess the overall impact of this literature on public opinion. Those who are ecologically minded or involved in environmental causes are more likely to be opposed to food irradiation and more likely to read publications supporting their opposition. As Bruhn *et al.* point out, these people are unlikely to change their attitudes in response to any information.

The focus again shifts to the large number of undecideds: what are they likely to be exposed to and what might their reaction be. The average person is likely to get their information on new technologies from newspapers and television. Because the most attention-getting information in the news reports from these media is opposition arguments, it is reasonable to assume that anti arguments will have a disproportionate impact in shaping attitudes. However, survey data trends do not indicate large public opinion shifts in the anti direction. This absence of a noteworthy effect is probably due to a continued lack of public attention to this issue rather than a lack of impact of the media.

In summary, while articles on food irradiation tend to exhibit the full range of information and opinion, opposition arguments tend to dominate by virtue of their emotional impact and the disproportionate attention granted them. It appears very difficult for a consumer of the print media to come away with a very positive image of food irradiation. It is easy, however, to get the impression from newspaper reports that the general public is opposed to this technology. Although the public opinion polls do not indicate noteworthy shifts in opinion to the anti side, it is probably safe to agree with the Brand Group's conclusion that, on this issue, it is easier to shift people in the anti than in the pro direction.

The crucial issue in consumer acceptance may not be what the uncommitted public actually thinks about irradiated food, but what major decision makers in industry and government believe public opinion to be. In the absence of an intensely committed pro-food-irradiation constituency to fight for expanded use, the anti-irradiation groups may be able to convince decision makers that they are the true representatives of the average consumer. The perception of public opinion, by elite decision makers, will guide their policy-making and eventually determine the fate of food irradiation.

The final section of this paper takes the position that it is easier for those opposed to food irradiation to convince elite decision makers that the public is opposed to food irradiation than it is for proponents to convince them that the public is not opposed.

4 FACTORS INFLUENCING DECISION MAKERS

First, the food irradiation issue is burdened by its links to a technology that has risks that arouse dread, that are involuntary, uncertain, and long-term. Those who examine how ordinary people define risk argue that technologies with these kinds of outcomes generate disproportionate fear and hostility.[34] The ability of opponents to raise the possibility that food irradiation may depend on radioactive wastes from weapons production is a substantial hurdle to overcome. The inability of the federal government to solve the radioactive waste disposal issue is widely known and has intensified the 'NIMBY' (not in my backyard) syndrome. If people are convinced that food irradiation is an alternative radioactive waste disposal technology, and opponents have explicitly so argued, it will be much more difficult to generate support.

Second, previously discussed research indicated the importance of trust in industry and government agencies in accounting for opposition to food irradiation. Given the burdens inherently carried by food irradiation it is absolutely essential that promoters of that technology enjoy impeccable performance records. Unfortunately, that is hardly the case. The Department of Energy, which has plans to set up food irradiation facilities, has recently been in the news for being less than forthcoming about public exposure to radiation during nuclear bomb tests and for mismanaging the Hanford, Washington nuclear facility site. Furthermore, one of the principle providers of data to the Food and Drug Administration for its ruling, 21 CFR part 179, which permits the irradiation of poultry, is a convicted felon who has been involved in mismanaging nuclear technology with resulting significant contamination to the environment.[35] A less than fully committed public cannot be reassured by these revelations. More importantly, politicians will find it very difficult to support a technology that appears to have so many social deficits.

Third, in any contentious issue, the persuasiveness of the arguments available to each side is crucial. The food irradiation issue suffers from an imbalance in the urgency of the messages presented by opponents and supporters. The key arguments available to proponents are the potential to ease world hunger, and the ability to provide a safer and higher quality product for some types of food. Americans, however, are generally not hungry and have easy access to relatively safe and good quality food. Opponents, on the other hand, raise the specter of cancer, nuclear weapons, greedy and irresponsible industrialists, and government regulators who are controlled by industry: all very powerful images that tie deeply into basic elements of American culture.

Finally, food irradiation is an issue that did not have to wait long for organized opposition to form. Well organized, dedicated, anti-nuclear groups were in place well before food irradiation became an issue. These groups did not have to worry about expending resources on mobilization, about accumulating the necessary knowledge, or about recruiting experts to their cause: these essentials were readily available. The National Coalition to Stop Food Irradiation simply emerged like a leaf on a stem and, as of March, 1987, represented 57 grass roots groups across the country. Other organizations, such as Food and Water, Inc, a non-profit public interest group based in Denville, New Jersey, and Floridians for Nuclear Free Food are dedicated to stopping food irradiation.[36]

The validity of the above arguments rests on evidence that opponents are carrying the day. The recent Food and Drug Administration permission to irradiate poultry products could be construed as evidence that proponents are winning the battle for the hearts and minds of US consumers. That, however, hardly appears to be the case. Anti-food irradiation bills have been introduced in at least 15 states. New Jersey banned irradiated foods for 2 years. Other states, including New York and Maine, have either imposed restrictions or are considering such restrictions on the sale of irradiated food.[37]

Equally telling, it has been virtually impossible to get major food producers behind irradiated food. Companies such as Quaker Oats, H.J. Heinz and Company, and Ralston-Purina have announced anti-irradiation policies. Campbell Soup Company dropped out of an industry coalition on irradiation because of fear that consumers interpreted their membership as a sign of support.[38] Even the University of California Berkeley School of Public Health *Wellness Letter* has called for further research on irradiated food because of the possibility it may carry carcinogenic substances and also cautions consumers to limit the amount of irradiated food they eat.[39] It is highly unlikely that food irradiation, regardless of FDA approval, can progress without this elite support.

5 SUMMARY

Survey data, and marketing studies, do not indicate 'overwhelming' opposition to food irradiation in the United States. Depending on how middle category responses are interpreted, US consumers are either moderately accepting of the irradiated food concept or quite uncertain. In either case, it appears that majorities or near majorities are willing to try irradiated food. On the other hand, there is substantial concern and doubt expressed as well. The fact that such large percentages cluster in middle categories and express substantial doubt indicates that few people have committed attitudes on this issue. Furthermore, consumers tend to assume that other people are more opposed than they are. Both of these findings indicate that people are open to information on this issue and perhaps easier to move in the anti than in the pro direction. It is highly likely that if irradiated food were free from organized criticism some types of foods would find market niches.

Studies that have examined the impact of information on attitudes have suffered from problems such as very limited samples and very cursory information. Their findings indicate that change in response to information is difficult to predict: some become more positive and some become more negative while the majority remains uncertain. There is some indication that information about a history of safe use by special populations may be more effective in molding supportive opinions than information about the process and its outcomes.

Assuming that information is important, the type of information easily available to the general public was canvassed. Organic food, health and sport, and counter-culture magazines and news letters are very hostile to the idea of food irradiation. However, its readers are likely to already be opposed so it is unlikely that these news sources will have much impact on the issue. While both pro and con issues are covered by popular newspapers and magazines, opposition arguments tend to be covered more extensively and in greater detail. Opposition arguments are inherently more dramatic and 'newsy'. Within the last 2 years newspapers have paid particular attention to the victories of the anti food irradiation groups. This, combined with the tendency to view others as opposed to food irradiation, may create a powerful normative impact. It may create the impression that knowledgeable elites, and caring people in general, oppose this technology.

Perhaps the most telling facts are the victories achieved by anti food irradiation groups. These groups emerged quickly out of existing anti-nuclear groups and had a ready cadre of expertise and experience to lead them. Anti food irradiation legislation is growing across the country and major food companies are taking anti stances. This sends a powerful message to consumers: if food producers and politicians are against food irradiation, there must be major problems.

In the end, this issue will be decided in legislative chambers and the board rooms of major food producers, not in sterile surveys and decisions by the Food and Drug Administration. Proponents of food irradiation have five fundamental problems: establishing a real need for the process; dissociating food irradiation from the Department of Energy's waste disposal dilemma; establishing safety more convincingly; producing companies that can inspire trust; and, convincing major decision makers that the public is not adamantly opposed. Unless these problems can be addressed successfully, the average consumer may never get the opportunity to make a choice between food that is, or is not, irradiated.

REFERENCES

1. Slovic, P., Fishhoff, B. and Lichtenstein, S., Rating the risks, *Environment,* 1979, **21**, 14–39.
2. Freudenburg, W.R., Waste not: the special impacts of nuclear waste facilities. In: *Waste Isolation in the U.S.*, R. G. Post (ed.), 1985.
3. Webb, T., Lang, T. and Tucker, K., In: *Food Irradiation: Who Wants It?* Thorsons Publishers, Inc., Vermont, 1987.
4. Brand Group. In: *Irradiated Seafood Products: A Position Paper for the Seafood Industry,* Chicago, Illinois, January, 1986.
5. Bullard, R.D. and Wright, B.H., The politics of pollution: implications for the Black community, *Phylon,* **47**, 71–78.
6. Ajzen, I. and Fishbein, M. In: *Understanding Attitudes and Predicting Social Behavior,* Prentice-Hall, Inc., Englewood Cliffs, New Jersey, 1980.
7. Bruhn, C.M. and Schutz, H.G., Consumer awareness and outlook for acceptance of food irradiation, *Food Technol.,* July, 1989, 97.
8. Gidwani, B., Isomedix Inc., Kidder, Peabody and Co. Company Analysis, November 2, 1984.
9. Brand Group, In: *Irradiated Seafood Products: A Position Paper for the Seafood Industry,* Chicago, Illinois, January, 1986, p. 24.
10. Schutz, H.G., Bruhn, C.M. and Diaz-Knauf, K.V., Consumer attitude toward irradiated foods: effects of labeling and benefit information, *Food Technol.,* October, 1989, 80–86.
11. Bord, R.J. and O'Connor, R.E., Who wants irradiated food? Untangling complex public opinion, *Food Technol.,* October, 1989, 87–90.
12. Gidwani, B., Isomedix Inc., Kidder, Peabody and Co. Company Analysis, November 2, 1984; Weise Research Associates, Consumer reaction to the irradiation concept. In: *Report prepared for the National Pork Producers Council and the U.S. Department of Energy,* Omaha, Nebraska, 1984, p. 24.

13. Opinion Research Corporation. In: *Trends: Consumer Attitudes and the Supermarket*, Food Marketing Institute, Washington, D.C., 1988.
14. Fazio, R.H., How do attitudes guide behavior? In: *The Handbook of Motivation and Cognition*, R.M. Sorrention and E.T. Higgens (eds.), Guilford Press, New York, 1986, p. 289.
15. Weise Research Associates, Consumer reaction to the irradiation concept. In: *Report prepared for the National Pork Producers Council and the U.S. Department of Energy*, Omaha, Nebraska, 1984, p. 30.
16. The Polling Report. *Thinking Green*. May 7, 1990.
17. Opinion Research Corporation. In: *Trends: Consumer Attitudes and the Supermarket*, Food Marketing Institute, Washington, D.C., 1990, 58.
18. Good Housekeeping. In: *Women's Attitudes Toward New Food Technologies*. Good Housekeeping Institute for Consumer Research Department Report, February, 1985, 26 pp.
19. Johnsrud, J.H., Food irradiation: its environmental threat, its toxic connection, *The Workbook*, 1988, **13**, 47–58.
20. Bruhn, C.M. and Schutz, H.G., Consumer awareness and outlook for acceptance of food irradiation, *Food Technol.*, July, 1989, 97.
21. Bruhn, C.M., Schutz, H.G. and Sommer R., Attitude change toward food irradiation among conventional and alternative consumers, *Food Technol.*, January, 1986, 86–91; Bruhn, C.M., Sommer, R. and Schutz, H.G., Effect of an educational pamphlet and posters on attitude toward food irradiation, *J. Ind. Irrad. Technol.*, 1986, **4**, 1–20; Wiese Research Associates, Consumer reaction to the irradiation concept. In: *Report prepared for the National Pork Producers Council and the U.S. Department of Energy*, Omaha, Nebraska, 1984, p. 30; Bord, R.J. and O'Connor, R.E., Who wants irradiated food? Untangling complex public opinion, *Food Technol.*, October, 1989, 87–90.
22. Bord, R.J. and O'Connor, R.E., Risk communication, knowledge, and attitudes: explaining reactions to a technology perceived as risky, *Risk Analysis*, 1991, **10**, 499–506.
23. Bord, R.J., Epp, D.I. and O'Connor, R.E. In: *Achieving Greater Consistency between Subjective and Objective Risks*. Final Report to the U.S. Environmental Protection Agency. 110 pp.
24. McStay, J.R. and Dunlap, R.H., Male-female differences in concern for environmental quality, *Int. J. Women's Stud.*, 1983, **6**, 291–301.
25. Mark Clements Research, Inc. What America eats: shopping, preparation, eating, nutrition, *Parade Magazine*, April 5, 1988.
26. Bruhn, C.M. and Noell, J.W., Consumer in-store response to irradiated papayas, *Food Technol.*, September, 1987, 83–85.
27. Diehl, J.F. In: *Safety of Irradiated Foods*. Marcel Dekker, Inc., New York and Basel, 1990, p. 291.
28. Gibbs, G., Zap, crackle, pop, *Progressive*, September, 1987, 22–24.
29. Rosenberg, B., A diner's guide to irradiation, *Science Digest*, 1986, **94**, 94.
30. Dworkin, P., Irradiated food: is it safe? *U.S. News and World Report*, 1986, **101**, 58.
31. Sugarman, C., Food irradiation is still a subject of much debate, *The Atlanta Constitution*, Wednesday/Thursday, November 4–5, 1987, Section W.

32. Halverson, G., Food irradiation process opposed, *The Christian Science Monitor,* Tuesday, December 19, 1989, p8.
33. SerVaas, P., The promise of irradiated foods, *The Saturday Evening Post,* May/June, pp. 10 and 108–109, 1988.
34. Slovic, P., Perception of risks, *Science,* 1987, **236**, 280–286.
35. Colby, M. In: Comments to the Food and Drug Administration (FDA) in Regards to: Irradiation in the Production, Processing, and Handling of Food, Food and Water, Inc., New York, June 1, 1990, 7pp.
36. Glickman, P., Battle rages over the future of food irradiation, *The Chrisian Science Monitor,* Tuesday, March 24, 1987, pp. 5 and 6.
37. Sugarman, C., Food irradiation is still a subject of much debate, *The Atlanta Constitution,* Wednesday/Thursday, November 4–5, 1987, Section W, p. 8.
38. Halverson, G., Food irradiation process opposed, *The Christian Science Monitor,* Tuesday, December 19, 1989, p. 8.
39. Puzo, D.P., Questions over irradiation safety, *Los Angeles Times,* February 5, 1987, p. 22.

Chapter 4

CURRENT STATUS OF FOOD IRRADIATION IN EUROPE

D.A.E. EHLERMANN

Institute of Food Engineering, Federal Research Centre for Nutrition, Engesserstrasse 20, D-7500 Karlsruhe, Germany

1 INTRODUCTION

Since the discovery of radiation (by Röntgen in 1895) and of natural radioactivity (by Becquerel in 1896) research into nature and effects of ionizing radiation has had a long and rich tradition in Europe. The bactericidal effects of radiation were discovered early; however suitable and intense sources of ionizing radiation for any practical application on a larger scale were not available and even not imaginable. Despite such obstacles the first of all patents was granted in 1906 in Britain.[1] It was based essentially on the radiation emitted from radon which at that time was called 'radium emanation'. The patented process cycles the radioactive gas over a thin layer of powdered products like flour or semolina; the goods are steadily moved through the chamber and during that period exposed to the short-range radiation.

When more intense radiation sources became available, other application were studied in France and Sweden. However, the scale of such studies was still far from any commercial size application. Beginning with the 1950s, X-ray machines and van de Graaf-generators became powerful radiation sources for industrial and commercial applications. The first of all commercial scale applications was radiation processing of spices in Germany in 1958.[2] However, after a change of the food law it had to be abandoned. At the same time, nuclear power reactors were introduced which could be used to breed radioactive Cobalt-60 which is now a common source material for many radiation processing facilities.

The powerful US research program 'Atoms for Peace' also stimulated similar research programs in Europe. Many European countries inaugurated research institutes devoted to radiation applications. The worldwide coordinated research culminated in the finding that radiation processing up to an overall average dose of 10 kGy is a safe process.[3] Consequently, international standards were developed.[4] However, these rules were not transferred to national laws.

2 LEGISLATION

When radiation processing of food became practically available the legislative systems were not prepared for any provision. Most countries introduced some provisions from time to time.[5] The solutions of the legislative problems ranged from a strict and total ban (example, Federal Republic of Germany) over new laws providing an easy system of rules and clearances (example, the former German Democratic Republic) to a casual granting of several clearances through direct ministerial orders (example USSR). As a consequence, free trade in irradiated foods is not yet possible even between member countries of the European Community. Even more paradoxical, an individual country may have issued a clearance for a certain food item and radiation treatment, but within that country a suitable irradiation facility is not available and any import of irradiated food is strictly prohibited.

Within the European Community a number of national clearances have been granted by individual countries (Table 1). The situation is not very clear for member countries of the former Council of Mutual Economic Assistance. It may be reconstructed from information available elsewhertr.[5,6] It is only within the European Community that in a

TABLE 1
APPROVED APPLICATIONS OF FOOD IRRADIATION IN MEMBER COUNTRIES OF THE
EUROPEAN COMMUNITY

Country	Items
Belgium	Potato, onion, garlic, strawberry, spices, gums, dehydrated vegetables
Denmark	Potato, spices
France	Potato, onion, garlic, mechanically deboned poultry meat, gums, dehydrated vegetables, cereals
Italy	Potato, onion, garlic, pepper
Netherlands	Potato, onion, rice, fishfillet, shrimps, chicken, frog legs, mushrooms, spices, dehydrated vegetables, egg powder

common effort a harmonization of the food laws of member countries is envisaged for 1993.[7] The list of individual foods or groups of foodstuffs is rather comprehensive (Table 2) and provides also for imports from tropical and developing countries. However, this list is still under discussion in the member countries, in the Council of the European Community and in the European Parliament.

TABLE 2

FOODSTUFFS TO BE AUTHORIZED WITHIN THE EUROPEAN COMMUNITY[7] FOR AN IRRADIATION TREATMENT AND MAXIMUM RADIATION DOSES

Foodstuffs	Maximum average radiation dose (kGy)
1.* Strawberries, papayas, mangoes	2
2. Dried fruits	1
3. Pulses (legumes)	1
4. Dehydrated vegetables	10
5. Cereal flakes	1
6. Bulbs and tubers	0·2
7.* Aromatic herbs, spices and vegetable seasonings	10
8. Shrimps and prawns	3
9. Poultry meat	7
10.* Frog legs	5
11. Arabic gum	10

*Crossed out as of 17 November 1989.

The proposed rules of the European Community[7] demand also a strict labelling of any irradiated foods and of any irradiated ingredients. It is still under discussion whether minor ingredients will not need a labelling as irradiated. Furthermore, these rules demand a special licensing of irradiation facilities for food processing which will be in accordance with the provisions of Codex Alimentarius.[4] At the international level a register of licensed food irradiation facilities is elaborated by the International Consultative Group on Food Irradiation (ICGFI) and will be maintained by the Joint FAO/IAEA Division of Radiation and Isotope Applications, Vienna serving as secretariat to ICGFI. Reference to such national or internationally agreed registers could help to facilitate international trade and to ensure national authorities of a correct radiation treatment.

3 COMMERCIALIZATION

It is rather difficult to collect representative and reliable data on the commercial utilization of radiation processing of food. Established facilities for radiation processing hesitate to publish figures on the amount treated as these are considered to be confidential to their customers. However, from figures available elsewhere and from direct inquiries the best available data show considerable extent of commercial utilization.[8,9] The tendency is considered increasing in total amount and in number of individual items treated. In most European countries only pilot-scale or research irradiation facilities are available and the amount of irradiated products may serve for test marketing only (Table 3). With spices, however, the total amount treated may serve the purpose to provide raw materials with sufficiently low microbial counts for critical applications in the food industry.

Within the European Community food irradiation at a commercial scale can be observed in the Netherlands and in France. In Eastern Europe, Hungary and USSR seem to have commercial scale installations. The facilities in the former GDR are decommissioned today. However, it is difficult to reliably judge extent and progress. Some details are presented below (cf. Table 3).

3.1 Netherlands

A large commercial irradiator using Cobalt-60 as a source is situated at Ede.[10] It processes mainly medical disposables. However, the irradiator is designed flexibly and can handle products of many sizes and densities. The amount of foodstuffs processed is not large with regard to its capacity. The irradiated food is mainly used by the food manufacturing industry and, consequently, is not directly available to the consumer. There had been some test marketing of mushrooms labelled as irradiated. However, opposed in the retail shop to mushrooms labelled as fresh, the consumers got the wrong view that irradiated mushrooms are not fresh and the test marketing failed. Already very early, Dutch authorities granted clearances for test marketing of individual food items in limited amounts.[11] Radiation processing was done in a pilot plant. When the commercial irradiator became available, unconditional clearances were granted and radiation processing was transferred to the commercial facility.

3.2 France

A mobile irradiator was available in France until mid-1970s. It contained initially 150 kCi of Cesium-137 providing a maximum throughput of

TABLE 3
COMMERCIALIZATION OF FOOD IRRADIATION IN EUROPE[8,9]

Country	City	Item (capacity)[a]	Starting date[a]
Belgium	Fleurus	Spices (8 000 t/y) Dehydr. vegetables (700 t/y) Deep-frozen foods (2 000 t/y)	1981 (1)
Finland	Ilomantsi	Spices (n.a.)	1986 (1)
France	Lyon	Spices (2 500 t/y)	1982 (1)
	Orsay	Spices and poultry (500 t/y)	1986 (1)
	Berric	Mechanically deboned Poultry meat (2 000 t/y)	1988 (2)
	Nice	Dehydr. vegetables (200 t/y)	1988 (1)
German Dem. Rep.[b]	Weideroda	Spices (600 t/y) garlic (4 t/y)	1983 (1)
	Spickendorf	Onions (5 000 t/y)	1986 (2)
Hungary	Budapest	Spices (400 t/y)	1982 (2)
Netherlands	Ede	Frozen products, poultry, spices, dehydr. vegetables, rice, egg powder (n.a.)	1978 (2) (3)
Norway	Kjeller	Spices (500 t/y)	1982 (1)
USSR	Odessa	Grains (400 000 t/y)	1983 (2)
Yugoslavia	Zagreb	Black pepper (n.a.)	1985 (1)
	Belgrade	Spices (100 t/y)	1987 (1)

[a]Explanations/remarks: n.a., capacity/utilization data not available; t/y, tons per year; (1) pilot or research installation; (2) large scale or commercial installation; (3) estimated total amount processed 18 000 t/y in 1988.
[b]Decommissioned since 1990.

about 600 kg kGy/h. The facility was used for demonstration purposes (Action IRAD of the Bureau EURISOTOP) and served also for the production of limited quantities of radiation processed foods to be used in pilot studies with the food industry or to be test-marketed. From the positive results a plan was derived very early to cover France in a network of regional irradiation centers.[12] Also test-marketing of irradiated strawberries[13] was successful and the plans for regional irradiators became reality.[14] Most recently, an irradiation facility utilizing an electron linear accelerator came into operation at Berric (Bretagne).[15] It is devoted to processing deep-frozen meat from mechanically deboned poultry. The dimensions of the meat-blocks were chosen in order to fit the penetration depth of the 7 MeV-electrons. A Cobalt-60 large scale facility commenced processing at Marseille in 1989.

3.3 The former German Democratic Republic
The national legislation had provided for radiation processing of food since 1984; however, neither practical utilization (cf. Table 3) nor the availability of large scale irradiators show commercial scale applications. For pilot scale studies on onions a special facility was set up, utilizing 30 kCi of Cobalt-60 and a turn-table for the conveying of the bulbs.[16] Successful sprout inhibition and a quality acceptable to food manufacturing industry is reported. The irradiation facilities were decommissioned in 1990 and after accession of GDR to the Federal Republic of Germany the ban on food irradiation is also effective there.

3.4 USSR
The grain irradiator at the harbour of Odessa uses two lines of electron accelerators (1.4 MeV).[17] The theoretical throughput is 2 times 200 kg/h at a dose of 200 Gy. As the facility is only used when infestation of the grain is detected, the average amount treated per year is about 400 000 tonnes only.[18] The grain is used for milling and flour production; labelling of the product on retail level is not provided.

4 CONSUMER ACCEPTANCE

Growing concern of the public about environmental pollution and contamination of food also affected the acceptance of radiation processing of food. The International Organisation of Consumer Unions (IOCU) at its 1986 meeting called for a moratorium before any commercial scale application should be allowed. Also on a national level, consumer organisations expressed concern about the safety of the process.[19] Strong opposition to radiation processing of food is observed in Great Britain and the Federal Republic of Germany. As a consequence the European Parliament also called for a moratorium in order to allow comprehensive report on a number of raised questions.[20] As the British Government intended to authorize some applications a national expert committee was set up to re-evaluate all available data on safety and feasibility of food irradiation.[21] Clearances for seven food items were granted in UK in January 1991.

It is also of great concern to consumers that irradiated foods are not

easily identified and that a solely administrative control of correct processing and of adherence to labelling prescriptions may not be sufficient
to ensure the consumer a free choice between irradiated and non-
irradiated products on the market. The possibilities of identification
methods were reviewed recently.[22] Only in a few cases, especially some
spices, reliable methods for official inspections are available. This fact is
one reason for the delay in clearances in Great Britain. It is also considered one of the main obstacles before any broader introduction of food
irradiation and international trade in irradiated food.[23] For this reason,
the European Parliament required, that reliable identification shall be
established before any clearance on food irradiation is granted.[20]

Usually, consumer organisations express their concern about food irradiation and not too much information on reaction of individuals is
available. Successful test marketing[13] cannot answer such questions.
Several studies into consumer attitudes are available from Canada and
the United States. In Europe only one study from the Netherlands is
available.[24] Not surprisingly, this study showed that consumers' knowledge of facts on food irradiation is rather low. When highly perishable
products or the use of chemical preservatives are concerned, a decision
in favour of food irradiation may be understood as a choice of the lesser
of two evils. This is considered a negative choice and governments are
called to take such negative attitudes of the consumers into account
before they grant any licence on food irradiation.

5 CONCLUSIONS

Food irradiation has not yet become a commercial process in Europe.
Concern of consumers about the safety and the need for radiation processing of food is the main obstacle for any broader authorization in
member countries of the European Community. A rule on the harmonization of food laws shall come into force in 1993 and shall provide only
for a small number of applications. France is considerably promoting
food irradiation by constructing and planning several regional irradiation centers.

REFERENCES

1. Appleby, J. and Banks, A.J., Improvements in or relating to the treatment
   of foodstuffs, more especially cereals and their products, *British Patent No.
   1609* (1905).

2. Maurer, K.F., Zur Keimfreimachung von Gewürzen, *Ernahrungswirtschaft*, 1958, no. 3, 45–47.

3. Anon., Wholesomeness of irradiated food. In: *Technical Report Series 659* WHO, Geneva, 1981.

4. Anon., In: *Codex Alimentarius Standard for Irradiated Foods and Recommended International Code of Practice for the Operation of Radiation Facilities Used for the Treatment of Foods,* FAO and WHO, Rome, 1984, CAC/VOL. XV-Ed.1.

5. Anon., In: *Legislation in the Field of Food Irradiation,* IAEA, Vienna, 1987, IAEA-TECDOC-422.

6. Anon., List of Clearances (as of 1988-03-22), *Food Irrad. Newsl.*, 1988, **12** no. 1 supplement.

7. Anon., In: *Proposal for a COUNCIL DIRECTIVE on the approximation of the laws of the Member States concerning foods and food ingredients treated with ionizing radiation,* Commission of the European Communities, Brussels, 1988, COM(88) 654 final SYN 169.

8. Anon., Commercialization of food irradiation, *Food Irrad. Newsl.*, 1986, **10** no. 2, 48–52.

9. Anon., Status of practical application of food irradiation (as of June 1988), *Food Irrad. Newsl.,* 1988, **12** no. 2, 42–43.

10. Leemhorst, J.G., Industrial application of food irradiation. In: *Food Irradiation Now,* Martinus Nijhoff/Dr W. Junk Publishers, The Hague, 1982, 60–68.

11. Ulman, R.M., 14 years clearing irradiated foods in the Netherlands. In: *Food Irradiation Now,* Martinus Nijhoff/Dr W. Junk Publishers, The Hague, 1982, 77–83.

12. Laizier, J. and Vuillemey, R., Present status and prospects of food irradiation in France, *Food Irrad. Newsl.,* 1985, **9**, no. 3, 43–50.

13. Laizier, J., Test market of irradiated strawberries in France, *Food Irrad. Newsl.,* 1987, **11**, no. 2, 45–46.

14. Laizier, J., Status report on food irradiation, *Food Irrad. Newsl.,* 1987, **11**, no. 2, 47–48.

15. Sadat, T. and Cuillandre, C., A linear accelerator in a chicken factory, *Food Irrad. Newsl.,* 1988, **12**, no. 1, 61–62.

16. Döllstädt, R., A new onion irradiator, *Food Irrad. Newsl.,* 1984, **8**, no. 2, 40–43.

17. Zakladnoj, G., Pertsovskij, E., Men'shenin, A. and Cherepkov, V., Radiatsionnij metod dezinsektsii zerna, *Mukomol'no- Elevatornaya i Kombikormovaya Prom-st',* 1981, no. 6, 29.

18. Anon., FAO/IAEA study tour on radiation disinfestation of grain, *Food Irrad. Newsl.,* 1987, **11**, no. 1, 19–25.

19. Anon., In: *Discussion Document on Irradiated Food* prepared for the 7th European Consumer Forum, Berlin, Bureau Europeen de Consommateurs, Brussels and Arbeitsgemeinschaft der Verbraucher e.V., Bonn, 1983, AgV/BEUC/127/82.

20. *Entschliessung zur Bestrahlung von Lebensmitteln (Resolution on Irradiation of Food),* European Parliament, 10 March 1987, Council of the European Community, Brussels, 5203/87.

21. Anon., In: *Report on the Safety and Wholesomeness of Irradiated Foods by the Advisory Committee on Irradiated and Novel Foods,* Ministry of Agriculture, Fisheries and Food, London, 1986.
22. Bogl, K.W., Regulla, D.F. and Suess, M.J. (eds.), In: *Health Impact, Identification, and Dosimetry of Irradiated Foods,* Institut für Strahlenhygiene des Bundesgesundheitsamtes, Neuherberg, 1988, ISH-Heft 125.
23. Anon., In: *Acceptance, Control of, and Trade in Irradiated Foods,* Proceedings of a Conference at Geneva, 1988, International Atomic Energy Agency, Vienna, 1989.
24. Feenstra, M., Schep, G.J. and Spijkerman-Van Zon, I., In: *Irradiation, a Long-life Method,* Institute for Consumer Research, Den Haag, 1988, Working Paper 7.

Chapter 5

CONSUMER ACCEPTANCE OF IRRADIATED FOODS

M.H. FEENSTRA & A.H. SCHOLTEN
SWOKA, Institute for Consumer Research, Kon. Emmakade 192–195, 2518 JP, The Hague, Netherlands

1 INTRODUCTION

Although the first experiments on food irradiation were carried out in 1916 in Sweden,[1] food irradiation is for consumers a relatively new technology. From the sixties food irradiation has been applied more and more, so that the consumer movement became alert on this technology. Since then a lot of controversies arise in the literature about wholesomeness, safety, effects, etc.[2,3] It looks much like the debate on pasteurization of milk at the turn of the century in the United States, among scientists, consumers and milk producers.[4]

Food irradiation is currently permitted on a small scale in about 30 countries[2]; in some countries or states food irradiation has been put under a ban (e.g. Australia, New Zealand, New Jersey)[5]. The World Health Organization (WHO) and the Food and Agriculture Organization of the United Nations (FAO) have, however, chosen food irradiation as a safe and sound method for preserving and improving the safety of food.

Reactions on the part of the consumer organizations of many countries are however not favourable for or are even opposed to food irradiation. But what do the consumers think about food irradiation? Do they accept it? Public acceptance, or consumer acceptance, is an important condition for the further application of food irradiation.

But food irradiation is not an unique issue concerning acceptance. Many new technologies give rise to problems with acceptance. And food irradiation is only one of the many questionable subjects arising from new technologies. For that reason, firstly the acceptance of new technologies in general from consumer point of view are described (section 2). This is important for understanding the issue of acceptance of

97

irradiated foods. Then we come in section 3 to food irradiation itself and the point of view of the consumer movement on food irradiation, as a factor of influence over the public opinion, is described. After that consumer acceptance on food irradiation is elaborated (section 4). And finally in the discussion (section 5) some explanations of and recommendations for a better public acceptance are given.

2 CONSUMER ACCEPTANCE RELATED TO TECHNOLOGICAL DEVELOPMENTS

Consumer acceptance will probably remain an important issue in the near future, since consumers will encounter more and more new technologies. It is not easy to determine whether consumers will accept the products of these new technologies or for what reasons products such as irradiated foods are accepted or rejected.

Acceptance has to do on the one hand with a positive attitude towards new technological developments and on the other hand with a negative attitude because of unknown side effects of new technologies and their applications. Both attitudes, the positive and the negative, arise often from different sources, inside as well as outside the consumer.

From research it can be concluded that a majority of the (Dutch) society has a positive attitude towards the meaning of technology for society.[6] A small majority (55%) has the opinion that todays problems will be solved by technology itself; besides this a majority (69%) also did agree with the hypothesis that 'technical experts take account of the real demands of people to too small a degree'. This was 1988 in the Netherlands. Most interesting for the subject of consumer acceptance, specifically consumer acceptance of food irradiation, are the European public's attitudes to scientific and technical developments. For general attitudes to technologies could influence attitudes to specific technologies and related subjects.

The European public's attitudes to scientific and technical developments were thoroughly investigated in 1977 and 1979.[7,8] Although the results are more then 10 years old, they are still in agreement with results of later years, in Europe as well as in the United States. This is shown by Daamen et al.[9] in an interesting comparison of these regions.

'The work carried out in 1977 clearly demonstrates that fundamentally there is no crisis of confidence in science amongst the people of Europe.

In the nine Community countries, the general public has a high regard for the contribution which science makes to human progress and expects still more of it in the future; at the same time it has an acute awareness of the risks involved in scientific and technical research'.[8] This sounds familiar with the results of the Dutch report[6] and in the United States.[9]

Since the report on attitudes to scientific and technical developments still holds true, it is useful to let the report speak for itself. The results have to be applied to the explanation of consumer acceptance of food irradiation.

'When faced with the problems of scientific and technical development, the general public in Europe is both modest and anxious to be involved. People are aware that they know too little about science and are not sufficiently in contact with the facts concerning scientific and technical development, but at the same time they express a widespread desire to be more involved in research policy. This is an European attitude that is found with slight differences of emphasis in all the peoples of the Community Member States.

This desire is rooted in the ambivalent nature of the confidence Europeans have in science, as revealed by the first survey conducted in October 1977 and fully confirmed by the latest poll: the image of a science that will be as beneficial in the future as it has been in the past is widely accompanied by anxiety about the growing risks that it may involve for society. This survey bears new witness to public anxiety as demonstrated in the general image of the way in which science is put into practice and of the relationship between the public and those who make decisions.

Not only does the European general public endorse the distinction between a science that is good in itself and the way it is put into practice, often questionable and problematic, but it is also widely convinced that some discoveries are put to use too quickly before a sufficient study has been made of their possible consequences. What is more, its confidence in the 'self-correcting' power of a science that is always capable of finding new inventions to counteract the harmful consequences of its applications appears fairly limited and also varies considerably from one country to another.

This uneasy confidence in scientific and technical development, accompanied by a tendency to demand more involvement, is fostered by very real and widely shared fears about the future of the world. On the four subjects of the despoiling of nature, the increase in unemployment as a consequence of automation of jobs, the more and more artificial things coming into the life we lead and the risk that the use of new medical or pharmaceutical discoveries may affect the human personality, 53–80% of the replies expressed personal concern.

One of the main lessons to be learnt from this survey is that the anxiety shown by Europeans about some consequences of scientific and technical development is neither undifferentiated nor blind to reasoning, especially to consideration of the risks that may sometimes be run by playing safe.

The replies obtained to the set of questions designed to discover whether the general public supports or opposes eight research areas and whether it believes the statements made about the issues at state and thinks these issues serious are absolutely clear: to support a research project, Europeans need to believe that the issue at state is real and that the matter is serious.

There are, however, revealing exceptions to this apparent logic which indicate the need for a closer investigation of the importance of the subjective perception of technological risk.

The most striking example is the European reaction to the development of nuclear power stations. The widespread belief that there is a real and serious risk of having to restrict electricity consumption if nuclear power is not developed does not mean that support for the project is equally widespread, far from it. Its supporters (44%) only just outnumber its opponents (36%) and the fear of nuclear power appears to be at least partly blind to the logic that the issue at stake is both real and serious.

Finally, support is expressed most readily in the context of research areas where the risks, rightly or wrongly, appear the most remote from the individual: there is little obvious risk for me as an individual in supporting research into organ transplants, new sources energy synthetic materials or even the increase in the number of observation satellites.

In contrast, the projects evoking the strongest opposition or rejection are those which conjure up the most direct risks to the individual: the possibility that his natural biochemical equilibrium will be harmed by eating synthetic food, infringement of his freedom and privacy by a single computerized information file, harm to his biological identity and that of his offspring by genetic experiments, danger to his life in the proximity of a nuclear power station.

Obviously this variability in European reactions as soon as precise aspects of scientific and technical development are mentioned complicates the job of those who are trying to develop a Community policy on information for the general public on this subject (i.e. scientific and technical development, MF and AS). However, they should also be encouraged by the many reasonable aspects of European public opinion that appear in this survey: widespread in the potential benefits of science, although not blind to the increasing risks that it may carry with it and the difficulties of putting it into practice; various fears for the future of the world, a temptation to reject automation and to dream of going back to nature, but also reactions that differ according to the type of research and a realization of the risk that may sometimes be run in playing safe; desire to be more closely involved in the thinking on research policy but also an awareness of not knowing enough about science'.[8]

This was and still is the opinion of the public. But not only the public shows these opinions. The parliaments of many countries have the same attitude. In order to exemplify this, it is necessary to make an excursion to the United States of America in the early 1970s.

In those days the parliament of the United States had the opinion that technologies have their own dynamics, i.e. that it was difficult to influence the direction of technology. The technologies developed from within and not out of a demand of society. This is called the technology push character of technologies versus the demand pull. The technology push character resulted in unforeseen impacts. So, at the beginning of the seventies an early warning system was introduced in the United States on behalf of the parliament. This early warning system had to give information on impacts of new technologies before decision making.

At the same time, the year 1972, the OTA, Office of Technology Assessment, was founded. This office carried out all kind of studies of technology assessment, e.g. impact studies. The attention of the TA-studies was not only focussed on the developments and applications of new technologies, but much more on the socio-economic, cultural and ethical aspects arising from new technologies.

Although a lot of studies have been performed, also in countries other than the United States, an early warning system or TA-studies were sometimes thought to be limited. The parliament, for instance, was not satisfied with only forecasts and risk assessment studies, they also wanted to have influence on new technologies at an early stage of development. The concept of influencing new technologies at an early stage of product development is called Constructive Technology Assessment, or CTA. In the CTA-concept attention is focussed on the aspect 'constructive'. This means that the point of view of social parties is taken into account in the development and application of new technologies. Thus CTA is a concept of broadening the process of decision making with the result that applications on basis of new technologies are in agreement with wishes and needs of social parties.

In order to realize this CTA-concept, several activities could be, or should be, performed; the list of activities is shown in Table 1.[10]

A conclusion from what is mentioned above on attitudes and on Technology Assessment, is that policy on science and technology is focussed on industry and research institutes. Social parties, e.g. consumer organizations, are not taken into account; they have neither a direct nor an indirect influence on policy itself. This is strange, especially since consumers are confronted with the pros and cons of the result of these policies: products and services. Probably one has the opinion that producers know consumers wishes and translate them well to products. In a lot of cases, this is true. On the other hand failures of innovations, concern about technological applications and the catastrophy with environment show that products and services are not always produced which are in agreement with consumers wishes and social acceptability.

TABLE 1

ACTIVITIES OF CONSTRUCTIVE TECHNOLOGY ASSESSMENT

- Early warning: focussed on giving information as early as possible about problematic and undesirable technological developments,
- Contribution to the development of the long term government policy,
- Supporting the short and medium term government policy (exploration of alternatives, evaluation, legitimation),
- Searching for, formulating and developing socially desirable and useful technological applications,
- Broadening information and decision making process by supporting social parties in the formulation of their strategic position in regard to technological applications,
- Strenghtening the position of social actors in the decision making process,
- Giving attention to the public acceptance of technological developments,
- Stimulating responsibility of scientists for social impacts of technology.

For that reason it could be very important and even necessary to integrate the opinions of social parties, such as environmental and consumer movements, in the process of decision making. They ought to have the real opportunity to participate on equal terms at an early stage of the development and applications of new technologies. And this means that they can have influence on and can determine the direction of new technologies. The concept for that is Constructive Technology Assessment.

An important point of CTA concerning new technologies is, that the use of a technology can have different emotional and rational aspects in it, which makes the broadening of the process of decision making not easily realized. Food irradiation, for instance, is for one party a sound technique, while it isn't for the other party. The problem is that parts of society cannot simply declare each others arguments irrational. As a society we are obliged to develop and apply the most beneficial technologies. This means that 'beneficial' sometimes must be translated in terms of decreasing anxiety of the public, although the anxiety may not be a real one.

Acceptance or rejection is realized by arguments, opinions, knowledge, feelings, etc. Acceptance of consumers is realized on the one hand by their own perception of the subject and on the other hand by the opinions and perception of those who defend the rights of consumers. In most cases, especially at the beginning or the development of a new technology, it is the consumer movement which is the voice of the consumer.

It expresses their feelings, although it may not know exactly what the consumer feeling is. But this is already a CTA-activity, as is shown in Table 1: early warning (the activities described in Table 1 are primarily activities performed by the consumer movement as a representative of the consumer).

For this reason, speaking of consumer acceptance of irradiated food, it is not only important to describe the acceptance of consumers as a public of individuals, but it is likewise important to give an impression of the arguments for acceptance of the consumer movement. Since, eventually, these arguments could be the expression of consumers on the subject. So, both are highly relevant in discussing consumer acceptance of food irradiation.

3 THE CONSUMER MOVEMENT AND FOOD IRRADIATION

As stated before, the consumer movement can be seen as the representative of the individual consumer. In fact, in the process of decision making, in the activities arising from the CTA-concept, it is the consumer movement which acts and speaks for the individual consumer. It can do that in two ways: (1) by representing the consumers without exactly knowing the consumer perception on the subject and (2) by speaking for consumers with the knowledge of consumer opinions on the specific issue.

The second way is unfortunately often not the case. This has several reasons. One reason is, of course, that it is not possible to know precisely what each individual consumer thinks about a certain subject. If you have the money, however, you can try to investigate the opinions of a representative sample, but even then it is sometimes difficult to know opinions of consumers (see section 4) and to use them in your policy. As a representative, it is however important to try to find out consumer opinions, since then the base for action could be much broader. But, and then we come to one of the most important reasons why it is difficult for the consumer movement to follow the second way, they normally lack the money to have consumers' opinions investigated. Then it is hard to perform CTA-activities, since you need men and means for that.

Means are also needed to investigate the technical aspects. Also, the consumer movement needs this research in order to verify technical details or to do safety studies and risk assessments to counterbalance the information of industry, research institutes and government. Heretofore, in many countries comparative testing takes place, but no verification of research data from consumer point of view is done. Again, this is an important aspect in broadening the process of decision making.

An interesting aspect of the above is, that even if the consumer movement would know the opinions of consumers by way of research, it is sometimes too late for them to use that information. They have then already performed activities in an early phase of development in order to get social acceptable applications of a new technology, which is one of the CTA-activities; in this phase of development consumers probably have no opinion at all, since they are not aware of this development. So, if the consumer movement can not participate in an early stage of product-development, they and the consumers are confronted with the results of these developments. If they then do not agree with aspects of the development and take action, the industry and government are confronted with this resistance. Since industry or government, and often together, have already chosen their strategy, these parties and the consumer movement can become in a short time diametrically opposed to each other. And then much effort is needed for both parties to realize a dialogue and to change their point of view.

So, reality is that the consumer movement often cannot do anything but follow the first way: representing consumers without exactly knowing what their opinion is. And that means: take action, be against it or pro, ask questions, etc. etc. One can ask whether this is a correct handling, for instance for a subject such as food irradiation, but probably there is up to now no other way.

Generally speaking, the consumer organizations are almost worldwide against food irradiation. This could, of course, strongly influence consumer opinions, especially if consumer organizations give information on food irradiation, which they do.

3.1 The IOCU Resolution on Food Irradiation

In 1987, the General Assembly of IOCU, the International Organization of Consumers Unions, adopted a resolution on food irradiation; in the resolution a worldwide moratorium on food irradiation is demanded. For understanding the point of view of IOCU, the whole resolution is shown hereafter.

The IOCU General Assembly:

recognising that there are widespread consumer concerns about food irradiation with regard to need, safety, abuse, food quality, nutrition, hygiene, labelling, detection, control of facilities, enforcement and the impact on national and international economies: calls upon IOCU to develop a comprehensive policy paper on food irradiation;

urges IOCU, as a priority, to clarify with the World Health Organization, the International Atomic Energy Agency and other UN Agencies their role in promoting the acceptance of food irradiation by member governments;

demands a worldwide moratorium on the further use and development of food irradiation until there is a satisfactory resolution of issues of nutrition, safety, labelling and detection;

urges scientists, governments and the food industry, the world over, to research more desirable methods of food preservation; and

realising that food irradiation is permitted in some countries: calls upon IOCU to lobby in the meantime for clear and explicit labelling of irradiated products taking account of practical, and potential problems in date marking, inspection and certification.

Adopted by the 12th World Congress of the International Organization of Consumers Unions (IOCU), Madrid, Spain, 15–20 September, 1987

The activities of IOCU resulted, among others, in the IAEA/FAO/WHO/ITC-UNCTAD/GATT International Conference on the Acceptance, Control of and Trade in Irradiated Food in Geneva from December 12–16, 1988. Unfortunately, the different parties couldn't get agreement on promoting acceptance of irradiated food.

3.2 FIN

Then, at the end of the International Conference, the Food Irradiation Network (FIN) was founded. The FIN 'is an informal global coalition of like-minded groups and individuals opposed to food irradiation unless and until all outstanding issues are fully resolved'. Looking at the aims of the FIN, several arguments can be seen, which have to do with firstly a technology which has been introduced without participation of social parties in an early stage of development, and secondly CTA-activities as shown in Table 1.

The aims of FIN can give a clear insight in for them relevant aspects for reaching acceptability of irradiation; only a part of them is directly mentioned by individual consumers (section 4).

The FIN Aims

- call for a global moratorium on food irradiation unless and until all issues relating to its need, appropriateness, safety, detectability, labelling, wholesomeness, control and overall costs to society and the environment have been fully evaluated and subjected to public scrutiny.

- call upon the World Health Organization to reopen investigations into and thoroughly examine all aspects of safety associated with irradiated food.

- call upon the World Bank and other financial institutions/aid agencies not to include food irradiation in their programmes and projects.

- support the development of appropriate technologies which improve the quality, quantity and safety of the world's food supply.

- call for public participation in all decision making processes related to food irradiation.

In the meantime, many organizations, not only consumer organizations, have not accepted or even have banned food irradiation, a part of them shown in Table 2.

3.3 The PAN-declaration on Food Irradiation

To give an idea of the arguments used — of which safety, wholesomeness and the lack of a method of detection are important ones — a declaration on food irradiation is shown; the declaration has been adopted by some 45 participants from 26 countries at the third Pesticide

TABLE 2

(CONSUMER) ORGANIZATIONS WHICH DO NOT ACCEPT OR EVEN REJECT FOOD IRRADIATION (NOT LIMITATIVE)

Country	Consumer organization	Environmental organization	Parliament/ state	Health organization	Other type of organization
Australia	o				
European Community	o				
Germany			o		
India	o		o		
Japan	o		o	o	o
Malaysia	o				
Netherlands		o			
New Zealand			o		
Sweden	o		o		
Switzerland	o	o			
United Kingdom	o		o	o	o
USA		o			
International	o				o

Action Network (PAN) International meeting in Penang, Malaysia, January 24–28, 1989. The declaration looks as follows.

1. 'Believing that the potential benefits of food irradiation, particularly its need and appropriateness, have been generally overplayed, while its potential problems, particularly its safety and control, have been grossly underplayed;

2. Realising that controls on the use and abuse of food irradiation cannot be adequately enforced of monitored in the absence of reliable tests to detect irradiated foods;

3. Concerned that the effects of irradiation on pesticide residues in food have not been sufficiently studied;

4. Reiterating that the bulk of agricultural pesticides are applied pre-harvest — as such, a post-harvest technology like food irradiation would not reduce the total amount of agricultural pesticides manufactured, traded or used in any significant manner;

5. Noting that in many Third World countries, the improper transportation and storage of grains and other foods, coupled with the scourge of rodents, contribute substantially to post-harvest food losses — a situation that will not be solved by food irradiation;

6. Understanding that food irradiation will involve the use of other chemical additives to maintain or enhance the organoleptic properties of irradiated foods;

7. Recalling that food irradiation is being aggressively promoted, despite widespread opposition by consumer, environmental and other citizens' groups;

We the undersigned, representing some 300 citizens groups worldwide working on pesticide and food-related issues:

1. Declare that all attempts to foist food irradiation, particularly on Third World countries, without the active and informed participation of independent, public interest, citizens groups be thwarted;

2. Urge that WHO reopen and thoroughly investigate all aspects of safety associated with food irradiation;

3. Urge that FAO intensify research into alternative methods of food preservation and treatment;

4. Urge that the World Bank, other International Financial Institutions and Aid Agencies do not include food irradiation projects in

their foreign aid programmes, particularly in the case of Third World countries.

5. Call for a global moratorium on the further use and development of food irradiation, unless and until all issues relating to its need, appropriateness, safety, detectability, labelling, wholesomeness, control and overall costs to society and the environment have been fully evaluated and subjected to public scrutiny;

6. Support the activities of the Food Irradiation Network International (FINI) in fostering closer cooperation and coordination amongst citizens' groups working on the issue of food irradiation, using the resources that FINI will make available'.

Now we have seen how consumer organizations react upon food irradiation and have insight into their arguments, it is interesting to know if their reaction differs from that of consumers.

4 THE CONSUMERS AND FOOD IRRADIATION

Although there is a large non-acceptance of food irradiation among consumer organizations especially and an increasing non-acceptance by states and governments and other organizations (see Table 2), relatively little research has been done among individual consumers upon acceptance. A reason for that is, that the exploration of consumers' views or attitudes on a (relatively) new technique of preservation, i.e. irradiation, supposes some knowledge and ideas on this new technique. When consumers lack knowledge, there are no views to study. Furthermore, there is the problem that the information given to the interviewees will influence their answers. Also, what interviewees say about their future behaviour is not always what they really do.

Large surveys upon this subject have been carried out in the Netherlands, the United Kingdom and the United States of America. Besides these surveys, a large number of market trials of irradiated food products has been carried out. They will not be taken into account because of the design of the market trial. The design of a market trial gives quickly a bias with taste, colour, smell etc. so that it is completely unclear whether you are measuring acceptance of irradiated food products (see section 4.5). So, although market trials are performed much more than consumer surveys, market trials will not be described.

4.1 Awareness and Knowledge of Food Irradiation

In the United States consumer awareness of food irradiation increased from 23% in 1984 to 66% who heard of the process in 1986.[11-13] In 1989 only 32% of 195 women interviewed had heard anything about the issue prior to participation.[14] In Great Britain only 48% of 2000 respondents were aware of the term 'food irradiation'.[15] So, most consumers seem not to be familiar with the subject. This was so in the Netherlands in 1973[16] and 1982[17], it still seems to hold at this time. In 1973 in the Netherlands, 63% out of 670 housewives had never heard of food irradiation, and 18% knew the correct reason for food irradiation,[18] whereas in 1988, 32% of the respondents had no associations with the term food irradiation at all, 41% had correct associations and 56% have chosen the correct reason.[19] In the study carried out by Wiese[12] about 56% of the respondents were not able to give any specific concern or advantage; when they could specify an advantage most of them mentioned the elimination or reduction of chemicals, preservatives and additives or providing longer shelf life.

All existing studies report that a majority of the respondents have not heard of food irradiation, is the conclusion of Bord and O'Connor.[14] It seems that it is currently not a hot issue with consumers; this in contrast with what we have seen in section 3 on the consumer movement.

Some research into the knowledge of consumers on food irradiation has been carried out. When respondents (even those who had not heard of the subject) were asked to describe food irradiation in their own words, 35% of the sample could not give any answer.[15] The other answers were a mixture of true, false and debatable descriptions; the descriptions most often given were true. Over a third of the respondents did believe food irradiation was practiced in other countries than Great Britain, 22% believed it was not. Over 40% did not know whether food irradiation is practiced; this high percentage reflects the level of awareness of the subject is the conclusion. There is also a great uncertainty about the possible effects of irradiation.

As stated before, Feenstra et al.[19] showed that 41% of the associations with food irradiation were on the right line. The question for the main reason for using food irradiation, was correctly answered by 56% of the respondents: extending shelf life. When these 56% ($n = 937$) were asked about the availability of irradiated foodstuffs in Dutch shops, 62% couldn't give the right answer. The 356 persons who thought that irradiated foodstuffs were available in Dutch shops (the correct answer), were subsequently asked whether possibly irradiated products were avail-

TABLE 3

IDEA OF THE USE OF IRRADIATION IN FIVE PRODUCTS
AVAILABLE IN DUTCH SHOPS

Product	Use of irradiation		
	Yes	No	Don't know
Shrimps	**26**	12	62
Spices	**19**	28	55
Minced meat	31	**22**	47
Potatoes	26	**32**	42
Cheese	6	40	55

$N = 356$; per cent. Correct answers are printed in bold.
Source: Feenstra et al.[19]

able. The products included those which are likely to be irradiated (shrimps and spices) and products that are definitely not irradiated before being offered for sale (cheese, minced meat and potatoes). These questions about a specific product proved hard to answer (Table 3). An average of about 50% of those who were presented with the questions failed to reply, and the others did not by any means answer all questions correctly.[19]

In order to discover exactly how much the interviewees knew, the answers to all knowledge questions about food irradiation were combined into one knowledge scale. Then the respondents were grouped according to four levels of knowledge; see Table 4.

TABLE 4

RESPONDENTS GROUPED ACCORDING TO LEVEL OF
KNOWLEDGE

Knowledge level	% of correctly answered questions	% respondents
I	<25	35
II	25–50	23
III	50–75	31
IV	>75	11

$N = 1.685$. Source: Feenstra et al.[19]

Out of all respondents 588 persons (35%) were found to know little or nothing about irradiation; only a small number of them (11%) answered correctly more than 75% of the questions about irradiation. This is in agreement with the results of the Consumers' Association:[15] 11% of the respondents were classified as 'very knowledgeable'. Also Ford and Rennie[20] found that people are fairly ignorant of the irradiation process. As an example, 44% of the respondents did know that irradiation effects the vitamin content and 45% that irradiation destroys bacteria. Bord and O'Connor[14] tested the knowledge of respondents with a few questions. They concluded that about 35% choose the right answer, about 60% the don't know answer and 5% the wrong answer. People felt that they needed to know more about the process.[12]

It was found that the knowledge of irradiation depends on sex and the level of education: women are found to know less about food irradiation than men and the higher the education level, the higher the level of knowledge.[19,20] In the study carried out by the Consumers' Association[15] men, younger age groups (those under 45 years) and respondents in the higher social classes were more knowledgeable about food irradiation.

4.2 Concern on Food Irradiation

Although the majority of the people may not know the concept 'food irradiation', they may have associations to the word.[15,20] In the Netherlands the associations with 'bestralen ' (irradiate) were substantially more negative than with 'doorstralen' (transradiate).[17] 'Doorstralen' is now the commonly used name.

Associations with food irradiation did not change much over the years. Normally, the spontaneous reaction is a negative one: dislike, fear of unknown, fear of disease, aversion because of unnaturalness, cancer, X-rays and unfamiliarity[17] and aversion because of potential harm to people, safety and effectiveness of the process, side effects and long-term effects, taste and nutrition.[12] This holds also for respondents who are employed in the retail food business even after explanation of the process of irradiation; the initial negative reactions deal with: safety of the process, residual activity of the products, taste and appearance of the product.[21]

Directly related to associations are expectations and concern. Around half of the informants had a negative opinion about the likely effects of food irradiation.[15] Most people have concern about food irradiation, although the concern in the United States seems to have decreased.[11] Almost 25% of the respondents in the United States show major and 21%

a minor concern with regard to irradiation; according to Schutz *et al.*[11] this is in agreement with most results of the United States (from 27% in the final report of the Brand Group[13] to 42% in the Wiese report[12]), and also with results from the Netherlands. On the question 'Are you concerned about the use of irradiation to extend the shelf life of food?' 26% of the respondents answered that they were very and 24% that they were somewhat concerned.[22]

The concern on food irradiation depends on sex, age and education. In many literature is found that women and less educated respondents show a higher concern level.[11,12,23] This compares with results from the Consumers' Association:[15] men and those in the higher social classes express a more favourable opinion about the effects of food irradiation.

There is no agreement on age. In the United States it is shown that younger people are more concerned than the others,[11,12] while in Great Britain and the Netherlands the younger people were significantly less concerned.[15,23] Dutch respondents in the 10–29 years age group are on average less concerned about the topic of food irradiation than the elderly. This can be concluded because young people more frequently answered all the questions with 'no opinion'.

4.3 Food Irradiation and Other Methods of Food Preservation

In all literature but the report of the Consumers' Association[15] it is shown that consumers express a higher concern for chemical sprays and pesticide residues and for use of preservatives than for food irradiation.[11,14,19,22] The Consumers' Association[15] reports that approximately half of the sample preferred food preserved with chemical preservatives to irradiated food.

Wiese[12] found that 55% of the respondents had a major concern for chemical sprays and 43% for preservatives in comparison to 38% for food irradiation. Also Schutz *et al.*[11] found a high degree of concern for both pesticides and herbicides, and additives and preservatives: 85% of the respondents for the first group, 80% for the second one. Cramwinckel[22] investigated the relationship between concern about pesticides and preservatives on the one hand and food irradiation on the other. The results are shown in Table 5.

Feenstra *et al.*[19] investigated the views of respondents on three methods of preservation: food irradiation, sterilization and preservatives. The respondents were asked which effect these three methods of preservation have on a number of aspects of foodstuffs that are in general important to customers: appearance, storage life, naturalness, safety,

TABLE 5

CONCERN ABOUT PRESERVATIVES, PESTICIDES AND IRRADIATION (IN %)

	N	Very concerned	Somewhat concerned	Little concerned	No concern	No answer
The use of preservatives in food for longer shelf life?	990	8·0	26·2	41·6	20·7	3·5
The use of pesticides for the production of vegetables and fruit?	990	16·6	37·8	31·7	12·2	1·7
The use of irradiation for a longer shelf life of food?	987	26·2	24·3	22·9	21·4	5·2

The question asked: 'Are you personally concerned about ... .
Source: Cramwinckel[22]

TABLE 6

ASSESSMENT OF THE EFFECT OF THREE METHODS OF PRESERVATION ON SEVEN
FOODSTUFF ASPECTS

Aspect	Method of preservation		
	Sterilization	Preservatives	Irradiation
Safety	2·57	1·70	1·72
Vitamin content	1·77	1·98	2·25
Wholesomeness	2·47	1·68	1·85
Naturalness	2·21	1·69	2·10
Taste	1·66	1·93	2·41
Storage life	2·04	1·96	2·00
Appearance	1·71	2·04	2·25
Total assessment for preservation method	2·17	1·78	2·05

Source: Feenstra et al.[19]

vitamin content, taste, and wholesomeness. In Table 6 the answers are
worked into an assessment of the effect of each of the three methods of
preservation on the seven foodstuff aspects. The figures show the mean
rank order number per method of preservation for each of the seven
foodstuff aspects. A high figure represents a relatively good opinion, a
low figure a relatively bad opinion of the effect of the relevant method
on a foodstuff aspect.

As regards safety and wholesomeness sterilization is considered more
favourable compared to the use of preservatives and food irradiation.
According to the interviewees, foodstuffs to which preservatives are added are less natural than foodstuffs treated with the two other methods of
preservation.

A striking point is that food irradiation scores better than other methods of preservation as regards appearance, taste, and vitamin content of
treated products. Food irradiation never emerges as the least favourable
of the three methods, although as regards safety the respondents are as
unfavourable to irradiation as to the use of preservatives.[19]

As a conclusion, the use of preservatives is considered the most unfavourable method as regards the effects on foodstuffs. Food irradiation
and sterilization are considered more favourable, sterilization even more
than irradiation.[19]

4.4 Acceptance of Food Irradiation and Irradiated Food Products

The percentages of acceptance, or better rejection, differ largely between surveys. According to Bord and O'Connor[14] the Brand Group classified the respondents as follows: 5–10% rejectors, 55–65% undecideds and 25–30% acceptors. This does not agree with Feenstra et al.[19]. On the basis of the acceptance of irradiation of various products (see also below) they constructed a scale in order to show the degree of acceptance of food irradiation in general. Among those who gave their opinion on the acceptance of food irradiation are obviously more opponents than advocates, that is, 728 opponents (56%) against 435 advocates (34%); 10% remained neutral.[19]

The degree of acceptance of food irradiation was examined by asking about the degree of acceptance of a number of specific irradiated products. These products were: potatoes, raisins, chicken, shrimps, spices, potato crisps, and strawberries. The prevailing opinion proves to be that food irradiation is unacceptable for all seven products. When the degree of acceptance of irradiation for each product is compared to the total of the other products, respondents prove to have fewer objections to irradiation of particularly shrimps and chicken, whereas irradiation of strawberries is considered less acceptable.[19] The same trend has been observed by Schutz et al.[11] They found a higher interest in purchase of irradiated raw poultry which may contain *Salmonella* (59% for irradiated and 18% for non-irradiated) than in the purchase of irradiated soft fruit (43% for irradiated and 33% for non-irradiated).

Feenstra et al.[19] give the following explanation for the difference in acceptance. The fact that irradiation as a method of preservation for shrimps and chicken meets with comparatively fewer objections, is not surprising in view of the (recent) news of lack of hygiene in shrimp processing in the Netherlands and the risks of the presence of pathogens. As regards chicken similar notes of warning have been sounded from time to time. Moreover, decay in products such as shrimps and chicken is often not clearly visible. Consumers are comparatively less favourable to irradiation of strawberries, perhaps because strawberries are considered a truly fresh, natural product. For such a product the use of any method of preservation is thought undesirable; besides, decay is easily recognizable.

Not surprisingly, young Dutch respondents (under 22 years) are comparatively less unfavourable to irradiation. Unexpected, however, is that persons of 65 and above also have fewer objections.[19]

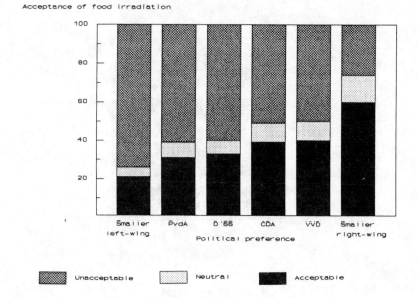

FIG. 1. Acceptance of food irradiation in relation to political preference. (Source: Feenstra et al.[19]).

Consumers' views on acceptance of food irradiation are clearly related to their political preferences (see Fig. 1): the more their sympathies are for the left-wing, the greater their objections. With the exception of supporters of the smaller right-wing parties, the majority of supporters of all political parties considers food irradiation unacceptable. The majority of the smaller right-wing parties endorses food irradiation.[19]

The degree of acceptance appears to be related to the knowledge of food irradiation, although not all agree with that. Feenstra et al.[19] show that, although irrespective of the level of knowledge there are more opponents than advocates, respondents knowing most about this method of preservation (level IV) have comparatively fewer objections (see Fig. 2).

Also Bord and O'Connor[14] found that those who gave more correct answers on the asked questions on food irradiation, are significantly more accepting it. They showed also that the information people have about irradiated food is a more predictive factor of attitude than other characteristics of the respondent. The same is shown by Feenstra et al.[19] Schutz et al.[11] described this in another way: they say that the safety of

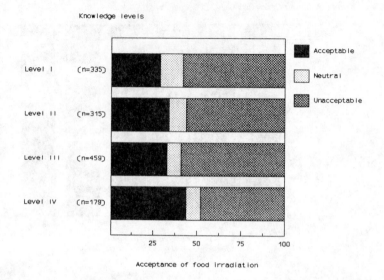

Fig. 2. Acceptance of food irradiation in relation to knowledge level. (Source: Feenstra et al.[19]).

the process rather than the food characteristics determines acceptance or hesitancy. In the study carried out by the Consumers' Association[15] the very knowledgeable seem to have a more negative opinion about the likely effects of food irradiation than some others and seem less willing to buy irradiated products.

When respondents know something about food irradiation, age, sex, education or food shopping frequency turn out to be irrelevant to the degree of acceptance of irradiation.[19] However, when consumers know little or nothing about food irradiation these characteristics turn out to be related to the acceptance of food irradiation. Women, the principal food shoppers for a household, appear to have comparatively more objections to irradiation than men. The higher the education level of the persons with little or no knowledge, the larger in proportion is the percentage that considers food irradiation unacceptable.

4.5 Willingness to Buy Irradiated Food Products

Above, something is said about acceptance of irradiation and irradiated food products. It is interesting to see, what people say they would do: are they willing to buy, or not? Unfortunately, in spite of much research,

even in spite of many market trials, this question is hard to answer unequivocally. Market trials seem to show a choice for irradiated food. This
is possible, but from the available literature it cannot be concluded for
certain. Even the often mentioned paper of Bruhn and Noell[24] leaves
questions, since in their experiment picked green, double-dipped papayas
were compared with picked ripe, irradiated papayas. One thing should be
clear: the results of the experiment are probably biased by the colour, the
smell and the taste of the papayas.

From literature it can be seen that the percentage of people who reject
irradiation is not in correspondence with the percentage who is willing
to buy products: it seems that people are more willing to buy than could
be expected on the basis of their non-acceptance.

Anyway, Schutz et al.[11] state that some researchers have found that
over 66% of the consumers would buy irradiated food, while others found
that about half of the consumers wanted more information on food irradiation before purchase. Only a quarter of the respondents in Great
Britain said they would buy irradiated food at the same price. The attitude to the purchase of irradiated food didn't primarily seem to be dependent on the type of food nor the benefits gained by irradiation, but
rather on the price.[15]

Bord and O'Connor[14] showed that a solid majority of the respondents
'will probably try' which according to them indicates a considerable lack
of intensity of feeling. After giving information about food irradiation
the number willing to try increased however significantly. These results
don't agree with results of Ford and Rennie.[20] They stated that there
was no clear evidence that those who had more knowledge on irradiation
were any more or less willing to buy irradiated food.

In general, Ford and Rennie[20] found a large number of people (70%)
who didn't want to buy food that had been deliberately treated with
radiation to extend shelf life; 12% said yes and 18% didn't know. In these
results it is clearly shown, and has been suggested above, that willingness to buy depends on much factors: information, description or definition, and labelling (see also hereafter). A description as 'deliberately
treated with irradiation to prolong its shelf life' influences attitude and
perception and thus willingness to buy immediately. Also labelling could
be such a factor. Schutz et al.[11] showed that about 45% would be (very)
likely to buy approved and labelled irradiated food in the market place;
still 34% of the respondents were uncertain. Also, Cramwinckel[22] found
that 58% of the people who consider buying irradiated products take
(very much) account of safety of the irradiated products, 57% of hygiene,

41% of unnaturalness and 63% of wholesomeness. These results seem in agreement with Ford and Rennie[20] who investigated the reasons for buying, for not buying and for uncertainty about buying. Reasons for buying irradiated foods ($n = 23$) were safe on health (35%), no strong feelings for or against (22%) or improves shelf life (17%); reasons for not buying ($n = 139$) are health risks and cancer (53%), fear and mistrust (27%), concern about nutrient value (26%) or insufficient knowledge (26%); reasons why one is uncertain about buying ($n = 36$) are insufficient knowledge (61%) and effect on food value (22%) or health risks (44%).

Males and older consumers have a greater willingness to buy irradiated products than female and young people.[11,15,20] The 'very knowledgeable' informants seem less willing to consider buying irradiated products than others.[15]

4.6 Impact of Information on Acceptance and Willingness to Buy

As mentioned before, the concern on irradiation is influenced by the way of explaining the concept or by choosing the word. The reaction to a specific word or concept can be more positive, as is shown for the word 'transradiate'[17] or for the definition of 'ionization':[12] in the Wiese report the percentages of respondents who have major concern about four initial descriptions of 'irradiation' (irradiation, gamma waves, ionization and combined) vary between 32% for ionization up to 42% for irradiation. Schutz et al.[11] found a significant influence of the label statement.

Information has a definite effect upon attitude, acceptance and willingness to buy irradiated food products. Most researchers state that the effect of information, especially on the benefits of irradiation, is positive except for those who are strongly opposed.[11,12,14,19-21,23,24,26]

The effect of information is not that all respondents get a positive attitude, it is more positive in general. Wiese[12] and also Cramwinckel[22] investigated the effect of presenting attributes of the process or a short or lengthy introduction. Wiese[12] showed that over 1/3 of the people who showed an initial major concern shifted to a minor concern; that from the people with an initial minor concern 25% changed to a major concern while the majority stayed minor concerned; the people who where initially undecided shifted for 1/3 to major concern and 6/10 to minor concern; the people who were initially without concern stayed in majority with no concern, although even 40% got at least a minor concern. Cramwinckel[22] finds a difference between concerned and unconcerned respondents. The non-concerned respondents have more faith in experts arguments in favour of food irradiation; the experts arguments against food irradiation are more agreed upon by the very concerned respondents.

The above is important for information or education programmes to consumers. Wiese says it this way: '... indications might imply that additional and more effective consumer information will be necessary to reduce the significant concern that exists presently within the population'.[12] Titlebaum et al.[21] have the same opinion and suggest that a straightforward and complete education program is necessary to promote consumer acceptance. They state that the consumer should have the possibility of being fully informed about the process and feeding experiments. Sapp and Harrod[26] go much further and concur with what is said in section 2 of this chapter. For them the adoption of an innovation is a social process and therefore they examined acceptance of food irradiation in group settings subject to the influence of social persuasion. They state as a result of their study, that consumer adoption of food irradiation cannot depend upon impulse of favourable information alone. Word-of-mouth discussion should therefore be taken into account. Another important instrument of information is labelling of irradiated food products.

4.7 Labelling of Irradiated Food Products

One of the consumer rights is the right to be informed. All the consumers claim this right as can be seen on many subjects. This is also the case for food irradiation.

American consumers, for instance, uniformly think the clear labelling of irradiated food to be necessary.[14] In addition to this, Ford and Rennie[20] state that, since people don't know exactly what labelling means and since people accept a familiar context unquestioningly, labelling is not enough: people must be made aware of what labelling means so that they are able to make an informed choice. Schutz et al.[11] and Wiese[12] showed that the description on the label influenced the consumer reaction.

Feenstra et al.[19] investigated the need for labelling a bit more extensively. A vast majority of respondents thinks food labelling is necessary, regardless of which method of preservation is used (see Fig. 3).

Those who believe irradiation should always or just in certain cases be labelled, also want a labelling statement indicating presence of irradiated ingredients, whether with regard to irradiated strawberries in jam (90%) or irradiated spices in tomato ketchup (85%).

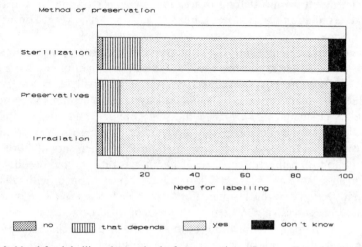

Fig. 3. Need for labelling the method of preservation. (Source: Feenstra et al.[19]).

The results of Feenstra et al.[19] compare with those of the Consumers' Association:[15] 86% of the sample think labelling is absolutely essential in case of irradiated ingredients, 90% in case of irradiated products.

When respondents are asked why they want products to be labelled as irradiated, 98% of the respondents reply; only 2% does not give a reason[19]. More than half of them (55%) say they simply want to know what has happened to the food; 19% of the consumers want to have this information on the label so they can choose between irradiated and non-irradiated products. Other respondents go even further: they, too, want to have the product labelled to be able to choose, that is, to avoid irradiated foodstuffs. Other reasons for wanting product labelling are: health risks (5%), irradiation consequences are not sufficiently well known (3%), certainty of long shelf life (1%).

As was to be expected, especially those who consider irradiation unacceptable want to be able to tell from the label whether a product was irradiated or not[19] (see Fig. 4).

The demand for food labelling is greater among women than among men; the demand increased with an increase in the shopping frequency and with an increase in general knowledge of foodstuff composition and manufacture. The frequency of being 'confronted' with food labelling (women are the principal shoppers) apparently increased the need for labelling. And more general knowledge of foodstuffs will partly result

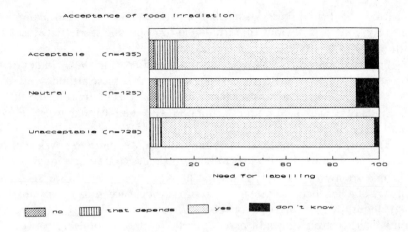

Fig. 4. The need for labelling irradiation in relation to the degree of acceptance of irradiation. (Source: Feenstra et al.[19]).

from reading labelling information. Very knowledgeable or higher educated consumers comparatively more often want to have the method of preservation labelled,[15,17] possibly because they are (able) to read this information sooner. A striking result is that young respondents (under 22 years) more often do not hold strong views on labelling. They answer more often that they don't know.[19]

5 DISCUSSION

It is shown clearly that the consumer organizations and other organizations are against food irradiation, while consumers itself don't know yet. It seems that at least half of the consumers are concerned about irradiation, although they have a much higher concern about chemicals in food. The concern depends on knowledge and information: the more information and knowledge, the less the concern. In spite of the large concern, 25–36% of the consumers will accept food irradiation. But if they will buy irradiated food products is not completely clear, although probably half of the consumers would try irradiated food products. The willingness to buy is, however, strongly influenced by the way irradiation is presented (e.g. 'deliberately treated with food irradiation') and by information about the process of irradiation. Also, concerning information, for everyone it is obvious that irradiated food products, irradiated as a whole or as a part, should be labelled clearly.

For what reason do consumers' attitudes differ from the attitudes of the consumer movement? Also, isn't it strange, that these attitudes differ so strongly? In order to answer these questions let us firstly compare the attitude of consumers on food irradiation with their attitude on science and technology.

The attitudes of the consumers on food irradiation agree well with what is said on public attitudes to scientific and technical developments. People are aware that they know too little about science. However, the image of a beneficial science in the future is widely accompanied by anxiety about the growing risks for society. The anxiety arises from the way in which science is put into practice and from the relationship between the public and those who make decisions. It is very clear that people want to get more involved. People are for instance widely convinced that some discoveries are put to use too quickly without knowing the possible consequences.

For the subject of food irradiation it is important to mention, that the anxiety of consumers about the consequences of scientific and technical development is neither undifferentiated nor blind to reasoning, especially regarding risks. The developments with the largest risks to the individual (for instance harming the natural biochemical equilibrium) evoke the strongest opposition or rejection.

The attitude of the consumer movement on food irradiation could be explained partly by the consumers' opinions and partly by the process of decision making, or CTA. Although about a third of the consumers accept food irradiation, still another third has a major concern; another third doesn't know what to think about food irradiation. Concern or anxiety, sometimes expressed in anxiety to get diseases like cancer, accompanies food irradiation, or new technologies. The consumer movement has to take care of the, right or wrong, feelings.

The problem with food irradiation (and a lot of other new technologies, e.g. biotechnology) is that on the one hand a lot of people have serious doubts about the technique, while on the other hand experts and a health organization such as the World Health Organization (WHO) say there are no risks. The consumer movement doesn't believe the reports of experts. According to the consumer movement the experts are all associated with the food irradiation industry. So the consumer movement put the safety of the process in doubt, although many consumers would accept it.

At this point it is interesting to question what should the consumer movement do if it knows for instance only half of the consumers' opinions? Should it say: alright, half of the consumers see no problems, so we don't see any problem either? Or should it say, no, since half of the consumers are not concerned, the other half is and that is reason enough to be careful? This dilemma is not easily answered. Short and long term effects have influence on the problem. Consumers in general only look at personal, short term effects, while the consumer movement should observe and take care of social, long term effects.

The dilemma is probably solved to a large degree in the second part of the explanation of the attitude of the consumer movement: the CTA-activities. The consumer movement was not involved in the development of the technology, they have no means to carry out the desired research, so they try to get influence in the current stage of technology. Since the consumer movement has not been involved in the early development and applications of food irradiation, they put in doubt the safety of the process and irradiated products, as stated before. Now they want to get involved in a phase of technology development in which products are coming into the market. Since their questions momentarily can only slow down market introduction, industry is not very enthusiastic about their wish of involvement. Then industry tries to push on public acceptance of food irradiation. The reaction of the consumer movement is then found in Table 1: firstly giving information about problematic and undesirable technological developments — in the case of food irradiation a way of late warning — and secondly searching for, formulating and developing socially desirable and useful technological applications, such as looking for more desirable and acceptable methods of preservation. Again, the consumer movement sees food irradiation momentarily not as socially desirable. Setting up research programmes in which both parties participate on equal terms can change the viewpoint of the consumer movement. If, then, the results of the research program are acceptable to both, a large step to a better understanding of each other point of view is taken. Also, this can result in an agreement on what to do with food irradiation: accept it worldwide for all products or for only a few or accept it only for special applications in some countries, or for some target groups, or don't accept it at all because of lack of safety and wholesomeness.

Then the broadening of the information and decision making process is reached and attention, in an overall view, is given to the public opinion on food irradiation. Also, probably the responsibility of scientists for social impacts of a technology will increase. CTA is then a means for involving social parties in (an early stage of product) development: it immediately improves the way science is put into practice (one of the concerns of consumers) and the relationship between the public and those who make decisions, and it prevents the introduction of applications too quickly.

As a conclusion, it can be stated, that a dialogue on basis of research which has been set up by all (social) parties is at present extremely necessary in order to get a better understanding of each others point of view, and possibly a better acceptance of food irradiation. The technology has been developed without involvement of the consumer movement. It is no use for the industry to go on with campaigning without consumer involvement. Probably it works contradictingly. Cooperation is the only solution. Since the consumer movement feels responsible for the future well being of the consumer as a whole, they cannot accept an, according to them, unsafe technology. Although consumers may be willing to buy irradiated food products, there is some fear, some anxiety and some concern. And this anxiety will finally win in a large campaign against food irradiation by for instance FIN as can be seen in several bans from government. If parties will stay diametrically opposed to each other and if the consumer movement, in their opinion, sees no evidence for the safety of the process and the willingness of the industry to listen to their arguments, then both parties become loosers. The industry looses its concept, the consumer looses a, according to experts, sound method of preservation.

So, labelling of all irradiated products and products with irradiated ingredients should be started. Secondly, a research program in which all parties are involved must be set up. And finally a dialogue on basis of the results should be started. If the results are positive, the consumer acceptance of irradiated food products will probably increase. This must be a clear point to governments. They are primarily responsible for protecting the public welfare, physically and psychologically, and therefore they must consider inputs from the consumer movement as well as scientific interests.

REFERENCES

1. Webb, T. and Lang, T., *Food Irradiation. The Facts,* Thorsons Publishing Group, Wellingborough, 1987.
2. WHO, *Food Irradiation. A Technique for Preserving and Improving the Safety of Food,* World Health Organization, Geneva, 1988.

3. Webb, T. and Lang, T., *Food Irradiation. The Myth and the Reality*, Thorsons Publishing Group, Wellingborough, 1990.
4. Kuchler, F., McClelland, J. and Offutt, S.E., The demand for food safety: an historical perspective on recombinant DNA-derived animal growth hormones, *Policy Stud. J.*, 1988, **17**, 125–135.
5. Consumer Currents, March and December, 1989.
6. Knulst, W. and van Beek, P., *Publiek en techniek: opvattingen over technologische vernieuwingen*, Cahier nr. 57, Sociaal en Cultureel Planbureau, Rijswijk, 1988.
7. Riffault, H. and de la Beaumelle, S., *Science and European Public Opinion*, on behalf of the EC, Brussels, 1977.
8. de la Beaumelle, S., The European public's attitudes to scientific and technical development. *Opinion Poll in the Countries of the European Community (XII/201/79-EN)*, Brussels, February, 1979.
9. Daamen, D.D.L., Biegman, M., Midden, C.J.H., van der Pligt, J. and van der Lans, I.A., *Individuele oordelen over technologische vernieuwingen.* ESC-34, Energie Studie Centrum, Petten, April, 1986.
10. Leyten, A.J.M. and Smits, R.E.H.M., *De revival van Technology Assessment. De ontwikkeling van TA in vijf Europese landen en de VS.* NOTA-rapport V2, Apeldoorn, 1987.
11. Schutz, H.G., Bruhn, C.M. and Diaz-Knauf, K.V., Consumer attitude toward irradiated foods: effects of labelling and benefits information, *Food Technol.*, 1989, **43**, 80–86.
12. Wiese Research Associates, *Consumer Reaction to the Irradiation Concept. A Summary Report.* 1984.
13. Brand Group, *Irradiated Seafood Products: A Position Paper for the Seafood Industry*, Chicago, Illinois, January, 1986.
14. Bord, R.J. and O'Connor, R.E., Who wants irradiated food? *Food Technol.*, 1989, **43**, 87–90.
15. Consumers' Association, *Consumer Attitudes to Food Irradiation*, London, 1987.
16. de Bekker, G., *Consumentenonderzoek bestraalde verse kip*, Association Euracom-Ital, External Report no. 11, Wageningen, The Netherlands, 1973.
17. Defesche, F., *Consumer attitude towards irradiation of food (A pilot study in the Netherlands)*, Marketing and Consumer acceptance of irradiated foods, p. 47 f., Joint, FAO/IAEA. Division of Isotope and Radiation Application. International Atomic Energy Agency, Vienna, 1982.
18. NIPO (Nederlands Instituut voor de Publieke Opinie en het Marktonderzoek), *Onderzoek over bestraalde voedingsmiddelen*, Amsterdam, 1973.
19. Feenstra, M.H., Schep, G.J. and Spijkerman-Van Zon, I., *Irradiation, a long-life method? Dutch consumers' views on food irradiation*, SWOKAtern nr.7, The Hague, 1988.
20. Ford, N.J. and Rennie, D.M., Consumer understanding of food irradiation, *J. Consumer Stud. Home Econ.*, 1987, **11**, 305–320.
21. Titlebaum, L.F., Dubin, E.Z. and Doyle, M., Will consumers accept irradiated foods? *J. Food Safety*, 1983, **5**, 219–228.

22. Cramwinckel, A.B., *Opvattingen over voedseldoorstraling, een enquête bij het thuispanel,* Rapport 88.24, RIKILT, Wageningen, 1988.

23. Cramwinckel, A.B. and van Mazijk-Bokslag, D.M., Dutch consumer attitudes toward food irradiation, *Food Technol.,* 1989, **43**, 104 and 109–110.

24. Bruhn, C.M. and Noell, J.W., Consumer in-store response to irradiated papayas, *Food Technol.,* 1987, **41**, 83–85.

25. Bruhn, C.M., Schutz, H.G. and Sommer, R., Attitude change towards food irradiation among conventional and alternative consumers, *Food Technol.,* 1986, **40**, 86–91.

26. Sapp, S.G. and Harrod, W.J., Consumer acceptance of irradiated food: a study of symbolic adoption, *J. Consumer Stud. Home Econ.,* 1990, **14**, 133–145.

Chapter 6

PHYSICOCHEMICAL METHODS FOR THE DETECTION OF FOOD IRRADIATION

C. HASSELMANN

Laboratoire de Chimie Analytique, Faculté de Pharmacie, 74 route du Rhin, 67400 Illkirch-Graffenstaden, France

&

E. MARCHIONI

Association d'Etudes et de Recherches pour l'Ionisation en Alsace, C.R.N., 23 rue du Loess, 67037 Strasbourg Cedex, France

1 INTRODUCTION AND GENERAL ASPECTS

1.1 Why Try to Detect Irradiated Food?

In 1980, the Joint FAO/WHO/IAEA Expert Committee[1] concluded that the irradiation of foodstuffs at doses not exceeding 10 kGy presented no risk whatsoever to the health of human beings. Consequently, toxicological expertise was no longer necessary to obtain the authorisation to irradiate food at such doses. This affirmation was based on a number of scientific studies which allowed a better understanding of the mechanism of food radiolysis[2-7] and which showed that the toxicological risks induced by the products of radiolysis were non-existent.[8] Such studies were never required at the time of approval of other preservation treatments (appertisation, lyophilisation, treatment by microwaves, freezing, etc...). In spite of this, up to now, irradiation techniques have not had the developments that they merit, mainly because of psychological reasons linked to the use of radioactive sources and high energy electromagnetic radiations.

Certain countries, such as the Federal Republic of Germany, totally refuse irradiation. For them, the only way to protect themselves against the illicit importation of irradiated food is to set up some methods of detection for this type of food. This is without doubt one reason why the studies carried out in the laboratories of Diehl at Karlsruhe (Federal

Research Centre for Nutrition, Karlsruhe, FRG) and of Bögl at Munich (Institute for Radiation Hygiene, Neuherberg, FRG) represent an essential contribution to this field of research.

Other countries, such as the United Kingdom, are not very favourable towards irradiation at the moment. Some powerful English Consumer Associations are not totally convinced of the absence of toxicological risks, and do not wish to allow the commercialisation of foodstuffs which have undergone a treatment which cannot be controlled *a posteriori*. It is therefore only by the setting up of detection tests that these associations will accept the use of irradiation.

Finally, other countries, amongst them the Netherlands, Belgium, and France, are rather favourable towards irradiation, perhaps because their Consumer Association are less powerful or better informed than elsewhere. Nevertheless, for regulation reasons (labelling, consumer information) and economic reasons (international trades) the establishment of detection tests is required for irradiated food. As a result, whatever their position in respect to the irradiation of food, all the countries would like to establish a control for this type of food.

1.2 The Problem to Resolve and its Difficulties

Twenty three years ago, when the irradiation technique was still at the experimental stage, Lafontaine and Bugyaki[9] already deplored the absence of detection tests. The first important studies in this field were published at the beginning of the 1970s.[10,11] They expressed great optimism, based on the hypothesis that the irradiation should induce very characteristic physical or chemical modifications in the food itself. A number of research workers have especially thought that a unique radiolysis product (URP) could be found in the irradiated food, thus allowing its detection. These hopes have always been later disappointed.

In other respects, the treatment by radiation, which provokes practically no temperature variation within the food, is often less damaging than the other preservation treatments, especially at the doses usually applied: from 0.02 kGy to 0.15 kGy for sprout inhibition, from 0.2 kGy to 1.0 kGy for disinfestation and from 1 kGy to 10 kGy for radurisation and radicidation.

This matter of fact, which is a positive characteristic of this preservation technique, however represents a drawback when one wishes to develop a detection test based on the modifications of a physical or chemical parameter within the food.

If we add to this that the purely scientific interest in this work is not obvious now that the mechanism of radiolysis of foodstuffs is better known, then we understand that, up until the beginning of the 1980s, very little work had been carried out in this field, apart from that presented at the International Colloquium at Karlsruhe in 1973. After 1980, following the conclusions of the Joint FAO/WHO/IAEA Expert Committee concerning the wholesomeness of irradiation treatment, the development of this technique stimulated the research into detection tests. This research was less empirical and used to the best advantage the most up to date knowledge of radiolysis mechanisms and the progress achieved in analytical methods.

1.3 The Mechanisms of Food Radiolysis

Principally, there are two methods of irradiation used: high energy electrons and γ-rays produced by a ^{60}Co or ^{137}Cs radioactive source. The latter, in contact with the substance, generate high energy electrons by the Compton effect. In both cases, it is therefore the high energy electron which will be the primary active species in the process of food radiolysis. This electron decays in energy while ionizing and activating the molecules of the foodstuffs.

The high energy electrons can act directly on the different components (proteins, lipids, carbohydrates, etc...) with the production of primary solute radicals.

They can also cause the radiolysis of water with the production of primary water radicals ($OH^{\cdot}$, $H^{\cdot}$), reactive species (solvated electrons (e^-_{aq}) and other molecules (H_2, H_2O_2). The solvated electrons and the OH and H radicals thus produced can lead in an indirect way to the formation of secondary solute radicals, possibly different from the primary solute radicals. Owing to the generally higher content of water in the foodstuffs, these reactive species play an essential role in the food radiolysis.

The solvated electrons react very rapidly with electrophilic compounds, either by a dissociation reaction:

$$R-CH_2-X + e^-_{aq} \rightarrow R-CH_2^{\cdot} + X^- \qquad (1)$$

in which X can be for example an SH group, or, more frequently, by an addition reaction with an S–S group, a conjugated double bond, an aromatic group or a carbonyl function:

$$R-CH_2-\underset{\underset{O}{\|}}{C}-CH_2-R' + e^-_{aq} \rightarrow R-CH_2-\underset{\underset{O}{\|}}{\overset{\cdot}{C}}-CH_2-R' \qquad (2)$$

These solvated electrons can also react on the oxygen present in the food, giving the superoxide radical:

$$O_2 + e^-_{aq} \rightarrow O_2^{-\cdot} \tag{3}$$

The OH radicals are very strong oxidizing species capable of extracting a hydrogen atom from a molecule. This reaction takes place rapidly with

$$\text{H}$$
$$|$$
$-\text{SH}, -\text{NH}_2 \text{ and } -\text{C}-\text{R groups:}$
$$|$$
$$\text{R}'$$

$$\begin{array}{cc} \text{R}' & \text{R}' \\ | & | \\ \text{R}-\text{C}-\text{R}'' + \text{OH}^\cdot \rightarrow \text{R}-\overset{\cdot}{\text{C}}-\text{R}'' + \text{H}_2\text{O} \\ | & \\ \text{H} & \end{array} \tag{4}$$

When the molecules are unsaturated, the predominant reaction is an addition reaction:

$$R-CH=CH-R' + OH^\cdot \rightarrow R-\overset{\cdot}{C}H-CH-R' \tag{5}$$
$$|$$
$$OH$$

The H radicals are good reducing species capable of extracting a hydrogen atom:

$$R-CH_2-R' + H^\cdot \rightarrow R-\overset{\cdot}{C}H-R' + H_2 \tag{6}$$

They can also lead to addition reactions with unsaturated compounds:

$$R-CH=CH-R' + H^\cdot \rightarrow R-CH_2-\overset{\cdot}{C}H-R' \tag{7}$$

predominant when we are dealing with aromatic compounds.

All the mechanisms proposed therefore lead to the formation of reactive radicals which will evolve towards the formation of chemically stable compounds. The modes of evolution are varied:

(a) *Radical-radical reactions,* which can lead to dimerisation reactions:

$$
\underset{\underset{R''}{|}}{\overset{\overset{R}{|}}{R'-\overset{\bullet}{C}}} \cdot + \underset{\underset{R''}{|}}{\overset{\overset{R}{|}}{\cdot C-R'}} \rightarrow \underset{\underset{R''\,R''}{|\ \ |}}{\overset{\overset{R\ \ R}{|\ \ |}}{R'-C-C-R'}} \tag{8}
$$

or to disproportionation:

$$
\underset{\underset{H}{|}}{R-\overset{\bullet}{C}-OH} + \underset{\underset{H}{|}}{R-\overset{\bullet}{C}-OH} \rightarrow \underset{\underset{H}{|}}{\overset{\overset{H}{|}}{R-C-OH}} + \underset{\underset{H}{|}}{R-C=O} \tag{9}
$$

or to cross reaction between different free radicals:

$$
R^{\bullet\prime} + R^{\bullet} \rightarrow R'-R \tag{10}
$$

(b) *Oxidation reactions of radicals,* in particular carbonyl, aromatic and heterocyclic radicals (indole, imidazole, cytosine), hence their importance in the radiolysis reactions of proteins and nucleic acids. The oxidizing agent can be the oxygen molecule — triplet oxygen — which therefore has a strong affinity for radicals:

$$
R-CH_2-\underset{\bullet}{CH}-R' + {}^3O_2 \rightarrow \underset{\underset{OO^{\bullet}}{|}}{R-CH_2-CH-R'} \tag{11}
$$

or

$$
R-CH_2-\underset{\bullet}{CH}-R' + {}^3O_2 \rightarrow R-CH=CH-R' + O_2^{-\bullet} + H^+ \tag{12}
$$

These reactions intervene especially during radiolysis of unsaturated fatty acids.

(c) *Radical-solute reactions,* leading to the formation of tertiary radicals, important in gamma radiolysis and in the presence of S–H bonds:

$$R^{\cdot} + R'-SH \rightarrow R-H + R'-S^{\cdot} \qquad (13)$$

All the radicals formed as a result of the reactions (b) and (c) stabilise finally by radical-radical reactions.

1.4 The Lines of Research Envisaged to Determine a Detection Test

The different mechanisms considered in the section 1.3 are not strictly specific to radiolysis. They are of the same type as those observed during other technological treatment or more simply during changes in storage.[12] In these conditions, it is hardly surprising that the research for a 'unique radiolysis product' has not yet been successful.

A more realistic path of study is to try and discover among all the modifications engendered by the radiolysis within the foodstuff, those which would be, qualitatively or quantitatively, different from the modifications resulting from other preservation treatments or simply from the effects of a more or less prolonged storage.

Radiolysis has for example the particularity, which we saw in the preceding paragraph, to create, within the foodstuff, high quantities of free radicals. These chemical species, which are very reactive, are unstable in the presence of water. However, their detection is possible in very dry foodstuffs. The study of these free radicals, in order to establish a detection test, has been the subject of a large number of recent publications.

A number of authors have also tried, for a longer period of time, to demonstrate that certain chemical or physical modifications of the principal components of the foodstuffs (proteins, lipids, carbohydrates, vitamins, nucleic acids), induced by irradiation, could be used as a detection test.

A certain number of bibliographical studies have been recently published on the subject and can be usefully consulted.[13-16]

2 THE DETECTION OF FREE RADICALS

2.1 The Field of Use

One of the first experiments concerning the research on free radicals for the detection of irradiated foodstuffs was carried out on potatoes in 1971 by Mehringer.[17] The research by ESR (electron spin resonance) of radicals in the skins of potatoes irradiated at 0.2 kGy proved futile, even when these were previously dried. The instability of the free radicals in moist foodstuffs was confirmed by Onderdelinden and Strackee.[18] These authors have shown that the research of free radicals was equally fruitless in foodstuffs with low levels of humidity such as fats.

Finally, the stability of free radicals is only effective in dry foodstuffs or in dry parts of foodstuffs, and it is therefore only in this field that the tests based on radical research have developed.

2.2 The Detection Methods of Free Radicals

To date, three analytical methods have been used for radical research in foodstuffs: chemiluminescence, thermoluminescence and ESR. Let us recollect concisely the principles of these three methods.

2.2.1 Chemiluminescence

Some oxidizing substances like hydrogen peroxide and oxidizing radicals are formed in the irradiated foodstuffs (see section 1.3) and remain relatively stable if the latter is dry (or may appear upon dissolution, if Ċ radicals are present). In an alkaline medium and in the presence of a photosensitizer like luminol or lucigenin, an intense light emission (chemiluminescence) is then produced. The mechanism of this reaction has been proposed by White et al.:[19]

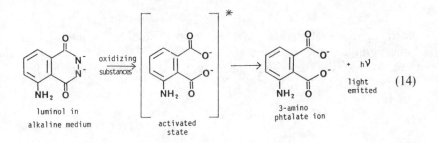

$$(14)$$

The chemiluminescence intensity is noticeably proportional to the dose absorbed.

2.2.2 Thermoluminescence

When a solid substance is submitted to ionizing radiation, three sorts of defects can appear in the structure:

- electronic defects which lead to changes in the valence states,
- ionic defects resulting in the displacement of ions in the network,
- serious imperfections, such as dislocation loops and voids.

The existence of these defects gives rise to discrete energy levels in a forbidden region between the valence and the conduction bands, into which electrons can be transferred. This energy, stocked at ambient temperature, can therefore be partially liberated as a result of heating in the form of light (thermoluminescence). The luminescence intensity, which is proportional to the dose absorbed, can be measured as a function of the heating temperature. We thus obtain what is known as a glow curve, the shape of which depends on the speed of heating and the maximum temperature. If these parameters are fixed, the shape of the glow curve is characteristic to the substance studied.

2.2.3 Electron spin resonance

Electron spin resonance (ESR) is a spectroscopic method applicable to compounds possessing an unpaired electron, thus to free radicals, but also to odd molecules, triplet electronic states and transition metal and rare earth ions. All the chemical species possess a non-zero electronic magnetic moment. Under the action of an external static magnetic field H_0, two energy levels are established, the energy separating the two levels being:

$$\Delta E = \frac{\mu_e H_o}{M_s}$$

where μ_e is the electronic magnetic moment and M_s is the angular momentum quantum number.

In this way the compound under study can absorb radiation of frequency:

$$\gamma = \frac{\Delta E}{h}$$

where h is Planck's constant. ΔE can be given under the form:

$$\Delta E = g \beta H_o$$

where β is a constant (Bohr's magneton) and g the splitting factor. This g-factor has a value of 2.0023 for free electrons and varies slightly (a few %) depending on the nature of the free radical. Therefore, the ESR spectrum enables us to differentiate the radicals. Depending on the environment of the unpaired electron and by coupling with the protons, the signal observed can transform into a multiplet.

The three analytical methods so far described allow the detection of radical species. However they differ sufficiently in their principle such that they do not necessarily have the same field of application and can, in certain cases, be complementary.

ESR detects all the radicals present, not only those produced by irradiation, and eventually allows their separation if they are of a different type. It is the only method which enables the possible detection of a unique radical product. The other methods — chemiluminescence and thermoluminescence — do not provide any information on the types of radicals formed.

The luminescent methods do not detect the same thing (presence of oxidizing species for chemiluminescence, electronic and ionic defects for thermoluminescence). One can therefore assume that they do not necessarily give the same results for the same foodstuffs. They can thus be complementary. Their main disadvantage is their lack of specificity, especially for the chemiluminescence method because any oxidizing substance can induce a luminescent signal in the presence of luminol.

2.3 The Different Foodstuffs Tested

2.3.1 Meat (with bone), fish, shellfish

In these different categories of foods, the radicals produced by irradiation can be detected in the dry parts of the foodstuffs, i.e. in the bones of meat and fish and in the cuticle of shellfish.

Apart from a single study by chemiluminescence,[20] without follow-up, carried out on frozen chickens irradiated at 10 kGy, which indicated that detection was still possible after 2 weeks of storage on condition that the cartilage was analysed and not the bone, all the other research works carried out with this type of foodstuff have been by ESR.

In 1974, Onderdelinden and Strackee[18] demonstrated that irradiation, at a high dose (50 kGy) and at ambient temperature, of the marrow bones of beef and pork, ribs of beef and pork and chicken bones, induced relatively time-stable signals after a fairly rapid decrease the first day after irradiation.

After this work, although very promising, no publication was made on the subject until 1985. That year, in a short report, Dodd et al.[21] studied the radicals produced in chicken bones, lemon sole bones and prawn cuticles at ambient temperature and irradiation doses of 4–10 kGy. In the chicken bones, the induced signal, considered to be characteristic of the irradiation, and clearly different to the endogenous signals, is not significantly reduced after 14 days of storage at +4°C (14 months for isolated bones). Soaking of the bone in water for 30 min does not alter the signal. Heating by flame produced a large signal, very different to the irradiation signal. The radicals formed are probably CO_2 radicals incorporated in the hydroxy apatite matrix[22,23] and not radicals formed in the collagen as thought by Onderdelinden and Strackee.[18] Generally speaking, for any type of foodstuffs studied, detection is still possible after 3 weeks storage at 4°C. The detection limit dose is ≈ 0.2 kGy.

Desrosiers and Simic[24] have shown that the radiation induced ESR signal in chicken bones — with or without surrounding meat — at a 1-kGy dose was clearly distinguishable from the weak endogenous signal. According to these authors, the detection limit is 0.1 kGy. The signal intensity of the excised bone remains stable for at least 4 months, even during storage at 20°C. The results obtained are valid for all chicken bones, but not for cartilage, even dried. The long-term stability of the radiation-induced ESR signal and the linear relationship observed between its intensity and the absorbed dose (1–5 kGy) should allow the feasibility in parallel of postirradiation dosimetry.

Dodd et al.[25] confirmed and expanded these results by the study of the effects of 2-kGy irradiation (10 MeV electron linear accelerator) on chicken, pork and cod bones. The observed ESR signals, specific to the irradiation, are stable for at least 3 weeks at +4°C (and for several months at −21°C). Their intensity increases with the bone calcification. In bones which are washed, reduced to powder and dried, the radiation-induced ESR signal remains stable for several years. According to these authors, the detection limit is 0.05 kGy. Also, Swallow[26] and Lea et al.[27] have shown that the ESR signal in chicken bones remains stable after cooking. However, according to Goodman and McPhail,[28] the ESR signal disappears very quickly in scampi cuticle after boiling.

In a recent publication, Desrosiers[29] has studied the feasability of identifying γ-irradiated seafoods (shrimps, mussels, flounder) by ESR technique. Generally speaking, ESR signals are relatively stable and the detection of irradiation is still possible after 5 months storage. Concerning flounder bones, the linear relationship between the signal intensity and the irradiation dose allows the realisation of a postirradiation dosimetry, as mentioned by Desrosiers and Simic[24] and Lea et al.[27] for chicken bones. Because of the complexity of the ESR spectra, this is unfortunately not true for shrimps and mussels (in spite of very intense signals induced by irradiation in the mussel shell). In other respects, the radiation-induced ESR spectra of the shrimp cuticle reported in this publication differ markedly from those observed by Dodd et al.,[25] probably because two different types of shrimps were used for each study. This observation indicates the difficulty for a general application of the ESR method to the detection of irradiated shrimps.

2.3.2 Cereals

Raffi et al.[30] have tried to use the ESR method to characterise the radicals produced during γ-irradiation (0.3 kGy, 5 kGy, 10 kGy) in wheat, barley and maize in the presence of air at ambient temperature. Two signals appear, one rather fine, which can also be noted very faintly in non-irradiated samples, the other much broader which seems, when it appears, characteristic of irradiation. The detection limit being 0.05 kGy, this technique could therefore be used for cereals which have undergone a disinfestation (the most common treatment). The radicals, apparently produced in starch, are however not very stable, more particularly in irradiated wheat and maize. This is why the authors show a cautious optimism, especially for the lowest radiation doses, and exclude any possibility of measurement of the absorbed dose by this technique, because of the heterogeneity of the radical distribution in the grains and the possible creation or disappearance of radicals during the grinding.

Dodd et al.,[25] taking into account the non-specificity of the ESR signals and the instability of the radicals, consider that, for cereals, this method does not show any advantages compared to thermoluminescence.

2.3.3 Fresh fruits

In the pulp of fruit, the radicals are very unstable and cannot be demonstrated. However, they can possibly be detected in the parts with low water content, such as the achenes and the stones.

Dodd et al.[21] studied the radicals produced in the achenes of strawberries. In those of unirradiated fruits, two ESR signals are seen, one produced by free radicals, of which the intensity increases with irradiation, the other due to the presence of Mn^{2+} ions and which remains stable after irradiation. After treatment at 10 kGy (which is very high for this fruit) and storage for 14 days at 4°C, the ratio of the two signals is still largely superior to that found in an unirradiated sample. This ratio can however vary, depending on the ripeness of the fruit and the storage conditions. The development of a test based on the measurement of the two ratios and valid for the normal irradiation doses on these fruits (1–5 kGy) appears therefore very unlikely.

The study of this fruit was continued by Raffi et al.,[31] who, after γ-irradiation at ambient temperature and at doses varying from 1 to 3 kGy, preferred to look, in the ESR spectrum of the irradiated achenes, for a signal which did not appear in the spectrum of the unirradiated sample. Such a signal, partly hidden by that produced by the Mn^{2+} ions, appeared after irradiation at 1 kGy and was still visible after 23 days storage at 5°C. According to the authors, this test could be used for other fruits such as the blueberry, the fig, the raspberry or the red currant.

Saint-Lebe and Raffi[32] have also studied by ESR the radicals produced by gamma irradiation in the stones of plums. These radicals which give a very complex signal, disappear very quickly. This test does not appear to be satisfactory.

2.3.4 Dried foodstuffs (spices, aromatics, fruits, milk, mushrooms)

Spices have been by far the subject of the most work at the present time, mainly because they are the category of foodstuffs most subjected to irradiation on an industrial scale.

The first work carried out in this area by Beczner et al.[33] was somewhat disappointing. After irradiation (5 kGy) and storage for 3 months of ground paprika, the ESR spectrum showed no difference whatsoever between the irradiated and the unirradiated samples. Also the grinding induced the production of a considerable quantity of free radicals.

Much more recently, still by ESR, Yang et al.[34] have studied the free radicals produced in powders of turmeric, black pepper, mustard, cinnamon and paprika, by high doses of gamma irradiation (30 kGy) at 30°C and in the presence of air. The spectrums obtained, due to their poor resolution, did not allow the attribution of the observed signals to particular radical types. These radicals, which appear to already exist in small quantities in unirradiated spices, are relatively unstable, since only 8% of those formed after irradiation in ground paprika are still in evidence after 36 days storage at 25°C (confirmation of results obtained by Beczner et al.[33]). The best results were obtained with dried mustard (45% of the radicals remained after same storage time). Similar observations were made by these authors after irradiation of spray-dried fruit powder.

According to Wieser and Regulla,[35] the identification of irradiated spices by conventional ESR techniques is not possible due to the short life span of the radicals formed by irradiation, and the presence in unirradiated samples of free radicals of which the concentration can increase by heating or during storage. On the other hand, by the use of a phase sensitive ESR technique, these authors have been able to show, in the spectrum of ground paprika irradiated at 10 kGy, two stable signals which do not exist in the unirradiated spice. These have a half-life >1 year and allow the detection of doses as low as 1 kGy. These promising results merit to be developed.

However, at the present time, the many studies carried out for the development of a detection test on dried foodstuffs have used chemiluminescence and/or thermoluminescence rather than ESR.

The first study of Bögl and Heide[36,37] states that it was still possible to quickly detect by chemiluminescence, 2 months after the treatment, powdered cinnamon, curry and paprika irradiated at 10 kGy. The authors did however specify that the intensity of luminescence was very variable from one spice to another, that it was reduced by heating or in the presence of water, but that it increased under the action of ultraviolet rays, observations that seem relatively foreseeable. This study was expanded to a large number of different spices.[20,38,39] Of the twenty spices irradiated at 4–10 kGy, all, except the sage, could be recognised by chemiluminescence immediately after irradiation, and most of them still after 2 months storage (the detection limit corresponds to a luminescence intensity of the irradiated sample double that of the unirradiated sample). The authors have however at times obtained very variable luminescence intensities for the same spice but from different batches[39] and note a great importance for the quality of the luminol (pH, oxygen concentration, age). Parallel tests carried out on milk powder irradiated to 10 kGy (a high dose for this food) were disappointing, insofar as the oxidation of the milk in the presence of air provokes a rapid increase in the luminescence intensity, making the detection impossible.[20]

Meier and Zimmerli[40] and Meier et al.[41] used chemiluminescence to test spices imported into Switzerland, and considered that the very large variations in the intensity of luminescence observed for identical spices but of different origins (variation coefficients from 30 to 100%) made the identification of irradiated spices questionable if no reference sample is available.

Delincee[42,43] has also made a thorough study of the possibility of using chemiluminescence as a unique test for the detection of irradiated spices and has arrived, for those studied (curry, juniper berries, marjoram), to the same conclusions as Meier and Zimmerli. The variation coefficients of the luminescence intensity of spices of different origins can, in many cases, exceed 50%. Replacement of luminol by lucigenin slightly improves the reproducibility of the measurement but reduces its sensitivity. Sattar et al.[44] have also shown on a sample of black pepper that the grinding, the particle size and the irradiation dose affects the chemiluminescence response in an inconsistent manner.

It would therefore finally appear that the use of chemiluminescence as a unique test for the detection of irradiated spices is not as favorable as originally hoped by Bögl and Heide. These authors in fact fully understood this and quickly proposed the use of thermoluminescence in parallel with chemiluminescence for the detection of irradiated spices and aromatics.[45-49]

Considering the different natures of the transitory species looked for in the foodstuffs by the two methods (see sections 2.2.1 and 2.2.2), it would appear in effect that often when one method does not give satisfactory results, the other allows better ones. In this way, the chemiluminescence measurement is for example better for aniseed, celery and juniper berries, whereas that obtained by thermoluminescence is better for cinnamon, coriander, cumin, curcuma, paprika, caraway, black pepper, red pepper and tarragon.[47]

Although the two methods show the same shortcomings — decrease of the luminescence intensity during storage, with acceleration of this phenomenon by heat and humidity — it seems however that the thermoluminescence signal is less sensitive to these two external parameters than the chemiluminescence signal, and therefore thermoluminescence allows the detection of spices and aromatics after a longer period of storage than does chemiluminescence. Simultaneous use of the two methods, according to Heide and Bögl,[47] allows the characterisation of all the spices studied 2 weeks after irradiation at 10 kGy. Most can still be detected after 6 months storage, and some after more than 1 year (curcuma, juniper berries, basil, red pepper, paprika, celery). Mediocre results have however been obtained with chive, cloves, garlic and white pepper.

Using only thermoluminescence, Moriarty et al.[50] considered that irradiated paprika, carvi, black pepper, cinnamon, curcuma and cayenne pepper could still be detected after 2 months of storage. On the other hand, disappointing results have been obtained with nutmeg. All these works have however a major disadvantage: they were always carried out, for each spice, by comparison of an irradiated and an unirradiated sample of the same origin. Now, during prior studies carried out by chemiluminescence, very large fluctuations in the luminescence intensity were observed for various spices of different origins.[39,40,42,43]

Since, during routine checks, it is rare that an unirradiated control sample is available, it is obviously essential to know if statistically the thermoluminescence and/or chemiluminescence signals given by an unir-radiated sample and by an irradiated sample, which is subjected to a fading effect during storage, stay sufficiently different to be differentiated for any foodstuff origin. This work was carried out by Heide *et al.*[51] and Heide and Bögl[52] on more than 40 spices and aromatics, unirradiated or irradiated at 10 kGy. These authors admit that if the maximum signal intensity obtained statistically for an unirradiated foodstuff is x, the sample can be considered to be irradiated if its signal intensity is greater than $2x$. Insofar as x is defined (which supposes a vast statistical study for each type of foodstuff) there is therefore no risk of considering an unirradiated foodstuff as irradiated. The opposite is however possible, particularly as during storage, the signal fading can be very different from one sample to another, whether it is irradiated or not. It follows, obviously, that the results obtained are much less satisfactory than if unirradiated control samples are available, less and less satisfactory as the storage time lengthens. Overall, thermoluminescence appears more satisfactory than chemiluminescence, mainly because the signal values obtained from unir-radiated samples show much less dispersion and are more reliable with the first method than with the second.

The detection of irradiated spices and aromatics by thermolumines-cence and by chemiluminescence has been the subject of an inter-laboratory analysis involving seven German laboratories, five of which were unfortunately not equipped to carry out thermoluminescence analy-sis.[53,54] It was therefore principally the chemiluminescence technique which was tested. Nine different spices and aromatics were analyzed (be-tween 11 and 71 days after irradiation). The results were relatively satisfactory for curcuma, coriander, juniper berries and celery, very poor for cinnamon, paprika and parsley, somewhat variable for fennel and pepper, reflecting undoubtedly a lack of practice in certain laboratories. It is certain that the vast work carried out on the detection of irradiated spices and aromatics by ESR, chemiluminescence and thermo-luminescence, merits an interlaboratory analysis realised on a larger scale.

3 STUDY OF THE PRODUCTS OF RADIOLYSIS — MODIFICATIONS OF THE PHYSICAL PROPERTIES

For a long time, researchers have hoped to discover the unique radiolysis product (URP), which would allow the unambiguous characterisation of an irradiated foodstuff. Up to the present, every time that such a product was believed to have been demonstrated, the product in fact was either formed by other technological treatments or simply during storage, which is hardly surprising now that the radiolysis mechanisms are better known, very similar to those of thermolysis or oxidation.[12] Most work is therefore limited to the research of the appearance or disappearance in abnormal quantities of a chemical compound (or a group of chemical compounds) or of an unexpected modification of a physical parameter. These tests, which are not unambiguous tests, generally require a non-irradiated control sample. Practically all the constituents of foodstuffs have been investigated, but most attention has been given to proteins.

3.1 Modifications Induced in Proteins

The type and magnitude of chemical reactions observed during the irradiation of proteins depend on a large number of parameters such as their structure — fibrous for myofibrillar proteins like myosin or actimyosin, globular for sarcoplasmic proteins such as myoglobin — their state (native or denatured, in cooked foodstuffs for example), their water content and the irradiation conditions (dose, dose rate, presence or absence of oxygen, temperature (ambient or subfreezing), etc...).

The mechanisms of protein radiolysis are well known.[2,4,6] The main reactions are caused by the solvated electrons and the OH radicals resulting from the radiolysis of water, the direct effects of the irradiation on the protein structure being totally negligible.

The solvated electrons react principally by an addition reaction (see reaction (2)) on aromatic rings, sulphur groups (with final liberation of sulphur compounds) and carbonyl bonds (with as a result either a terminal deamination or an intermediate deamination followed by a main-chain scission).

The OH radicals easily abstract a hydrogen atom (see reaction (4)), followed by the possibility of dimerisation (see reaction (8)), but can also add to the protein, particularly to aromatic rings (see reaction (5)).

Apart from these reactions, which all concern the primary structure of the protein, destruction of the complex structure can also occur — mainly in the case of fibrous proteins — or, on the contrary, cross linking in the case of globular proteins, these aggregates being formed by a mechanism poorly defined, however, partly by the establishment of covalent bonds between two protein radicals or through non-covalent cross linking.[6] When the protein is an enzyme, it is evident that these structure alterations can lead to a modification of its activity.

All these lines have in fact been explored for the definition of a detection test. We have chosen to class the suggested tests in three groups, depending if they result in a modification of the amino acid residues, in an alteration of the complex protein structure or in a reduction of the enzyme activity.

It should be remembered that the effects produced by irradiation on the protein are practically negligible up to doses of 10 kGy,[6] therefore making the setting up of a detection test even more difficult.

3.1.1 Modifications of the amino acid residues

3.1.1.1 Meat, fish, shellfish. (a) Reduction of the SH group content: The sulphur containing amino acids appear to be very sensitive to irradiation (see section 1.3) and their partial disappearance in irradiated meat has been proposed as a detection test by Stockhausen *et al.*[55,56] These authors, by irradiating respectively chicken and pork meat (fillet, liver and heart), at ambient temperature, in the presence of air and at doses varying from 1 to 25 kGy, have effectively shown that the SH group content (as determined by photometry, using 6,6'-dithiodinicotinic acid as a reagent) of these irradiated foodstuffs was quickly reduced in relation to the absorbed dose. It is however probable that the reduction is related to the presence of oxygen during irradiation, inducing the formation of superoxide radicals O_2^-, liable afterwards to destroy the SH groups.[2]

This leads one to understand why Hamm *et al.,*[57] whilst irradiating lean pork in the absence of oxygen, observed no reduction in the SH group content up to doses of 50 kGy. As in general meat is always irradiated in the absence of oxygen, the suggested test — the specificity of which has not been studied since — appears to be of limited interest.

(b) Dimerisation reactions: Formation of amino acid dimers during irradiation of foodstuffs by radical-radical type reaction (see reaction (8)) is possible.[58] Boguta *et al.*[59] have effectively been able to prove the formation of bityrosine during irradiation of pure proteins, particularly collagen. Karam *et al.*[60] and Dizdaroglu *et al.*[61] have since shown that this cross linking can occur either between two carbons of the aromatic ring or between a carbon of this ring and the oxygen of the hydroxy group. Dimers have also been formed during irradiation of methionine peptides,[62] of alanine peptides[63] and of phenylalanine peptides.[64] These dimers which to date have not been shown in irradiated foodstuffs, are certainly not unique radiolysis products. Bityrosine can effectively be formed by photolysis of tyrosine[65] or by enzymatic oxidation of proteins.[59]

(c) Hydroxylation reactions: Irradiation of protein containing phenylalanine can lead by addition of OH radicals (see reaction (5)) to the formation of hydroxycyclohexadienyl radicals which, by disproportionation and oxidation reactions, yield to ortho-, meta- and paratyrosines.[66,67] If paratyrosine is naturally present in proteins, the two other isomers are not, and can therefore be considered as unique radiolysis products, allowing the development of a detection test. Later, however, small quantities of ortho- and metatyrosines were found in the unirradiated control sample, probably produced during homogenisation of the sample by carbon tetrachloride.[68] It appears however that the orthotyrosine formed in this way is only found in the aqueous fraction of the meat, and not in the water insoluble fraction. By gas chromatography and mass spectrometry, Karam and Simic[66,69] have demonstrated the presence of orthotyrosine in fresh chicken meat, irradiated in the presence of air, in quantities increasing linearly with the irradiation dose (1–80 kGy). For Hart et al.,[70] the validity of this test is questionable: using GC-MS with negative ion chemical ionisation detection technique, these authors have detected orthotyrosine in fibres of unirradiated chickens and prawns. It should also be noted that this compound, in the presence of air, is also likely to be formed by photolysis.[71]

(d) Formation of polar non-volatile carbonyl compounds: The presence of carbonyl compounds in meat is normal, whether it is irradiated or not. However, according to Schubert and Esterbauer,[72] the formation of polar non-volatile carbonyl compounds is characteristic of irradiation. This test, which involves isolating these compounds by thin-layer chromatography after derivatisation by 2,4-dinitrophenylhydrazine, would give positive results with beef and codfish, but negative with potatoes. This is the reason why the authors consider that these carbonyl compounds are formed from proteins, and not from carbohydrates, an observation which could be contested since the irradiation doses for potatoes (≤ 0.2 kGy) are much less than those for meat (1–10 kGy) and that in these conditions the quantities of carbonyl compounds formed in the potatoes could be negligible. In addition, Schubert and Esterbauer state that their test is not applicable with doses inferior to 2.5 kGy and that it only seems correct for sterilisation doses (50–80 kGy).

(e) Modifications of free peptides and amino acids: By high voltage electrophoresis and by ion exchange liquid chromatography, Partmann and Keskin,[73] during the irradiation of beef, pork, chicken and fish (carp) at 25°C, in the presence of air and at doses greater than 5 kGy, demonstrate the presence of two compounds reacting to ninhydrin, one in meat (compound Y), the other in fish (compound X) and differing from the free amino acids and peptides present in the unirradiated samples. These compounds have not been fully identified,[74] but the substance Y seems to form from the dipeptides carnosine and anserine, whereas the substance X originates from the degradation of histidine (the imidazole ring however staying intact). The formation of these compounds which only form in the presence of oxygen and at relatively high irradiation doses, therefore, in such conditions, cannot be retained as a detection test for meat and fish. This study has not given rise to any further research work.

3.1.2 Alterations of the complex protein structure

Diverse effects of irradiation on the protein structure — rupture of the links between subunits, rupture of the polypeptide chain, formation of aggregates — can be demonstrated, mainly by electrophoresis or by gel chromatography, and possibly followed by a detection test, even though their specificity can be questioned.

   3.1.2.1 Meat, fish, shellfish. In 1971, Adriaanse,[75] while studying, by disc electrophoresis on polyacrylamide gel, proteins of cooked shrimps irradiated at doses varying from 0.5 to 3 kGy, showed the appearance of certain bands and the disappearance of others, and proposed this result as a detection test. Van Der Stichelen Rogier,[76] using identical experimental conditions (irradiation dose 2 kGy), however noted no modification in the protein structure after irradiation, thus contradicting the results obtained by Adriaanse. Again in a study of cooked shrimps, with irradiation doses from 1 to 5 kGy, Deschreider and Vigneron[77] observed the modification of two bands in the infrared spectrum of the whole sample, which was, according to them, due to a disturbance of the protein structure. This test which appears very empirical and not very specific, has given negative results with cod.

   Van Der Stichelen Rogier[76] has studied by electrophoresis the behaviour of sarcoplasmic and myofibrillar proteins of raw beef irradiated at doses from 5 to 50 kGy. For the sarcoplasmic proteins, the electrophoresis in starch gel has shown no alteration by irradiation at 5 kGy. After 4 weeks storage at $+4°C$, the electrophoregram of unirradiated and irradiated samples can however be differentiated, mainly because of the reduction in myoglobin content, which, unfortunately, is not radiation specific. Less satisfactory results have been obtained by electrophoresis in polyacrylamide gel. The effect of irradiation on myofibrillar proteins (studied by isoelectric focusing in polyacrylamide gel) can only be observed at a dose of 50 kGy. These electrophoretic techniques have given disappointing results with cod fillet irradiated at 10 kGy. It would appear that irradiation at doses ≤ 10 kGy does not lead to noticeable modifications in the electrophoretic behaviour of proteins. This observation has also been made by Altmann et al.[78] (study by gel electrophoresis of proteins of chicken irradiated at doses ≤ 5 kGy) and by Bruaux et al.[79] (study by immunoelectrophoresis of proteins of meat and fish irradiated at 5 and 10 kGy).

Results which were more interesting and more surprising were obtained by Hamm et al.[57] in a study by SDS polyacrylamide gel electrophoresis on proteins of lean pork (sliced) irradiated at 2 kGy. These authors have effectively noted a marked reduction in the myosin content, but above all the formation of peptides of low molecular weight (without modification of actin). They have also shown that the effects were very different to those resulting from heating. Unfortunately, Rossler and Hamm,[80,81] by irradiating vacuum-packed pork and beef at 2, 10 and 50 kGy, were unable to confirm this fragmentation. They have indeed noted, contrary to that which had previously been reported, a similarity of the effects produced by medium heating (70°C) and by irradiation, this similarity being confirmed by Taub et al.[82]

The globular proteins, by irradiation, are likely to form high molecular weight aggregates, which can be isolated by gel chromatography.[6] By thin-layer chromatography on sephadex gel, Radola[83] has been able in this way to demonstrate such aggregates in small quantities from pork and beef sarcoplasmic proteins irradiated at −30°C and at doses from 10 to 50 kGy. By SDS gel electrophoresis, Taub et al.[82] did not obtain aggregates from irradiated myosin, at low temperature (−40°C) and at doses up to 80 kGy. Delincee and Paul[84] and Hajos and Delincee,[85] however, do not exclude the possibility of destruction of the aggregates by the sodium dodecyl sulphate (SDS). If this is the case, it would still have to be proven that the aggregate formation is specific to irradiation, which is not certain.

3.1.2.2 Eggs. According to Van Der Stichelen Rogier,[76] freezing has much more noticeable effects on egg proteins than has irradiation at doses from 5 to 10 kGy. The results are thus negative, whether the study is carried out by starch or polyacrylamide gel electrophoresis or polyacrylamide gel isoelectric focusing.

Morre and Janin,[86] have suggested to look for malonaldehyde as a detection test on eggs frozen in bulk and irradiated at 5—10 kGy, this compound being formed from glycoproteins present in the egg. Since malonaldehyde can also be formed by autoxidation of lipids, the specificity of this test is very uncertain.

3.1.2.3 Mushrooms. Proteins of unirradiated mushrooms and mushrooms irradiated at 2.5 kGy, studied by polyacrylamide gel isoelectric focusing, showed no significant difference[76] and cannot be used as a detection test.

3.1.2.4 Potatoes, onions. Irradiation doses to inhibit sprouting in these foodstuffs are low (0.10–0.15 kGy). The effects produced on the proteins are therefore very limited and it is hardly surprising that studies carried out by polyacrylamide gel electrophoresis or isoelectric focusing have been negative.[76]

3.1.3 Modifications of the enzymatic activity

3.1.3.1 Meat. The alterations produced by irradiation on the protein structure can induce a reduction in the enzyme activity. It is in this way that Altmann *et al.*[78] have noted a reduction of the proteolytic activity of chicken juice of around 50% after irradiation at 10 kGy. Briski[87] who did not find any reduction of the activity of a commercial protease irradiated at 10 kGy, has however obtained a reduction of 3–17% for an α-amylase irradiated at doses varying from 3 to 9 kGy.

This reduction of enzymatic activity by irradiation is however not characteristic and cannot be used, in our opinion, as a detection test.

3.1.3.2 Mushrooms. An original method, the mechanism of which is not fully understood, has been developed by Münzner.[88] In unirradiated mushrooms, which contain dehydrogenase enzyme, 2,3,5-triphenyltetrazolium chloride transforms into red formazan. Irradiation at 2.5 kGy seems to modify the activity of the dehydrogenase, and in the presence of the reagent, a brown coloring appears. This effect is at a maximum 36 h after the irradiation and disappears after 6 days of storage. For Zehnder,[89] this method is not suitable as routine food control because irradiated mushrooms recover too quickly from radiation damages.

3.2 Modifications Induced in Carbohydrates

In foodstuffs, carbohydrates are present essentially in the form of digestible polysaccharides (starch), non-digestible polysaccharides (cellulose, hemicellulose, pectin), but can also be found in the form of disaccharides (maltose, lactose,...) or monosaccharides (glucose, fructose). The radiolysis of these various compounds, very complex, has been widely studied,[2,4,7] especially that of starch.[90–93] It can be summarised in the following way: (1) rupture of the glycosidic bond with depolymerisation, (2) destruction of the monomer (essentially glucose, in starch and cellu- lose) under the action of OH radicals, with, in a first step, abstraction of a hydrogen atom (see reaction (4)):

$$
\begin{array}{c}
\text{HO OH} \\
| \ \ | \\
-\text{C}-\text{C}- \\
| \ \ | \\
\text{H} \ \ \text{H}
\end{array}
\ + \ \text{OH}^{\cdot} \ \rightarrow \
\begin{array}{c}
\text{HO OH} \\
| \ \ | \\
-\text{C}-\overset{\cdot}{\text{C}}- \\
| \\
\text{H}
\end{array}
\ + \ \text{H}_2\text{O}
\tag{15}
$$

the evolution of this radical leading to the formation of a deoxyketo- carbohydrate:

$$
\begin{array}{c}
\text{O} \\
\| \\
-\text{C}- \\
\\
\end{array}
\ \ \
\begin{array}{c}
\\
\overset{\cdot}{\text{C}}- \\
| \\
\text{H}
\end{array}
\ \rightarrow \
\begin{array}{c}
\text{O} \ \ \text{H} \\
\| \ \ | \\
-\text{C}-\text{C}- \\
| \\
\text{H}
\end{array}
\tag{16}
$$

then, by rupture of other bonds of the molecule, to the formation of com- pounds such as malonaldehyde.

All of the proposed detection tests, based on a modification of the structure of carbohydrates uses one or other of these two possibilities: the depolymerisation, characterised by a modification of physical properties, and the destruction of a monomer leading to the formation of carbonyl compounds.

The research for detection tests based on modifications induced in car- bohydrates has been obviously mainly carried out on foodstuffs which are rich in polysaccharides, such as cereals and potatoes. Since the radia- tion doses generally used in the two cases are relatively small, < 0.2 kGy, to avoid sprouting of potatoes, from 0.2 to 1 kGy to carry out disinfesta- tion in cereals the tests proposed must of necessity have an excellent sen- sitivity. Also it is probable that heating is equally capable of producing such effects, which at first sight could lead one to doubt the specificity of the tests which are presented below.

3.2.1 Formation of deoxy and carbonyl compounds

3.2.1.1 *Cereals.* The production of malonaldehyde was proposed first of all as a detection test for irradiated maize starch by Berger and Saint

Lebe[94] (colorimetric determination with thiobarbituric acid). Winchester[95] later confirmed that this test allowed him to detect starch irradiated at 0.1 kGy. But this experiment was carried out on recently irradiated starch. Later, Stewart and Winchester[96] confirmed the formation of malonaldehyde in irradiated maize starch, but found that this compound disappeared quickly on storage and that this posed a problem for the detection. Scherz[97] had previously shown that the malonaldehyde totally disappeared in wheat flour irradiated at 1 kGy after 8 days storage and that its production was not specific to irradiation.

This same author however thought that the research of more stable compounds such as deoxyketocarbohydrates would be of more interest for the development of a detection test in wheat and wheat flour.[98] The study of radiolysis products, by thin-layer chromatography and mass spectrometry-gas chromatography, has shown the presence of hydroxymaltol and dihydropyrones in irradiated wheat starch, in appreciable quantities only if the irradiation dose is > 5 kGy,[98] which is too high for a disinfestation. Also it would not appear that the deoxy compounds are totally absent in unirradiated foodstuffs, which reduces the interest of the method proposed. Another deoxy compound, 5-deoxyxylosehexodialdose, was proposed by Adam[99] as an indicator of irradiation. Unfortunately this compound has never been found in foodstuffs.

3.2.1.2 Potatoes. The tests for malonaldehyde[100] and for deoxy compound[98] have also been proposed as detection tests for irradiated potatoes. Given the doses necessary to prevent sprouting, the sensitivity of these tests is insufficient.

3.2.1.3 Spices. Study by thin-layer chromatography of the carbonyl compounds present in irradiated and non-irradiated ground paprika, up to doses of 50 kGy, has not allowed differentiation of the two samples.[33]

3.2.1.4 Fruits. Den Drijver et al.[101] by irradiating a fruit model (mango) have shown glucosone (D-arabino-hexos-2 ulose) which would be a unique radiolysis product. This compound has however never been found in a real fruit.

3.2.2 Rupture of glycosidic bonds and depolymerisation

3.2.2.1 Cereals. The soluble carbohydrates resulting from depolymerisation of the starch, among which are found dextrins, form cloudiness when ethanol is added to the clarified suspension of wheat flour irradiated at 0.25 and 1 kGy. This cloudiness is measured by turbidimetry.[10] This method, apparently not very specific, has not been used since its publication.

3.2.2.2 Spices and aromatics. The depolymerisation of non digestible polysaccharides in foodstuffs after irradiation causes changes in viscosity.[102] This characteristic was used for the first time for the detection of irradiated spices by Mohr and Wichmaan.[103] This initial study was later completed by the studies of Heide *et al.*[104,105]

By spice suspension in water (10%), heating at 80°C, then cooling, the viscosity diminishes when the spice in suspension is irradiated. Significant changes have been obtained with cinnamon, ginger, black pepper and white pepper. The results are much less satisfactory with cloves, onion and leek. In the authors own opinion, this method can give an indication, or can possibly complete results given by another method, but cannot alone constitute a reliable method for the detection of irradiated spices and aromatics.

3.3 Modifications Induced in Lipids

Radiolysis of lipids has been widely studied for some 20 years, as much on the reaction mechanism as on the nature of the radiolysis products.[3,5,106-108] The effects of irradiation on triglycerides are essentially a result of the action of solvated electrons and OH radicals produced by radiolysis of the water or, in hydrophobic regions, of direct excitation and ionisation.[3] In the absence of oxygen, hydrogen abstraction and radical splitting are mainly observed, preferably close to carbonyl groups, as indicated below:

$$
\begin{array}{cccc}
2 & 3 & 4 & 5 \\
\downarrow & \downarrow & \downarrow & \downarrow \\
\end{array}
$$

$$
H_2C-O-CO-CH_2-CH_2-R
$$
$$
1 \rightarrow \quad |
$$
$$
HC-O-CO-R'
$$
$$
|
$$
$$
H_2C-O-CO-R''
$$

A large number of volatile compounds can be formed — esters (1), organic acids (2), aldehydes (3), alkanes and alkenes ((4) and (5)), but also, by radical rearrangement, compounds of medium volatility (diesters, hydrocarbons, long chain aldehydes and esters — and non-volatiles (triglyceride dimers for example).[109]

The production of these compounds obviously depends on the chemical composition of the irradiated lipids (saturated or unsaturated), the irradiation (dose, temperature) and the storage conditions (atmosphere, temperature).

In the presence of air, the radiolysis mechanisms are different. But their study, in this case, is of little interest, in as much as the fat foodstuffs are always irradiated in the absence of air in order to slow down the development of rancidity.

Amongst the many radiolysis products formed, researchers have for a long time believed it possible to find a unique radiolysis product. In fact these products are practically the same as those formed by heating.[110] Nevertheless Letellier and Nawar,[111] by radiolysis of saturated triglycerides, have found a category of compounds — 2-alkylcyclobutanones — which is formed neither by heating, nor by oxidation. Unfortunately the formation of these compounds has never been observed in an irradiated foodstuff.

The study of low volatile and non-volatile radiolysis products is not finished, but there is very little chance, amongst these products, of discovering a unique radiolysis product.

In fact, from the start, the research work carried out to find a detection test has been essentially orientated towards the quantitative differences between the decomposition products depending if they are formed during storage, by heating or by irradiation. The gas chromatography study shows that the chromatograms are very different depending on the type of treatment undergone by the foodstuff.[110] Indeed, it appears that the volatile compounds found in the fats which have been heated normally have much shorter chains than the compounds obtained by irradiation. These recordings can constitute a fingerprint of the irradiated foodstuff. Based on this idea, Nawar and Balboni[112] proposed in 1970 a detection test for irradiated meat. In fact, all the studies realised in this field by Merritt and Navar have a major disadvantage: important quantitative effects have been noticed only for very high irradiated doses (from 10 to 60 kGy and sometimes even higher), far removed from the doses usually employed for the meat irradiation at ambient temperature ($\leq$ 5 kGy). This is undoubtedly the reason why, until now, no precise detection test involving lipids and applicable in normal irradiation conditions has been proposed.

3.4 Modifications Induced in Nucleic Acids (DNA)

Little attention has been given to the effects of radiations on the DNA in foodstuffs, given the lack of nutritional interest in this molecule. However it is the alteration of this very radiosensitive molecule which causes the inactivation of microorganisms or the inhibition of vegetal cells development. So the irradiation effects, in spite of the small amounts of this molecule in foodstuffs, must be discernible.

The action of γ-rays or accelerated electrons on DNA is quite well known.[113,114] The radiation effects, resulting from the action of the solvated electrons and of the OH radicals, can be classified in two main categories: (1) the alterations of the purine and pyrimidine bases; (2) the DNA strand break. DNA is present in all foodstuffs, but the only detection tests proposed at the moment concern meat, fish and shellfish.

3.4.1 Alteration of the nucleic bases

3.4.1.1 Meat, fish, shellfish. Some structures of bases modified by irradiation have been studied, such as 5-hydroxymethyl 2′-deoxyuridine[115] or 8,5′-cycloadenosin 5′-monophosphate and its deoxyanalog.[116] These molecules, which have never been found in foodstuffs, cannot be considered straight away as unique radiolysis products.

Only thymine glycol (5,6-dihydroxy-5,6-dihydrothymine) formed by addition of OH radicals on thymine according to the reaction:

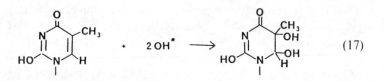

$$\text{(17)}$$

has up till the present been the object of more extensive studies, with the development of a detection test in view.

Schellenberg and Shaeffer[117] and Pfeilsticker and Lucas[118,119] have attempted, not to show directly the formation of this molecule, but to obtain, by chemical methods, a characteristic fragment of it. Schellenberg and Shaeffer have in such a way isolated the ^{3}H-labelled methyl ester of 2-methylglyceric acid after treatment of irradiated DNA by ^{3}H-labelled sodium borohydride followed by methanolic chlohydric acid. Pfeilsticker and Lucas[118,119] have chosen to degrade, by potassium hydroxide, thymine glycol in hydroxy acetone, a molecule which, in the presence of orthoaminobenzaldehyde, gives a fluorescent compound. This method has been applied to the detection of shrimps, mock salmon, chicken breast and calf thymus, irradiated at 5 and 10 kGy, at ambient temperature. On the whole, the results obtained are rather mediocre: in all cases, except for calf thymus, the fluorescence of the unirradiated sample represents more than half the fluorescence given by the sample irradiated at 5 kGy, which is a relatively high dose for these types of foods. This apparently slightly specific test will be rather hazardous in the frequent case where reference samples are not provided.

3.4.2 DNA strand break

3.4.2.1 Meat, shellfish. DNA is a very fragile molecule and rupture of the strands can occur, not only by irradiation, but also by enzyme action, after a chemical treatment or a variation of a physical parameter (for example temperature). This effect cannot therefore be considered a priori as a specific test; its use can only be envisaged in very special cases.

The first research in this field was carried out by Altmann et al.[78] These authors have shown that the ratio RNA/DNA in chicken meat rose considerably after irradiation at 5 and 20 kGy, because of significant breaks of the DNA strands. The duration and conditions of storage can however considerably modify the results obtained, thus removing any value for this test.

More recently, Flegeau et al.[120] and Copin et al.[121] have proposed a detection test for Norway lobsters irradiated at 3 kGy, at a temperature of −20°C and preserved afterwards at this temperature. This test is based on measurement by microfluorometry of the amount of DNA retained on a polycarbonate filter of controlled porosity. This amount is much less for an irradiated than for a unirradiated sample (respectively (0.21 ± 0.14) μg and (2.32 ± 0.25) μg). Given the irradiation and storage temperature (−20°C), this reduction is certainly due to the irradiation. The test, which has the merit of being simple to perform, however has a very limited application. In fact, in fresh foods irradiated at ambient temperature, and a fortiori in irradiated cooked foodstuffs, the degradation of cellular DNA is such that it would appear that no irradiation detection test could be based on the break of DNA strands.[26,122] Under these conditions, it would seem unlikely that the DNA study by agarose gel microelectrophoresis on single cells proposed by Östling and Hofsten[123] would be applicable to irradiated foodstuffs.

The problem caused by lysis, essentially enzymatic, of DNA in foodstuffs, irradiated or not, may possibly be resolved by studying the effects of irradiation, not on cellular DNA, which is a fragile molecule of high molecular weight, but on mitochondrial DNA, which is smaller and therefore more stable, and also protected from the action of cellular enzymes by the mitochondrial wall. By agarose gel electrophoresis, Hasselmann and Marchioni[122] have effectively shown that the mitochondrial DNA extracted of unirradiated beef slices, after storage at +4°C remained mainly in a supercoiled form. If beef slices, wrapped under reduced pressure, are irradiated at 1.5 and 3.0 kGy, the DNA in the supercoiled form disappears completely and transforms into circular DNA (break of one of the strands), then into linear DNA (second break of the strand less then 10 pairs of bases from the first).

Using this method, two slices of beef, one unirradiated, the other ir-radiated at 3 kGy, could still be distinguished after 37 days of preserva-tion at +4°C. The validity of this test, of which the major disadvantage lies in its technical complexity, must however still be confirmed by a number of experiments, concerning in particular the influence of the storage temperature on the stability of the supercoiled DNA.

3.5 Modifications Induced in Vitamins

Vitamins, with the exception of vitamin C, are always present in small amounts in foods and are only correctly determined by relatively complex methods. If we add to this that the radiolysis products of these molecules are generally not well known, it is easily understandable why these molecules have been of little interest as a means of detection test.

Only Thayer[124] has proposed a test based on the disappearance of thiamin in meat and fish products. Its degradation is both dose and temperature dependent. In these conditions, according to Thayer, it is possible at the same time to know if the food has been irradiated and to estimate the dose received (at least if this is > 1 kGy) if the initial content in thiamin and if the conditions of treatment are known.

This test, which does not take into account the instability of thiamin during storage and which necessitates a perfect knowledge of the irradia-tion conditions, does not appear appropriate for the detection of meat and fish products under usual irradiation conditions.

3.6 Modifications Induced by Irradiation but of Unknown Origins

3.6.1 Modification of the electrical conductivity

3.6.1.1 Potatoes. The γ-irradiation provokes a marked reduction of the electrical conductivity of potatoes. This peculiarity which till now has not been explained, is effective for very weak irradiation doses, and has been used by various authors for the development of a detection test.[125-128] The observed effect increases with the irradiation dose, at least up to 0.3 kGy (higher doses are never used to inhibit tubers sprouting). The varia-tions of the electrical conductivity depend mainly however on the species of potato irradiated and on the temperature of storage, and decrease noticeably during storage.[127]

The reliability of this test, which appears insufficient by simple measurements of the electrical conductivity has been improved by deter-mination of the impedance. Hayashi et al.[129] and Hayashi[130] have pro-posed the measurement, for each samples, of the ratio of impedances measured at two different frequencies (50 and 5 kHz). This ratio decreases in irradiated samples. According to these authors, the test employed does not depend on the species of potato irradiated and has a good reliability on condition that at least 10 measurements are performed, the differentiation between two samples, irradiated or unirradiated, is possible with a success probability of at least 95%. An estimation of the absorbed dose is also possible.[130]

3.6.1.2 Fish. The measurement of electrical conductivity has also been applied to the determination of fish irradiation, at doses between 1 and 2 kGy.[131] The first results were interesting but no further works have since been carried out.

3.6.2 Modifications of the composition in volatile compounds

3.6.2.1 Spices. The lack of specificity of the majority of tests proposed throughout this section 3 is due to the fact that they look for the concentration variations of a particular chemical compound, or possibly of a chemical function. This type of test can however become much more specific if it is concerned with simultaneous variations in concentration of a large number of different chemical compounds, allowing therefore the establishment of an actual fingerprint for the foodstuff. This possibility had already been proposed by Nawar and Balboni[112] concerning modifications to the composition of fatty acids of irradiated meat (see section 3.3). In fact it is absolutely unnecessary to identify the radiolysis products formed in order to use them as test compounds, nor, moreover to know from which component of the foodstuff they are formed.

Based on this idea, two very similar studies have been undertaken, both having the detection of irradiated black pepper as an objective.[13,26] The BBS extract obtained after treatment of the sample with hexane[26] or with petroleum ether[13] is analyzed by a capillary column gas chromatography. The chromatogram obtained constitutes the fingerprint of the irradiated black pepper. If the chromatograms obtained by Swallow,[26] contain an insufficient number of peaks to allow any conclusions concerning the validity of this test, those obtained by Hasselmann *et al.*[13] have a much greater number and indicate clearly the limited effect of irradiation at 10 kGy on the chemical composition of black pepper. A single significant peak, existing in both non-treated samples and ethylene oxide treated samples, disappears following irradiation. However, the few disturbances caused by the irradiation to the general appearance of the chromatogram make the application of this method to other samples of black pepper and *a fortiori* to other species, very hazardous.

4 CONCLUSION

The development of a detection test for an irradiated food (or a group of foodstuffs) must involve at least three successive stages. In the first place, a test should be proposed which appears, *a priori*, characteristic of an ionising treatment and sufficiently sensitive to allow a detection at usual irradiation doses. Secondly, it should be verified that this test, if it is not absolutely specific — and none of the tests proposed up till now can claim to be totally specific — only gives weakly positive results when the foodstuff has undergone another technological treatment or simply when it has been in prolonged storage. It must also be verified that this test gives a result which is approximately the same, whatever the origin of the food studied. Finally, a statistical work should be carried out on a large scale in order to demonstrate the validity of this test under actual control conditions, i.e. when a non-irradiated reference sample is not available, and thus to clearly prove its good specificity.

At the end of this bibliographical study devoted exclusively to the physicochemical methods of detection, it must unfortunately be stated that the majority of the tests proposed have not passed the first stage. More often than not, they have only been the object of one publication (or of very similar ones), have never been reused in another laboratory and are then generally forgotten, without doubt because they have an uncertain specificity or/and because they cannot be applied to the more usual conditions of irradiation. It should also be emphasised that the validation of the second stage, not only assumes that a satisfactory test has been found, but also the commitment to a long, methodical and tedious study in order to assure that this test is valid, whatever the origin of the foodstuff and its history before irradiation.

In fact very few tests have given rise to thorough research work in different laboratories. We can mention the detection of irradiation in potatoes by measurement of impedance (but the development of this test suffers from the lack of economic interest in the irradiation of this tuber in the developed countries), in meat by the study of induced modifications in the fatty acids, in dried foodstuffs (especially spices and aromatics), or in food of which at least some part is dehydrated, by the research of induced radicals, either by chemiluminescence and thermoluminescence or by ESR. Of course these methods only allow the control of part of the foodstuffs that we consume. Also none of the methods have really satisfied all the requirements previously mentioned. The effects of irradiation on fatty acids have been the object of a very large number of publications, but, to our knowledge, only one, that of Nawar and Balboni[112] in 1970, actually deals with the detection. The validity of this method for detecting irradiated meat at < 10 kGy is still to be demonstrated. The considerable work realised by Bögl and Heide on the detection of irradiated spices, the most complete on a foodstuff up to the present time, has nevertheless shown all the limits of the proposed method. The most promising test at the moment seems to be the one using ESR for the research of stable radicals formed by irradiation in meat and fish bones, and in the cuticle of shellfish. This test must however soon be investigated by an interlaboratory study at international level.

Generally speaking a number of research programs concerning the detection of irradiated foodstuffs have been elaborated in different countries of the European Community, using not only physicochemical methods, but also biological and microbiological methods, not reviewed in this study, but which have up to now not given very satisfactory results (see bibliographic study of Delincee et al.[16]). However, in spite of all the research work carried out at the present time in the detection field, it seems obvious that the development of the irradiation technique cannot await the establishment of a satisfactory test for each foodstuff.

REFERENCES

1. Joint FAO/IAEA/WHO Expert Committee on the wholesomeness of irradiated food. Geneva, 27 oct. 1980 (WHO/TRS 659).
2. Simic, M.G., Radiation chemistry of water-soluble food components. In: *Preservation of Food by Ionizing Radiation,* Vol. II, E.S. Josephson and M.S. Peterson (eds.), CRC Press, Boca Raton, 1983, pp. 2–73.
3. Nawar, W.W., Radiolysis of non aqueous components of food. In: *Preservation of Food by Ionizing Radiation,* Vol. II, E.S. Josephson and M.S. Peterson (eds.), CRC Press, Boca Raton, 1983, pp. 75–124.
4. Taub, I.A., Reaction mechanisms, irradiation parameters and product formation. In: *Preservation of Food by Ionizing Radiation,* Vol. II, E.S. Josephson and M.S. Peterson (eds.), CRC Press, Boca Raton, 1983, pp. 125–166.
5. Delincee, H., Recent advances in radiation chemistry of lipids. In: *Recent Advances in Food Irradiation,* P.S. Elias and A.J. Cohen (eds.), Elsevier Biomedical, Amsterdam, 1983, pp. 89–114.
6. Delincee, H., Recent advances in radiation chemistry of proteins. In: *Recent Advances in Food Irradiation,* P.S. Elias and A.J. Cohen (eds.), Elsevier Biomedical, Amsterdam, 1983, pp. 129–147.

7. Adam, S., Recent developments in radiation chemistry of carbohydrates. In: *Recent Advances in Food Irradiation*, P.S. Elias and A.J. Cohen (eds.), Elsevier Biomedical, Amsterdam, 1983, pp. 149–170.

8. Brynjolfsson, A., Wholesomeness of irradiated foods: a review, *J. Food Saf.*, 1985, **7**, 107–126.

9. Lafontaine, A. and Bugyaki, L., Etude sur les méthodes d'identification des denrées alimentaires irradiées. EURATOM-EUR 2402 f, Brussels, 1965.

10. Deschreider, A.R., Modifications des constituants de la farine irradiée mises en évidence par: spectrophotométrie, spectropolarimétrie et analyse thermodifférentielle. EURATOM-EUR 441 f, Brussels, 1970.

11. Drawert, F., Emberger, R., Westphal, N., Rolle, K. and Tressl, R., Messbare Veränderungen in bestrahlten Lebensmitteln. Anwendung chromatographischer, elektrophoretischer und spektralphotometrischer Methoden zur Untersuchung von Aromastoffen, Aminosäuren, Fettsaüren, Naturfarbstoffen und Enzymen. EURATOM-EUR 4617 d, Brussels, 1971.

12. Taub, I.A., Free radical reactions in food, *J. Chem. Ed.*, 1984, **61**, 313–324.

13. Hasselmann C., Grimm, P. and Saint Lebe, L., Mise en évidence de l'irradiation des aliments par des méthodes physicochimiques, *Med. Nutr.*, 1986, **22**, 121–126.

14. Hayashi, T., Detection of irradiated foods, *New Food Ind.*, 1986, **28**, 11–16.

15. Delincee, H., Identification of irradiated food. In: *Tilgängliga och tänkbara metoder för identifierung ar bestrolade livsmedel*, Statens Livsmedelverk, Uppsala, Sweden. *Rapport*, 1986, **4**, 7–23.

16. Delincee, H., Ehlermann, D.A.E. and Bögl, K.W., The feasibility of an identification of radiation processed food. An overview. In: *Health Impact, Identification and Dosimetry of Irradiated Foods*, K.W. Bögl, D.F. Regulla and M.J. Suess (eds.), (Neuherberg). ISH-Heft, 1988, **125**, 58–138.

17. Mehringer, W., Elektronenspinresonanzversuche an hochenergetisch bestrahlten und unbestrahlten Kartoffelschalen, *Z. Lebensm. Unters. Forsch.*, 1971, **147**, 278–281.

18. Onderdelinden, D. and Strackee L., ESR as a tool for the identification of irradiated material. In: *The Identification of Irradiated Foodstuffs*, CEC-EUR 5126, Luxembourg, 1974, 127–140.

19. White, E.H., Zafiriou, O., Kägi, H.H. and Hill, J.H.M., Chemiluminescence of luminol: the chemical reaction, *J. Am. Chem. Soc.*, 1964, **86**, 940–941.

20. Bögl, K.W. and Heide, L., Chemiluminescence measurements as an identification method for gamma irradiated foodstuffs, *Radiat. Phys. Chem.*, 1985, **25**, 173–185.

21. Dodd, N.J.F., Swallow, A.J. and Ley, F.J., Use of ESR to identify irradiated food, *Radiat. Phys. Chem.*, 1985, **26**, 451–453.

22. Ikeya, M. and Miki, T., Electron spin resonance dating of animal and human bones, *Science*, 1980, **207**, 977–979.

23. Geoffroy, M. and Tochon-Danguy, H.J., ESR identification of radiation damage in synthetic apatites: a study of the carbone-13 hyperfine coupling, *Calcif. Tissue Int.*, 1982, **34**, 99–102.

24. Desrosiers, M.F. and Simic, M.G., Postirradiation dosimetry of meat by electron spin resonance spectroscopy of bones, *J. Agric. Food Chem.*, 1988, **36**, 601–603.
25. Dodd, N.J.F., Lea, J.S. and Swallow, A.J., ESR detection of irradiated food, *Nature*, 1988, **334**, 387.
26. Swallow, A.J., Some approaches based on radiation chemistry for identifying irradiated foods. In: *Health Impact, Identification and Dosimetry of Irradiated Foods*, K.W. Bögl, D.F. Regulla and M.J. Suess (eds.), (Neuherberg). ISH-Heft, 1988, **125**, 128–138.
27. Lea, J.S., Dodd, N.J.F. and Swallow, A.J., A method of testing for irradiation of poultry, *Int. J. Food Sci. Technol.*, 1988, **23**, 625–632.
28. Goodman, B.A. and McPhail, D.B. Unpublished results (cited by Swallow[26]).
29. Desrosiers, M.F., γ-irradiated seafoods: identification and dosimetry by electron paramagnetic resonance spectroscopy, *J. Agric. Food Chem.*, 1989, **37**, 96–100.
30. Raffi, J., Agnel, J.P. and Kassis, S.R., Identification par résonance paramagnétique électronique de céréales irradiées, *Sci. Aliments*, 1987, **7**, 657–663.
31. Raffi, J., Agnel J.P., Buscarlet, L. and Martin, C.C., Electron spin resonance identification of irradiated strawberries, *J. Chem. Soc. Trans Faraday*, 1988, **80**, 3359–3362.
32. Saint-Lebe, L. and Raffi, J., A point of view about identification of irradiated foods by electron spin resonance. In: *Health Impact, Identification and Dosimetry of Irradiated Foods*, K.W. Bögl, D.F. Regulla and M.J. Suess (eds.), (Neuherberg). ISH-Heft, 1988, **125**, 139–154.
33. Beczner, J., Farkas, J., Watterich, A., Buda, B. and Kiss, I., Study into identification of irradiated ground paprika. In: *The Identification of Irradiated Foodstuffs*, CEC-EUR 5126, Luxembourg, 1974, 255–267.
34. Yang, G.C., Mossoba, M.M., Merin, U. and Rosenthal, I., An EPR study of free radicals generated by gamma-radiation of dried spices and spray-dried fruit powders, *J. Food Qual.*, 1987, **10**, 287–294.
35. Wieser, A. and Regulla D.F., Identification of irradiated paprika by ESR spectroscopy. In: *Health Impact, Identification and Dosimetry of Irradiated Foods*, K.W. Bögl, D.F. Regulla and M.J. Suess (eds.), (Neuherberg). ISH-Heft, 1988, **125**, 155–161.
36. Bögl, K.W. and Heide, L., Die Messung der chemilumineszenz von Zimt-, Curry-, Paprika-, und Milchpowder als Nachweiss einer Behandlung mit ionisierenden Strahlen. (Neuherberg). ISH-Berich 1983, **32**.
37. Bögl, K.W. and Heide, L., Nachweis der Gewurzbestrahlung-Identifizierung gamma bestrahler Gewürze durch Messung der Chemilumineszenz, *Fleischwirtsch.*, 1984, **64**, 1120–1126.
38. Heide, L. and Bögl, K.W., Die Messung der Chemilumineszenz von 16 Gewürzen als Nachweiss einer Behandlung mit ionisierenden Strahlen. (Neuherberg). ISH-Heft, 1984, **53**.
39. Heide, L. and Bögl, K.W., Chemiluminescenzmessungen an 20 Gewürz-sorten. Methode zum Nachweis der Behandlung nit ionisierenden Strahlen, *Z. Lebensm. Unters. Forsch.*, 1985, **181**, 283–288.

40. Meier, W. and Zimmerli, B., Experiments with chemiluminescence measurements. Preliminary results with imported spices. In: *Health Impact, Identification and Dosimetry of Irradiated Foods*, K.W. Bögl, D.F. Regulla and M.J. Suess (eds.), (Neuherberg). ISH-Heft, 1988, **125**, 266–268.

41. Meier, W., Konrad-Glatt, V. and Zimmerli, B., Nachweis bestrahlten Lebensmittel: Chemilumineszenz messungen an Gewürzen und Trockengemüsen, *Mitt. Geb. Lebensm. Hyg.*, 1988, **79**, 217–223.

42. Delincee, H., Ist die Bestrahlung von Gewürzen durch Chemilumineszenz nachweisbar? *Fleischwirtsch.*, 1987, **67**, 1410–1417.

43. Delincee, H., Use of chemiluminescence for identifying irradiated spices. In: *Health Impact, Identification and Dosimetry of Irradiated Foods*, K.W. Bögl, D.F. Regulla and M.J. Suess (eds.), (Neuherberg). ISH-Heft, 1988, **125**, 248–265.

44. Sattar A., Delincee, H. and Diehl, J.F., Detection of gamma irradiation pepper and papain by chemiluminescence, *Radiat. Phys. Chem.*, 1987, **29**, 215–218.

45. Heide, L. and Bögl, K.W., Die Messung der Thermolumineszenz ein neues Verfahren zur Identifizierung Strahlenbehandelter Gewürze. (Neuherberg). ISH-Heft, 1984, **58**.

46. Heide, L. and Bögl, K.W., Analysen Verfahren zur Identifizierung bestrahlter Trockenlebensmittel, *Fresenius Z. Anal. Chem.*, 1985, **320**, 682–683.

47. Heide, L. and Bögl, K.W., Identification of irradiated spices with thermo- and chemiluminescence measurements, *Int. J. Food Sci. Technol.*, 1987, **22**, 93–103.

48. Heide, L. and Bögl, K.W., Thermoluminescence and chemiluminescence investigations of irradiated food. A general survey. In: *Health Impact, Identification and Dosimetry of Irradiated Foods*, K.W. Bögl, D.F. Regulla and M.J. Suess (eds.), (Neuherberg). ISH-Heft, 1988, **125**, 190–206.

49. Albrich, S., Stumpf, E., Heide, L. and Bögl, K.W., Chemilumineszenz- and Thermolumineszenzmessungen zur Identifizierung Strahlenbehandelter Gewürze. Eine Gegenüberstellung beider Verfahren. (Neuherberg). ISH-Heft, 1985, **74**.

50. Moriarty, T.F., Oduko, J.M. and Spyrov N.M., Thermoluminescence in irradiated foodstuffs, *Nature*, 1988, **332**, 22.

51. Heide, L., Albrich, S., Mentele, E. and Bögl, K.W., Thermolumineszenz- und Chemilumineszenzmessungen als Routine-Methoden zur Identifizierung Strahlenbehandelter Gewürze, Untersuchungen zur Festlegung von Grenzwerten für die Unterscheidung bestrahlter von Unbestrahlten proben. (Neuherberg). ISH-Heft, 1987. **109**.

52. Heide, L. and Bögl, K.W., Thermoluminescence and chemiluminescence measurements as routine methods for the identification of irradiated spices. In: *Health Impact, Identification and Dosimetry of Irradiated Foods*, K.W. Bögl, D.F. Regulla and M.J. Suess (eds.), (Neuherberg). ISH-Heft, 1988, **125**, 207–232.

53. Heide, L., Delincee, H., Demmer, D., Eichenauer, D., Von Grabowski, H.U., Pfeilsticker, K., Redl, H., Schilling, M. and Bögl, K.W., Ein erster-

Ringversuch zur Identifizierung Strahlenbehandelter Gewürze mit Hilfe von Lumineszenzmessungen. (Neuherberg). ISH-Heft, 1986, **101**.

54. Heide, L. and Bögl, K.W., Routine application of luminescence techniques to identify irradiated spices. A first counter check trial with 7 different research and food control laboratories. In: *Health Impact, Identification and Dosimetry of Irradiated Foods*, K.W. Bögl, D.F. Regulla and M.J. Suess (eds.), (Neuherberg). ISH-Heft, 1988, **125**, 233–244.

55. Stockhausen, K., Bögl, K.W. and Weise, H.P., A new biochemical technique for detection of radiation treatment of meat, *Atomkerenergie*, 1978, **31**, 184–188.

56. Stockhausen, K., Bögl, K.W. and Weise, H.P., Nachweis der Strahlenbehandlung bei bestrahlten Geflügelfleisch anhand des strahleninduzierten Verlustes von Proteinsulfhydrylgruppen, *Z. Lebensm. Unters. Forsch.*, 1988, **167**, 256–261.

57. Hamm, R., Hofmann, K., Grunewald, T. and Partmann, W., Veränderungen von Aminotransferasen und Muskelproteinen bei der Behandlung von Schweinefleisch mit ionisierenden Strahlen, *Fleischwirtsch.*, 1975, **55**, 1105–1112.

58. Garrison, W.M., The radiation chemistry of amino acids, peptides and proteins in relation to the radiation sterilization of high-protein foods, *Radiat. Eff.*, 1981, **54**, 29–39.

59. Boguta, G. and Dancewicz, A.M., Radiolytic and enzymatic dimerisation of tyrosyl residues in insulin, ribonuclease, papain and collagen, *Int. J. Radiat. Biol.*, 1983, **43**, 249–265.

60. Karam, L.R., Dizdaroglu, M. and Simic, M.G., OH radical-induced products of tyrosine peptides, *Int. J. Radiat. Biol.*, 1984, **46**, 715–724.

61. Dizdaroglu, M., Simic, M.G. and Karam, L.R., Chemical analysis of irradiated foods, *Trans. Am. Nucl. Soc.*, 1985, **49**, 10–13.

62. Gajewski, E., Dizdaroglu, M. and Simic, M.G. OH radical-induced crosslinks of methionine peptides, *Int. J. Radiat. Biol.*, 1984, **46**, 47–55.

63. Dizdaroglu, M. and Simic, M.G., Isolation and characterisation of radiation-induced aliphatic peptide dimers, *Int. J. Radiat. Biol.*, 1983, **44**, 231–239.

64. Simic, M.G., Gajewski, E. and Dizdaroglu, M., Kinetics and mechanism of hydroxyl radical-induced crosslinks between phenylalanine peptides, *Radiat. Phys. Chem.*, 1985, **24**, 465–473.

65. Hasselmann, C., Etude photochimique des acides aminés aromatiques et de l'histidine en solution aqueuse. Photolyse directe et sensibilisée à 254 nm. Thesis (University of Strasbourg), 1976, 140 pp.

66. Dizdaroglu, M. and Simic, M.G., Radiation induced conversion of phenylalanine to tyrosine, *Radiat. Res.*, 1980, **83**, 437.

67. Dizdaroglu, M., Gajewski, E., Simic, M.G. and Krutzsch, H.C., Identification of some OH radical-induced products of lysozyme, *Int. J. Radiat. Biol.*, 1983, **43**, 185–193.

68. Karam, L.R. and Simic, M.G., Detecting irradiated foods use of hydroxyl radical biomarkers, *Anal. Chem.*, 1988, **60**, 1117–1119.

69. Karam, L.R. and Simic, M.G., Orthotyrosine as a marker in post irradiation dosimetry (PID) of chicken. In: *Health Impact, Identification and Dosimetry of Irradiated Foods*, K.W. Bögl, D.F. Regulla and M.J. Suess (eds.), (Neuherberg). ISH-Heft, 1988, **125**, 297–304.

70. Hart, R.J., White, J.A. and Reid, W.J., Technical note: Occurence of o-tyrosine in non-irradiated foods, *Int. J. Food Technol.*, 1988, **23**, 643–647.

71. Hasselmann C. and Laustriat G., Photochimie des acides aminés aromatiques en solution. I DL-Phénylalanine, DL-Tyrosine et L-Dopa, *Photochem. Photobiol.*, 1973, **17**, 275–294.

72. Schubert, J. and Esterbauer, H., Chemical and biological techniques for identifying irradiated foods and food constituents — carbonyls and 2-deoxygluconic acid. In: *The Identification of Irradiated Foodstuffs*, CEC-EUR 5126, Luxembourg, 1974, 75–95.

73. Partmann, W. and Keskin, S., Radiation-induced changes in the patterns of free ninhydrin-reactive substances of meat, *Z. Lebensm. Unters. Forsch.*, 1979, **168**, 389–393.

74. Partmann, W. and Schlaszus, H., Investigation on the origin of two radiation-induced compounds in irradiated meat, *Z. Lebensm. Unters. Forsch.*, 1980, **171**, 1–4.

75. Adriaanse, A., A technique for the detection of low dose gamma irradiation in cooked dutch shrimp (Crangon Crangon) by disc protein patterns, *J. Sci. Food Agr.*, 1971, **22**, 498–499.

76. Van der Stichelen Rogier, M., Utilisation de techniques électrophorétiques en vue de l'identification de certaines denrées alimentaires. In: *The Identification of Irradiated Foodstuffs*, CEC-EUR 5126, Luxembourg, 1974, 45–60.

77. Deschreider, A.R. and Vigneron, J.M., Recherches sur l'identification de crevettes et de cabillaud irradiés. In: *The Identification of Irradiated Foodstuffs*, CEC-EUR 5126, Luxembourg, 1974, 17–26.

78. Altmann, H., Klein, W. and Dolejs, I., Untersuchungen zur Identifizierung von Bestrahltem Fleisch. In: *The Identification of Irradiated Foodstuffs*, CEC-EUR 5126, Luxembourg, 1974, 61–73.

79. Bruaux, P., Bugyaki, L., Claeys, F., Lafontaine, A. and Van Der Stichelen Rogier, M., Electrophorèse et immunoélectrophorèse des protéines de la viande, du poisson et de l'oeuf irradiés, *Atompraxis*, 1970, **16**, 390–394.

80. Rossler, I. and Hamm, R., Veränderungen der Muskelproteine bei der Behandlung von Rind- und Schweinefleisch mit energiereichen Strahlen. 1 Mitteilung. Veränderungen in der Löslichkeit der Sarkoplasmaproteine, *Fleischwirtsch.*, 1979, **59**, 711–718.

81. Rossler, I. and Hamm, R., Veränderungen der Muskelproteine bei der Behandlung von Rind- und Schweinefleisch mit energiereichen Strahlen. 4 Mitteilung, Sensorisch wahrnehmbare Veränderungen des Schweinefleisches. Allgemeine Schlussfolgerungen, *Fleischwirtsch.*, 1979, **59**, 1325–1331.

82. Taub, I.A., Robbins, F.M., Simic, M.G., Walter, J.E. and Wierbicki, E., Effect of irradiation on meat proteins, *Food Technol.*, 1979, **33**, 184–193.

83. Radola, B.J., Identifizierung von Bestrahlten Fleisch mit Hilfe der Dünn-schicht — Gelchromatographie und Dünnschicht-Isoelektrischen Fokus-sierung. In: *The Identification of Irradiated Foodstuffs*, CEC-EUR 5126, Luxembourg, 1974, 27–44.

84. Delincee, H. and Paul, P., Protein aggregation in food models: effect of irradiation and lipid oxidation, *J. Food Proc. Preserv.*, 1981, **5**, 145–149.

85. Hajos, G. and Delincee, H., Structural investigation of radiation-induced aggregates of ribonuclease, *Int. J. Radiat. Biol.*, 1983, **44**, 333–342.

86. Morre, J. and Janin, F., Méthode de détection de la radiopasteurisation des oeufs congelés, *Bull. Acad. Vét.*, 1972, **45**, 191–193.

87. Briski, B. Qualitative and quantitative characterization of enzymatic changes for the identification of irradiated foods. In: *Health Impact, Identification, and Dosimetry of Irradiated Foods*, K.W. Bögl, D.F. Regulla and M.J. Suess (eds.), (Neuherberg). ISH-Heft, 1988, **125**, 320–341.

88. Münzner, R., Nachweis einer Strahlenbehandlung bei Champignons, *Z. Lebensm. Unters. Forsch.*, 1973, **151**, 318–319.

89. Zehnder, H.J., Nachweis bestrahlter Champignons (Agaricus bisporus) — Theorie und Praxis — *Mitt. Geb. Lebensm. Hyg.*, 1988, **79**, 362–370.

90. Raffi, J., Agnel, J.P., Dauberte, B., d'Urbal, M. and Saint-Lebe, L., Gamma radiolysis of starches derived from different foodstuffs. Part I. Study of some induced carbonyl derivatives, *Stärke*, 1981, **33**, 188–192.

91. Raffi, J., Frejaville, C., Dauphin, J.F., Dauberte, B., d'Urbal M. and Saint-Lebe, L., Gamma radiolysis of starches derived from different foodstuffs. Part II. Study of induced acidities, *Stärke*, 1981, **33**, 235–240.

92. Raffi, J. Agnel, J.P., Dauberte, B. and Saint-Lebe, L., Gamma radiolysis of starches derived from different foodstuffs. Part III. Study of induced hydrogen peroxide, *Stärke*, 1981, **33**, 269–271.

93. Raffi, J., Dauberte, B., d'Urbal, M., Pollin, C. and Saint-Lebe, L., Gamma radiolysis of starches derived from different foodstuffs. Part IV. Study of radiodepolymerization, *Stärke*, 1981, **33**, 301–306.

94. Berger, G. and Saint-Lebe, L., Un test d'irradiation de l'amidon de maïs, basé sur l'emploi de l'acide 2-thiobarbiturique, *Stärke*, 1969, **21**, 205–211.

95. Winchester, R.V., Detection of corn starch irradiated with low doses of gamma rays, *Stärke*, 1973, **25**, 230–233.

96. Stewart, A.B. and Winchester, R.V., Detection of corn starch irradiated with low doses of gamma rays. Part 3. Positive identification of malonaldehyde in irradiated starch by isolation and characterisation as the thiobarbiturate derivative, *Stärke*, 1975, **27**, 9–11.

97. Scherz, H., Über die Bildung von Malondialdehyde bei der Bestrahlung von Lebensmitteln, *Chem. Mikrobiol. Technol. Lebensm.*, 1972, **1**, 103–105.

98. Scherz, H., Evaluation of the structure of the deoxy compounds found in irradiated starch and determination of these products in irradiated wheat and wheat flour. In: *The Identification of Irradiated Foodstuffs*, J. Smeets (ed.), CEC-EUR 4695, Luxembourg, 1970, 27–36.

99. Adam, S. Radiolysis of α, α'-trehalose in concentrated aqueous solution; the effect of co-irradiated proteins and lipids, *Int. J. Radiat. Biol.*, 1982, **42**, 531–544.

100. Winchester, R.V. and Le Roux, H.C., A possible chemical method for the detection of irradiated potatoes, *Chem. Mikrobiol. Technol. Lebensm.*, 1976, **4**, 170–172.
101. Den Drijver L., Holzapfel, C.W. and Vander Linde, H.J., High performance liquid chromatographic determination of D-arabino-hexos-2 ulose (D glucosone) in irradiated sugar solutions. Application of the method to irradiated mango, *J. Agric. Food Chem.*, 1986, **34**, 758–762.
102. Dwight, C.H. and Kersten, H., The viscosity of sols made from X-irradiated apple pectin, *J. Phys. Chem.*, 1938, **42**, 1167–1169.
103. Mohr, E. and Wichmann, G., Viskositätserniedringungen als Indiz für eine Cobaltbestrahlung von Gewürzen, *Gordian*, 1985, **85**, 96.
104. Heide, L., Mohr, E., Wichmann, G., Albrich, S. and Bögl, K.W., Viskositätsmessung. Ein Verfahren zur Identifizierung Strahlenbehandelter Gewürze. (Neuherberg). ISN-Heft 1987, **120**.
105. Heide L., Mohr, E, Wichmann, G. and Bögl, K.W., Are viscosity measurements a suitable method for the identification of irradiated spices. In: *Health Impact, Identification and Dosimetry of Irradiated Foods*, K.W. Bögl, D.F. Regulla and M.J. Suess (eds.), (Neuherberg). ISH-Heft, 1988, **125**, 176–189.
106. Nawar, W.W., Reaction mechanism in the radiolysis of fat: a review, *J. Agric. Food Chem.*, 1978, **26**, 21–25.
107. Merritt, C. and Taub, I.A., Commonality and predictability of radiolytic products in irradiated meats. In: *Recent Advances in Food Irradiation*, P.S. Elias and A.J. Cohen (eds.), Elsevier Biomedical, Amsterdam, 1983, pp. 27–57.
108. Vajdi, M. and Merritt, C., Identification of adduct radiolysis products from pork fat, *J. Am. Oil Chem. Soc.*, 1985, **62**, 1252–1260.
109. Nawar, W.W., Analysis of volatiles as a method for the identification of irradiated foods. In: *Health Impact, Identification and Dosimetry of Irradiated Foods*, K.W. Bögl, D.F. Regulla and M.J. Suess (eds.), (Neuherberg). ISN-Heft 1988, **125**, 287–296.
110. Nawar, W.W., Comparison of chemical consequences of heat and irradiation treatment of lipids. In: *Recent Advances in Food Irradiation*, P.S. Elias and A.J. Cohen (eds.), Elsevier Biomedical, Amsterdam, 1983, pp. 115–127.
111. Letellier, P.R. and Nawar, W.W., 2-alkylcyclobutanones from radiolysis of triglycerides, *Lipids*, 1972, **1**, 75–76.
112. Nawar, W.W. and Balboni, J.J., Detection of irradiation treatment in foods, *J. Assoc. Off. Anal. Chem.*, 1970, **53**, 726–729.
113. Schussler, H. and Hartmann, H., Chromatographic studies on the radiolysis of DNA in aqueous solution, *Int. J. Radiat. Bot.*, 1985, **47**, 509–521.
114. Dizdaroglu, M. and Bergtold, D.S., Characterization of free radical induced base damage in DNA at biologically relevant levels, *Anal. Biochem.*, 1986, **156**, 182–188.
115. Frenkel, K, Cummings, A., Solomon, J., Cadet, J., Steinberg, J.J. and Teebor, G.W., Quantitative determination of the 5-(hydroxymethyl)-uracil. moiety in the DNA of gamma irradiated cells, *Biochemistry*, 1985, **24**, 4527–4533.

116. Fuciarelli A.F., Miller, G.G. and Raleigh, J.A., An immunochemical probe for 8,5' cycloadenosine-5' monophosphate and its deoxyanalog in irradiated nucleic acid, *Radiat. Res.,* 1985, **104**, 272–283.

117. Schellenberg, K.A. and Shaeffer, J., Formation of methyl ester of 2-methylglyceric acid from thymine glycol residue; a convenient new method for determining radiation damage to DNA, *Biochemistry,* 1986, **25**, 1479–1482.

118. Pfeilsticker, K. and Lucas, J., The fluorometric determination of thymine glycol as criterion of the treatment of foodstuffs (biological material) with ionizing radiation, *Angew. Chem.,* 1987, **26**, 340–341.

119. Pfeilsticker, K. and Lucas, J., Radiation-damaged DNA as dosis-correlated indicator for an ionizing irradiation of moisture containing food. In: *Health Impact, Identification and Dosimetry of Irradiated Foods,* K.W. Bögl, D.F. Regulla and M.J. Suess (eds.), (Neuherberg). ISH-Heft, 1988, **125**, 308–312.

120. Flegeau, J, Copin, M.P. and Bourgeois, C.M., Detection of irradiated Norway lobsters by DNA elution method. In: *Health Impact, Identification and Dosimetry of Irradiated Foods,* K.W. Bögl, D.F. Regulla and M.J. Suess (eds.), (Neuherberg). ISH-Heft, 1988, **125**, 453–460.

121. Copin, M.P., Bourgeois, C.M. and Le Grand M., Détection des produits alimentaires ionisés, *Bretagne Agro-Alim.,* 1988, **15**, 2–5.

122. Hasselmann, C. and Marchioni, E., La détection des aliments ionisés, *Ann. Falsf. Exp. Chim.,* 1989, **82**, 169–175.

123. Östling, O. and v. Hofsten, B., Radiation-induced DNA strand breaks in single cells. In: *Health Impact, Identification and Dosimetry of Irradiated Foods,* K.W. Bögl, D.F. Regulla and M.J. Suess (eds.), (Neuherberg). ISH-Heft, 1988, **125**, 453–460.

124. Thayer, D.W., Residual thiamin analysis as a method for the identification of irradiated foods. In: *Health Impact, Identification and Dosimetry of Irradiated Foods,* K.W. Bögl, D.F. Regulla and M.J. Suess (eds.), (Neuherberg). ISH-Heft, 1988, **125**, 313–319.

125. Scherz, H., Conductivity measurements as a method for the differentiation between irradiated and non irradiated potatoes. In: *The Identification of Irradiated Foodstuffs,* J. Smeets (ed.), CEC-EUR 4695, Luxembourg, 1970, 13–21.

126. Scherz, H., Nachweis einer erfolgten Bestrahlung bei Kartoffeln durch Messung der electrischen Leitfähigkeit. In: *The Identification of Irradiated Foodstuffs,* CEC-EUR 5126, Luxembourg, 1974, 193–202.

127. Van. Dongen, R., Onderdelinden, D. and Stackee, L., Additional measurements of conductivity changes in potatoes induced by ionizing radiation. In: *The Identification of Irradiated Foodstuffs,* CEC-EUR 5126, Luxembourg, 1974, 203–215.

128. Hayashi, T. and Ehlermann, D., Identification of irradiated potatoes by means of electrical conductivity, *Rept. Natl. Food Res. Inst.,* 1980, **86**, 91–97.

129. Hayashi, T., Iwamoto, M. and Kawashima, K., Identification of irradiated potatoes by impedance measurements, *Agric. Biol. Chem.,* 1982, **46**, 905–912.

130. Hayashi, T., Identification of irradiated potatoes by impedemetric methods. In: *Health Impact, Identification and Dosimetry of Irradiated Foods,* K.W. Bögl, D.F. Regulla and M.J. Suess (eds.), Neuherberg, ISH-Heft, 1988, **125**, 432–452.
131. Ehlermann, D., The possible identification of an irradiation treatment of fish by means of electrical (ac) resistance measurements, *J. Food. Sci.,* 1972, **37**, 501.

Chapter 7

COMPARATIVE EFFECTIVENESS OF GAMMA-RAYS AND ELECTRON BEAMS IN FOOD IRRADIATION

TORU HAYASHI

National Food Research Institute, Ministry of Agriculture, Forestry and Fisheries, Kannondai, Tsukuba, Ibaraki, Japan

1 INTRODUCTION

Ionizing radiations which can be used for the treatment of foods are gamma-rays from Co-60 and Cs-137, accelerated electrons from a machine at an energy of 10 MeV or lower and X-rays from a machine at an energy of 5 MeV or lower. Joint FAO/IAEA/WHO Expert Committee on the Wholesomeness of Irradiated Food held in 1980 concluded that the foods irradiated at overall average doses up to 10 kGy with the radiation listed above are wholesome for human consumption. While most of the commercial food irradiations are conducted with gamma-rays from Co-60, accelerated electrons are increasingly utilized for treating foods. For example wheat is disinfested with accelerated electrons at an energy of 1·4 MeV at Port Odessa in Russia and deboned poultry is pasteurized with accelerated electrons in France. Spices are also treated with electron beams in France and Denmark. X-rays are generated by converting electrons from an electron accelerator with the aid of heavy metal target. Although X-rays have several advantages such as high penetration capacity and high dose rate, they are not used for practical irradiation of foods and medical products because of the poor conversion ratio of accelerated electrons to X-rays.

Gamma-ray and X-ray are electromagnetic radiations, which are converted into fast electrons in the medium through which they pass by Compton scattering, photoelectric absorption and pair-production. The reactions in the foods irradiated with gamma-rays or X-rays are brought about mainly by the fast electrons thus formed, so the reactions caused by gamma-rays or X-rays and those by electron beams from an accelerator are essentially the same. Whether or not there is any difference in the

effect on foods and living organisms between gamma-rays and accelerated electrons is controversial, although a lot of data are available on the
comparative effectiveness of the two types of radiation.

An important difference between gamma-rays and accelerated electrons is the penetration capacity in materials. The penetration capacity
of gamma-rays is much higher than that of accelerated electrons. The
penetration capacity of accelerated electrons increases with their energy
and the electrons at 10 MeV can penetrate ~4 cm of material if its density is 1 g/cm^3 (Fig. 1). The energy which the electrons give to the material
is different depending upon the depth in the material. The energy
gradually increases with the distance from the surface and then rapidly
decreases, while the energy which gamma-rays give to the material
decreases with the depth. It is quite important to irradiate a sample at the
same dose by taking account of the penetration capacity, when the comparative study on the effects of gamma-rays and electron beams is carried
out. Gotoh et al.[1] compared the sterilization effects of gamma-rays and

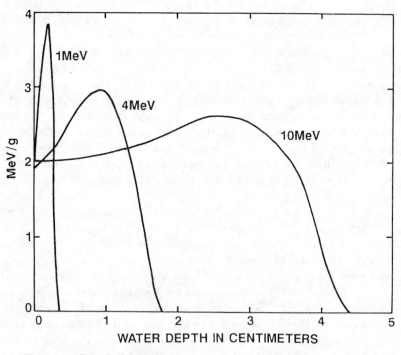

FIG. 1. Depth-dose curves of electrons in water.

electron beams and found that the sterility of irradiated ham was different between gamma irradiation and electron irradiation. The microbial growth during storage was faster in electron-irradiated ham than in gamma-irradiated ham. This difference was attributed to the low penetration capacity of electron beams resulting in insufficient sterility inside the sample, because the thickness of the sample was 1 cm, which is too thick to be efficiently irradiated with electron beams at 1·5 MeV. The study by Gotoh et al.[1] demonstrates the importance of irradiating a sample with radiation with a sufficient penetration capacity when the effects of gamma-rays and electron beams are compared.

Another important difference between gamma-rays and accelerated electrons is dose rate. The dose rates of gamma-rays from commercial Co-60 sources are 1–100 Gy/min, while those of electron beams from electron accelerators are 10^3–10^6 Gy/s. Ideally a comparison of the effects of different types of ionizing radiation should be carried out at the same dose rate, but this has been difficult due to the design of irradiators. Therefore very little work has been undertaken to compare the effects of different types of radiation at the same dose rate in food irradiation. Although dose rate is generally considered not to be a very critical parameter in food irradiation, there are many reports that dose rate can influence the effect of irradiation on foods and living organisms, and the difference in the effect in food irradiation between gamma-rays and electron beams has been ascribed to the difference in the dose rate rather than the type of radiation. Extremely high dose rate of accelerated electrons will bring about anoxic conditions in the reaction system, because the oxygen in the system is depleted at a rate greater than it can be replaced by diffusion process transferring atmospheric oxygen into the system. Oxygen facilitates indirect reactions through the formation of active oxygen groups including superoxide anion radical and hydrodioxyl radical, and the anoxic conditions brought about by extremely high dose rates of irradiation may reduce the rate of indirect reactions and alter the end result of irradiation. At higher dose rates, free radicals can be formed in such high concentrations that recombination of radicals rather than reaction of radicals with food components is favored, which reduces the amount of indirect reactions. Thus anoxic conditions and the increased radical-radical reactions can be responsible for the difference in the chemical and biological effects between gamma-rays and accelerated electrons.

Energy is also an important parameter which varies depending upon type of radiation. It is generally recognized that the energy does not

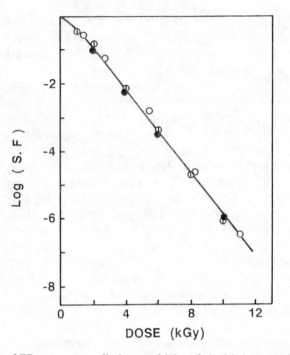

FIG. 2. Effect of EB energy on radiation sensitivity of air-dried *B. pumilus* spores irradiated in air. o, 0·5 MeV; ⊕, 1·0 MeV; •, 3·0 MeV. (From Watanabe *et al.*[2]).

directly influence the effect in food irradiation. For example, Watanabe *et al.*[2] reported that the energy of electron beams did not influence the D_{10} value of *Bacillus pumilus* (Fig. 2).

It is very difficult to draw a definite conclusion on the difference in the effectiveness in food irradiation between gamma-rays and electron beams based on published data. This chapter deals with as many reports as possible on the comparative effectiveness of gamma-rays and electron beams and on the effect of dose rate on chemical reactions and living organisms, whether or not they demonstrate any dependency of the effect of irradiation on dose rate and type of radiation.

2 EFFECT ON CHEMICAL REACTIONS

Chemical reactions caused by ionizing radiation such as gamma-rays, accelerated electrons and X-rays are dependent upon dose and the amount of radiolytic products increases with dose. It is well known that the chemical changes induced by radiation are influenced by temperature, water content and oxygen concentration. Dose rate also influences the chemical reactions caused by ionizing radiation.

2.1 Chemical Reaction in Simple Systems

Higher dose rate of irradiation resulted in a lower degree of chemical reactions in solution. Mead[3] investigated the rate of oxidation of linoleic acid in borate buffer (pH 9) caused by irradiation with X-rays and observed the decrease in the oxidation rate with the increase in dose rate in the dose-rate range between 10 R/min and 540 R/min. Tartrazine in aqueous solution degradated to lower extents at higher dose rates when it was irradiated with electron beams at various dose rates.[4]

The degradation of polyethylene film was also dependent upon dose rate and the formation of carboxylic acids increased with the decrease in dose rate when polyethylene film was irradiated with accelerated electrons at various dose rates.[5] The amount of carboxylic acids formed from polyethylene film irradiated with electron beams at 1 kGy/s was smaller than that from film irradiated with gamma-rays at 6 kGy/h, when the dose applied to the film was the same.[5] The dose-rate effects are generally discernible when electron irradiation at high dose rates and gamma irradiation at relatively low dose rates are compared as described by Taub et al.[6]

However there are reports that the chemical reactions in simple systems are independent of dose rate and type of radiation. Diehl[7] did not observe any dose-rate dependency of the destruction of thiamine when thiamine solution was irradiated with accelerated electrons at various dose rates. Losses of vitamin E in sunflower oil and in oil-water-emulsion induced by irradiation were independent of the type of radiation (electron, gamma-ray, X-ray) and of the dose rate.[8] Loss of tochopherol and tochopherol acetate was also independent of dose rate of electron beams.[9]

The contribution of dose rate to competitive pathways for reaction and to the relative yields of the products will be influenced by several parameters: the presence of other reactive constituents, additives or radiolytic products can influence the dose-rate effect in chemical reactions. The temperature, viscosity and water content can also influence the significance of dose-rate effect in radiation-induced reactions, because the diffusion of radicals in/from spurs is influenced by these parameters. Gopal[10] reported that the dose-rate effects in chemical reaction were contradictory. Although polypropylene and cellulose and aqueous solution of alkyl dimethyl benzalkonium chloride were more stable to radiation at high dose rates than at low dose rates, plasticized PVC tubes showed more discoloration when irradiated at higher dose rates than at lower dose rates.[10] Gopal concluded that dose-rate effects could not be predicted easily for a product which was a complex mixture of ingredients.

2.2 Chemical Reaction in Foods

Food comprises various kinds of compounds such as proteins, carbohydrates, lipids, minerals etc. The effect of dose rate on the radiation-induced reactions in complex systems such as foods may differ from those in simple systems such as solution of chemical compound and polymer. If dose-rate effect is attributed mainly to the increased radical-radical reactions at an extremely high dose rate, the dose-rate effect involving primary radicals is not expected to be significant in a system as complex as food, because pseudo-first order reactions with the main components will predominate at almost all dose-rate range. Reactions of primary radicals with proteins, carbohydrates and lipids are favored because of the high concentrations of these components and produce secondary radicals. Only a minor amount of combination of the primary radicals will take place. The secondary radicals thus formed will react with the major food components again and some of them will be preferentially converted to radicals of greater stability. Thus the radical reactions will proceed in irradiated foods. Consequently, it can be expected that there would be little or no difference in the yields of radiolytic products in foods between electron irradiation and gamma irradiation or between irradiation at a low dose rate and irradiation at a high dose rate, if only radical-radical reaction is taken into consideration. On the other hand, if only anoxia caused by high dose rates of radiation plays an important role in dose-rate effect in chemical reactions, reactions caused by high dose rates of irradiation will be smaller than those by low dose rates of irradiation and the dose-rate effect in foods will be as discernible as in simple systems.

The comparative effects of gamma-rays and electron beams on beef and pork have been extensively investigated, and in most of the studies no difference in the effect between the two types of radiation was observed. Josephson et al.[11] investigated the effects of various treatments of beef on the amino acid contents and found that there was no significant destruction of cystine, methionine and tryptophan, the three amino acids considered most sensitive to ionizing radiation, in beef immediately after the treatments and during storage for 15 months at room temperature

TABLE 1

EFFECTS OF DIFFERENT PROCESSING METHODS UPON CYSTINE, METHIONINE, AND TRYPTOPHAN CONTENT OF ENSYME-INACTIVATED BEEF (wt. %)

Amino acid	Storage (months)	Treatment			
		Frozen control	Heat sterilized $(F_0 = 5\cdot8)$	Co-60 $(4\cdot7-7\cdot1$ Mrad)[a]	Electron $(10$ MeV, $4\cdot7-7\cdot1$ Mrad)[a]
Cystine	0	0·28	0·29	0·26	0·28
	15	0·27	0·25	0·27	0·30
Methionine	0	0·53	0·58	0·57	0·59
	15	0·54	0·54	0·54	0·56
Tryptophan	0	0·25	0·26	0·25	0·26
	15	0·24	0·23	0·22	0·23

[a]Packaged beef air-evacuated to internal pressure (IP) of ~100 mmHg. IP at start and after irradiation and thawing was ~250 mm and 350 mmHg, respectively. Temperature of product was −40 to −5°C during irradiation. (From Josephson et al.[11]).

(Table 1). The contents of the three amino acids were slightly higher in electron-irradiated beef than gamma-irradiated beef but the difference in the contents of the three amino acids was not significant. It is also reported[12] that the tryptophan, histidine and cystine/cysteine in beef were destroyed to the same degree by gamma irradiation and electron irradiation.

Josephson et al.[11] did not get any consistent results on the contents of thiamine, riboflavin, niacin and Vitamin B_6 between gamma-irradiated beef and electron-irradiated beef (Table 2), and concluded that there was no significant difference in the degradation of vitamins between gamma-rays and electron beams. However there is a report[13] with the result that the loss of thiamine in pork was larger after gamma irradiation than after electron irradiation when the sample was irradiated at the same dose under frozen conditions (Table 3). Restricted diffusion of oxygen and mobility of radicals under frozen conditions might be responsible for the difference between the effect of gamma irradiation at a low dose rate and that of electron irradiation at a high dose rate.

The radiolytic products of fat were also investigated by irradiating beef with gamma-rays or electron beams.[14] The formations of heptane, dimethyl sulfide, alkenes, aldehydes and thioalkanes in gamma-irradiated beef and electron-irradiated beef were quantitatively the same.[14]

TABLE 2
EFFECT OF DIFFERENT PROCESSING METHODS UPON THIAMINE, RIBOFLAVIN, NIACIN AND VITAMIN B_6 CONTENT OF ENZYME-INACTIVATED BEEF (mg/kg)

		Treatment			
	Storage (months)	Frozen	Heat sterilized	Co-60 (4·7–7·1 Mrad)	Electron (4·7–7·1 Mrad)
Thiamin	0	0·97	0·63	0·83	0·77
	15	0·68	0·14	0·21	0·26
Riboflavin	0	2·80	2·63	2·83	2·60
	15	1·69	2·60	2·60	1·46
Niacin	0	48·6	48·1	48·8	46·8
	15	57·2	54·9	50·1	44·5
Vitamin B_6	0	2·50	2·13	3·93	5·20
	15	0·97	0·57	0·35	0·42

(From Josephson et al.[11]).

TABLE 3
EFFECT OF PROCESSING ON THIAMIN IN PORK

Treatment	Temp. (°C)	Dose Mrad	Retention (%)
Thermal,	116		12
F_0 = 6.0	121		9
Gamma rays	−45	1·5	72
(^{137}Cs)		3·0	50
		4·5	40
		6·0	35
		7·5	27
Electrons	−20	1·2	82
(LINAC)		2·4	68
		3·6	57
	−45	1·5	83
		3·0	75
		4·5	66
		6·0	58
		7·5	52
		9·0	50

(From Thomas et al.[13]).

The functional qualities of protein were not significantly different between gamma-irradiated and electron-irradiated dehydrated plasmas.[15]
Hayashi *et al.*[15] investigated the changes in the functional qualities of
dehydrated blood plasma caused by various treatments for decontamination and found that there was no difference in the solubility (Table 4) and
HPLC chromatogram (Fig. 3) between gamma-irradiated and electron-

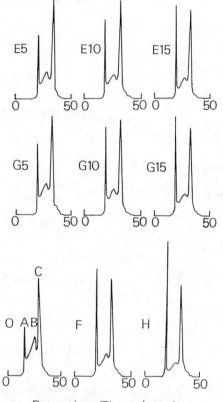

Retention Time (min)

FIG. 3. Liquid chromatograms of dehydrated plasma. O, non-treated plasma;
G5, plasma gamma-irradiated at 5 kGy; G10, plasma gamma-irradiated at 10
kGy; G15, plasma gamma-irradiated at 15 kGy; E5, plasma electron-irradiated
at 5 kGy; E10, Plasma electron-irradiated at 10 kGy; E15, Plasma electron-
irradiated at 15 kGy; F, fumigated plasma; H, heat treated plasma; A,
macromolecular peptide; B, gamma-globulin; C, serum albumin. (From Hayashi
et al.[15]).

TABLE 4
PROPERTIES OF TREATED DEHYDRATED PLASMA

Treatment	Solubility (%)	SH-groups (μmol SH/ g prot.)	EAI (m^2/g)	Hydrophobicity
Non-treated	98·8 ± 3·7	1·77 ± 0·28	59·17 ± 7·74	2500
Gamma-5 kGy	99·6 ± 7·9	1·56 ± 0·19	59·10 ± 6·94	2700
Gamma-10 kGy	95·2 ± 3·9	1·51 ± 0·24	55·41 ± 5·28	2850
Gamma-15 kGy	95·0 ± 4·4	1·47 ± 0·26	52·13 ± 7·29	2975
Electron-5 kGy	99·6 ± 4·4	1·73 ± 0·29	58·62 ± 6·96	2550
Electron-10 kGy	97·8 ± 6·0	1·70 ± 0·23	61·63 ± 7·05	2750
Electron-15 kGy	95·6 ± 5·6	1·67 ± 0·18	56·18 ± 8·78	2800
Fumigated	58·5 ± 3·9*	1·34 ± 0·24*	60·06 ± 2·83	2225
Heated (solution)	100·3 ± 6·8	0·68 ± 0·14*	44·18 ± 5·37*	2250
Heated (dry)	28·3 ± 4·3*	—	—	—

The mean values and the standard deviations are from 5 measurements.
*Significantly different from non-treated sample, $P > 0.05$ (From Hayashi et al.[15]).

irradiated samples. Electron beams influenced the SH-groups, emulsifying activity and hydrophobicity of the plasma to a greater degree than gamma-rays, but the difference between the plasma samples irradiated with the two types of radiation was not significant (Table 4). Fumigation with ethylene oxide gas and heat treatment affected these functional qualities of plasma to a significantly greater extent than radiation treatments.

The difference between gamma-irradiated and electron-irradiated foods can not be organoleptically detected. There was no significant difference in the acceptability of roast beef irradiated at doses of 4·7–7·1 Mrad at −30°C between the treatment with Co-60 gamma-rays and the treatment with accelerated electrons (Table 5).[16] Heiligman and Rice[17] compared gamma-irradiated and electron-irradiated codfish cakes and did not find any detectable differences in the acceptability that could be attributed to the type of irradiation.

Most of the reports on the chemical changes in irradiated foods demonstrated that there was no significant difference in the effect on chemical reactions in foods between gamma-rays and accelerated electrons. These data suggest that the chemical reactions caused by gamma irradiation and those by electron irradiation are qualitatively and quantitatively almost the same in food irradiation.

TABLE 5
CONSUMER ACCEPTANCE OF ROAST BEEF

No. of raters	Average acceptance rating[a]		
	Co-60[b]	Electron[b]	Control
32	5·5	5·4	5·5
32	6·2	5·6	6·1
32	5·4	5·8	6·0
32	6·6	5·9	6·2
32	5·6	6·3	6·0
30	5·0	5·6	4·8
30	5·8	6·3	5·4

[a]9-point hedonic scale: 9 = like extremely; 5 = neither like nor dislike; 1 = dislike extremely.
[b]Irradiation dose 4·7–7·1 Mrad at −30°C ± 10°C.
(From Josephson et al.[16]).

3 EFFECT ON MICROORGANISMS

Very little work has been done on the comparative effect of different types of radiation on living organisms by irradiating samples at the same dose rate. Williams and Hendry[18] compared the effects of gamma-rays, accelerated electrons and X-rays on the production of tail necrosis in mouse. The doses for the production of tail necrosis in 50% of mice (ND_{50}) were not significantly different among the radiations tested, when the samples were irradiated at the same dose rate of 200 rad/min (Table 6). These results suggest that the effect of radiation on living organisms does not directly depend on the type of radiation and that the difference, if any, in the biological effect among gamma-rays, accelerated electrons and X-rays is attributable to the difference in the dose rate.

TABLE 6
ND_{50} VALUES AND RATIOS

Primary source		$ND_{50} \pm S.E.$ (rad)	ND_{50} ratio $\pm S.E.$ (1/RBE)
Energy	Radiation		
300 kV	X	2820 ± 50	0·91 ± 0·05
1·1 and 1·3 MeV	γ(Co-60)	3100 ± 150	1·00
4 MV	X	3080 ± 210	0·99 ± 0·08
10 MeV	e^-	2970 ± 70	0·96 ± 0·05
3 MeV	e^-	3310 ± 50	1·07 ± 0·06
		(3160 ± 50)[a]	(1·02 ± 0·06)[a]

[a]Corrected doses.
(From Williams and Hendry[18]).

Whether there is any dose-rate effect in bactericidal effect of radiation or not is controversial.[19] Even if there is any dose-rate effects, whether the effect of radiation on microorganisms increases or decreases with increasing dose rate is dependent upon the region of the dose rate of radiation applied to microorganisms.

3.1 Dependency of Bactericidal Effect on Dose Rate in a Low Dose-rate Region

Ionizing radiation damages biomolecules, which contributes to the bactericidal effect. However some of damaged DNA and other biomolecules are repaired by the action of enzymes. This repair is of significance in the microorganisms irradiated at low dose rates, because the damaged biomolecules are repaired before the damage becomes lethal during a long period of mild irradiation. Therefore it is reasonable to expect that in a low dose-rate region the sensitivity of bacteria to radiation decreases with the decrease in the dose rate.

Decay of fruits could be prevented by irradiation at lower doses with dose rates of 20–40 krad/min than with dose rates of 1–3 krad/min.[20] For example, peaches were radurized at 125–137 krad and 157–182 krad when they were irradiated at dose rates of 40 krad/min and 20 krad/min, respectively, while irradiation at 182 krad did not effectively radurized peaches when the dose rate was 3 krad/min.[20] Within the dose range of 137–182·4 krad, irradiation at 7 krad/min controlled potato leak better than irradiation at 3 krad/min.[21] *Penicillium italicum* conidia that received a total dose of 100–200 krad formed fewer colonies after irradiation at 28 krad/min than at 1·9 krad/min.[20]

Serratia marcescens irradiated with X-rays at dose rates of ~2 krad/min was more sensitive to the radiation than that irradiated at dose rates of ~0·2 krad/min, when oxygen was present in the system.[22] However the difference in the sensitivity was not observed when the bacteria were irradiated under 100% nitrogen (Fig. 4). Dewey[22] explained that the increased survival of the bacteria irradiated at the lower dose rates under an aerobic condition could be due to normal multiplication and recovery from radiation damage of the surviving bacteria during irradiation because the irradiation at the lower dose rates took longer. Whereas in the presence of oxygen repair of radiation-induced damage of cells during the low-dose-rate treatment for a longer period resulted in the larger number of survivors, a long term of anaerobic condition prevented the multiplication and repair of the cells during irradiation, which did not lead to the difference in the bactericidal effect between irradiation at high dose rates and irradiation at low dose rates.

3.2 Independency of Bactericidal Effect of Dose Rate and Type of Radiation

Tarpley *et al.*[23] did not observe the dose-rate dependency of bactericidal effect in the almost same range of dose rate as that in the study of Dewey.[22] They[23] irradiated mixed bacterial suspension with gamma-rays at a series of dose rate varying from 13 to 116 krep/h and did not observe any difference in bactericidal effect among varied dose rates of radiation. Edwards *et al.*[24] also reported the independency of bactericidal effect of dose rate of electron beams. When *Bacillus subtilis* spores were irradiated

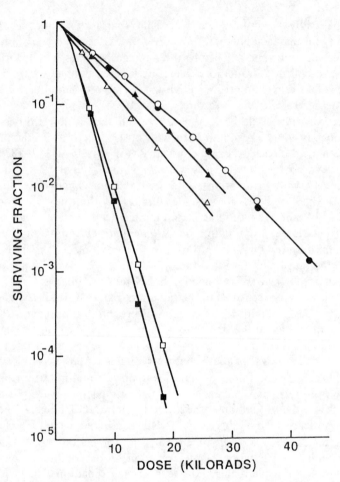

FIG. 4. The effect of X-ray dose rate on *Serratia marcescence* gassed under different conditions. ■, 2129 rads/min in 100% oxygen; □, 169 rads/min in 100% oxygen; ▲, 2120 rads/min in 0·11% oxygen; △, 168 rads/min in 0·11% oxygen; ●, 2142 rads/min in nitrogen; ○, 190 rads/min in nitrogen. (From Dewey[22]).

with accelerated electrons of 2 MeV at various dose rates by changing belt speed from 1·0 to 8·3 ft/min, there was no dose-rate dependency of the inactivation effect of accelerated electrons.[24] These results suggested that the dose-rate effect could not be clearly observed in a narrow range of dose rate. It is expected that clear dose-rate effect can be observed between irradiation at several kGy/h or lower and irradiation at several kGy/s or higher, and most of the reports on the comparative bactericidal effectiveness of irradiation at low dose rates and extremely high dose rates have demonstrated dose-rate effect as discussed in section 3.3.

However a few reports have demonstrated that there is no difference in the effect on microorganisms between high dose-rate irradiation and low dose-rate irradiation or between gamma irradiation and electron irradiation. For example, for both *Escherichia coli* and spores of *Bacillus thermoacidurans*, accelerated electrons, X-rays and gamma-rays had essentially the same bactericidal effectiveness.[25]

3.3 Dependency of Bactericidal Effect on Dose Rate and Type of Radiation

It is assumed that at extremely high dose rates oxygen is used up by radiation-chemical reactions at a rate faster than it can be replaced by diffusion from gas phase and the resulting anoxia reduces the sensitivity

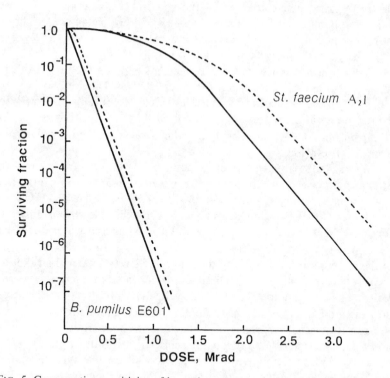

FIG. 5. Comparative sensitivity of bacteria under dry conditions to gamma-rays and electron beams. — , gamma-ray; - -, electron beams at 3 MeV. (From Ito and Tamura[26]).

of bacteria to radiation. Irradiation at very high dose rates produces a high concentration of radicals resulting in the recombination of radicals to a great extent, which reduces the radical reactions of biomolecules in microorganisms. Therefore it is expected that microorganisms are less sensitive to extremely high dose rates of radiation than to medium or low dose rates of radiation.

The sensitivity of dry preparation of *Streptococcus faecium* $A_2 1$ to gamma-rays was higher than that to accelerated electrons, where bacteria irradiated with electron beams showed a larger shoulder in the survival curve (Fig. 5).[26] A similar difference between the sensitivity to gamma-rays and that to accelerated electrons was observed for *Bacillus pumilus*, although the difference was not so significant as for *Streptococcus faecium*. Furuta *et al.*[27] also reported that the spores of *Bacillus pumilus* dried on a paper filter were a little more sensitive to gamma-rays than to electron beams, where the D_{10} of the spores for accelerated electrons was 1·9 kGy and that for gamma-rays was 1·7 kGy. Tabei *et al.*,[28] however, reported that the dry preparation of *Bacillus pumilus* showed a higher sensitivity to accelerated electrons than to gamma-rays (Fig. 6). Ito and Tamura[26] ascribed this disagreement between the results of the study by Ito and Tamura[26] and those reported by Tabei et al.[28] to the difference in the estimation of scattering of electron beams. Contrary to the dry preparation, Tabei *et al.*[28] reported that the sensitivity of *Bacillus pumilus* in aqueous suspension to gamma-rays was higher than to accelerated electrons (Fig. 7).

Emborg[29] compared the effects of gamma-rays and electron beams on *Streptococcus faecium* $A_2 1$ under various conditions. Gamma irradiation could inactivate dry preparation of *Streptococcus faecium* $A_2 1$ with serum broth at 10−50% R.H. to a greater degree than electron irradiation (Fig. 8).[29] While washing bacteria with 0·9% saline and water increased its sensitivity to both gamma-rays and accelerated electrons, the degree of the increase in the sensitivity to gamma-rays was greater than that to accelerated electrons (Fig. 9). These results indicated that serum was protective against radiation inactivation of bacteria and that the extent of protective effect of serum was different between gamma-irradiated and electron-irradiated bacteria. The dry bacteria was more resistant to radiation than the bacteria in aqueous suspension irrespective of the kind of radiation, and no significant difference between the two types of radiation was observed in the inactivation effect on the bacteria suspended in water or serum (Fig. 9; Table 7). The sensitivity of the bacteria to radiation increased with the increase in relative humidity and the extent of this

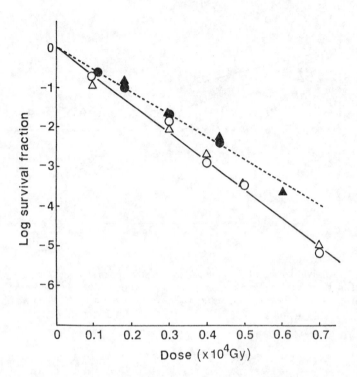

FIG. 6. Survival curves of *B. pumilus* (dry spores) irradiated by electron beam and gamma-ray. △, electron beam, 1.4×10^9 spores/strip; ○, electron beam, 2.9×10^8 spores/strip; ▲, gamma-ray, 1.3×10^9 spores/strip; ●, gamma-ray, 1.1×10^8 spores/strip. (From Tabei *et al.*[28]).

increase in the sensitivity was greater for gamma-rays than for electron beams (Table 7). The inactivating effect of gamma irradiation was higher than that of electron irradiation even under the irradiation condition of oxygen exclusion. When the bacteria were equilibrated to 100% R.H. after hard predrying their resistance to gamma irradiation was reduced by ~50%, whereas the resistance to electron irradiation remained almost unchanged. The results of the study by Emborg[29] indicated that the inactivation effect of gamma-rays from Co-60 for dried preparations of *Streptococcus faecium* $A_2 1$ was higher than the effect of accelerated electrons irrespective of the conditions of sample preparation and irradiation and that the parameters such as humidity and presence of organic com-

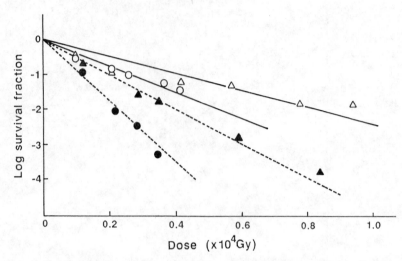

FIG. 7. Survival curves of *B. pumilus* (spores suspended in water) irradiated by electron beam and gamma-ray. $\triangle$, electron beam, $1{\cdot}7 \times 10^{10}$ spores/ml; $\circ$, electron beam, $4{\cdot}5 \times 10^7$ spores/ml; $\blacktriangle$, gamma-ray, $1{\cdot}4 \times 10^{10}$ spores/ml; $\bullet$, gamma-ray, $4{\cdot}6 \times 10^7$ spores/ml. (From Tabei *et al.*[28]).

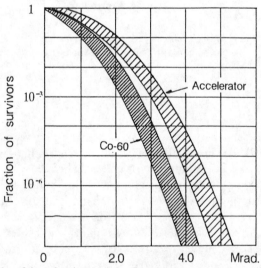

FIG. 8. Bands of inactivation curves for electron accelerator irradiation and Co-60 gamma irradiation of *S. faecium*, strain $A_2 1$, in the serum broth system at 10–50% R.H. (From Emborg[29]).

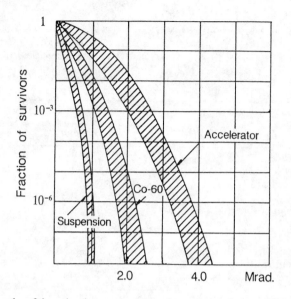

FIG. 9. Bands of inactivation curves for electron accelerator irradiation and Co-60 gamma irradiation of *S. faecium*, strain A$_2$1, in dry cleaned preparations and in aqueous suspension. (From Emborg[29]).

pounds were more influential to the sensitivity of bacteria to gamma-rays than to the sensitivity to electron beams.

Ohki *et al.*[30] reported that the circumstance of the spores at the time of irradiation influenced the sensitivity of spores to gamma-rays and electron beams. There was a significant difference in the D$_{10}$ value among gamma-rays, electron beams and X-rays when the spores were irradiated in the presence of peptone and glycerin, while the difference in the D$_{10}$ value of bacterial spores among different types of radiation was insignificant when the spores were irradiated on a glass-fiber filter (Table 8).

Hayashi *et al.*[15] reported that the sterilization of dehydrated blood plasma can be achieved at a slightly lower dose with gamma-rays than with electron beams.

The results of these studies[15,26-30] on the dependency of bactericidal effect on the type of radiation and on the dose rate suggest that whether or not there is any difference in the inactivation effect on bacteria between gamma-rays and accelerated electrons and the extent of the difference in the inactivation effect depend on several conditions for

TABLE 7

LD 99.99 IN MRADS FOR RADIATION INACTIVATION BY ELECTRON
ACCELERATOR IRRADIATION AND COBALT-60 GAMMA IRRADIATION OF *STR.*
FAECIUM, STRAIN $A_2$1, IN PREPARATIONS WITH SERUM BROTH (SB) AND IN
CLEANED PREPARATIONS (CL) AT DIFFERENT HUMIDITY LEVELS AND
PREPARATION TECHNIQUES

Preparation technique	Radiation source	
	Accelerator	Co-60
0% r.h. (SB)	3.7 ± 0.3	3.2 ± 0.3
10–50% r.h. (SB)	3.7 ± 0.2	2.8 ± 0.2
55–100% r.h. (SB)	2.6 ± 0.2	1.8 ± 0.2
30–100% r.h. (CL)	2.7 ± 0.2	1.7 ± 0.2
Nitrogen atmosphere (CL)	2.8 ± 0.2	1.5 ± 0.2
Hard predrying 100 percent r.h. (CL)	3.0 ± 0.6	0.7 ± 0.1
Hard predrying 100% r.h. (SB)	—	1.0 ± 0.1
Suspension (SB)	1.1 ± 0.2	1.2 ± 0.3
Suspension (CL)	0.7 ± 0.1	0.7 ± 0.1

(From Emborg[29]).

irradiation including oxygen concentration in atmosphere, the presence
of organic compounds in microenvironment around bacteria and relative
humidity.

Emborg[29] proposed the possibility that the reduction of damage of
biomolecules which was responsible for decreased inactivation effect of
high dose rates of radiation was brought about by increased radical-
radical reactions rather than anoxic conditions at high dose rates,
because the doses applied were very much higher than those known to
remove oxygen from water or biological tissues (40–200 krad). However
there are many reports[31–42] that suggest the role of oxygen in the different inactivation effects between gamma irradiation and electron irradiation as described in section 3.4.

Combination of irradiation and heat treatment is sometimes a useful
method to eliminate thermotolerant microorganisms from foods. Thermophilic bacteria, *Bacillus stearothermophilus*, can not be efficiently inactivated by treating them at 100°C. The D_{10} values of *Bacillus stearo-*
thermophilus in aqueous suspension were 1.86 kGy and 2.12 kGy for

TABLE 8

D VALUES ON *B. PUMILUS* AND *B. SUBTILIS* AT DIFFERENT CONDITION OF TEST PICES

	Dry condition	B. pumilus	B. subtilis	B. megaterium	B. brevis
r-rays	Glass fiber filter with non additives	1·5	1·4	1·9	1·6
	Glass fiber filter with pepton + glycerin	1·9	1·8	2·4	2·1
	Cellulose filter with pepton + glycerin	1·6	1·5	2·4	2·1
EB	Glass fiber filter with non additives	1·6	1·5	2·0	2·0
	Glass fiber filter with pepton + glycerin	2·1	2·0	2·8	2·4
	Cellulose filter with pepton + glycerin	1·7	1·6	2·8	2·4
X-rays	Glass fiber filter with non additives	1·6	1·6	2·0	2·0
	Glass fiber filter with pepton + glycerin	2·1	2·0	2·6	2·2
	Cellulose filter with pepton + glycerin	1·7	1·6	2·6	2·2

(From Ohki et al.[30]).

gamma-rays and electron beams respectively.[31] Both gamma irradiation
and electron irradiation enhanced their sensitivity to heat treatment
depending upon dose. The sensitivity of *Bacillus stearothermophilus* to
heat treatment was slightly higher after irradiation with gamma-rays than
after irradiation with electron beams (Fig. 10).[31]

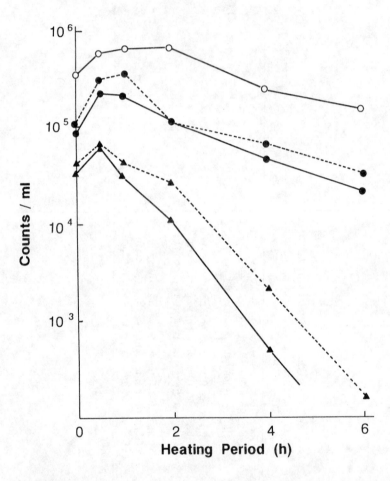

FIG. 10. Heat inactivation curves of irradiated *B. stearothermophilus*. —— ,
gamma-rays; - -, electron beams; o, 0 kGy; •, 1 kGy; ▲, 2 kGy. (From
Takizawa *et al.*[31]).

3.4 Oxygen Effect

Many living organisms, including bacteria, are more readily damaged when oxygen is available to the cells at the time of irradiation than when it is not available. Oxygen plays an important role in the inactivation of microorganisms by irradiation. For example, *Rhizopus stolonifer* is more sensitive to radiation under aerobic conditions than anaerobic conditions.[32]

Dewey and Boag[33] extensively investigated the dependencies of bactericidal effect of radiation on oxygen content and on type of radiation. *Serratia marcescence* increased sensitivity to X-rays of 1·5 MeV with the increase in oxygen concentration in the atmosphere. Irradiation with X-rays at a dose rate of 600 rad/min under a condition of 1% O_2 inactivated the bacteria to the same degree as irradiation with accelerated electrons at a dose rate of 10^9 rad/s under a condition of 5% O_2. The bacteria irradiated with electrons at 1% O_2 and 0% O_2 showed almost the same inactivation curve as those irradiated with X-rays at 0% O_2. These results indicate that oxygen plays a role in the inactivation of irradiated bacteria and that the influence of the oxygen concentration on the bactericidal effect is different between irradiation at high dose rates and irradiation at low dose rates. Oxygen effect was more distinct at low dose rates than at extremely high dose rates.

Greater importance of oxygen effect at lower dose rates was also reported by Powers and Boag.[34] The sensitivity of dry spores of *Bacillus megaterium* to X-rays in air at 20 krad/min was about 1·25 times that in the absence of oxygen, while no significant difference in the effectiveness of inactivation between the presence and absence of oxygen was observed when the spores were irradiated at 20 Mrad/min.

Epp *et al.*[35] reported that irradiation of *Escherichia coli* with electron beams at a high dose rate of 10^{12} rad/s demonstrated that irradiation at high dose rates inactivated the bacteria under anoxia after depletion of oxygen within cells. Survival curves for *Escherichia coli* were determined in the presence of various concentrations of oxygen. The curves for pure nitrogen and pure oxygen were exponential after a small initial shoulder and the ratio of slopes between these two curves was 3:1 (Fig. 11). At intermediate oxygen concentrations, before the depletion of oxygen the bacteria were inactivated by accelerated electrons in the same manner as they were inactivated in 100% O_2 and, after the depletion of oxygen, the cells were inactivated in the same manner as in N_2. The break-points of the survival curve of *Escherichia coli* between aerobic inactivation and

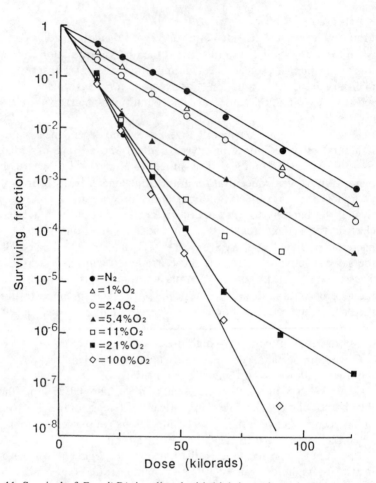

FIG. 11. Survival of *E. coli* B/r irradiated with high-intensity pulsed electrons in the presence of various concentrations of oxygen. Dose delivered in single pulses of time duration of about 30 ns. (From Epp *et al.*[35]).

anaerobic inactivation varied depending upon oxygen concentration and the break-points were observed at higher doses as oxygen concentration increased.

Berry *et al.*[36] observed such a break-point in the survival curve of Chinese hamster cells irradiated with X-ray at a high dose rate but not in gamma-irradiated cells (Fig. 12). The cells irradiated with X-rays

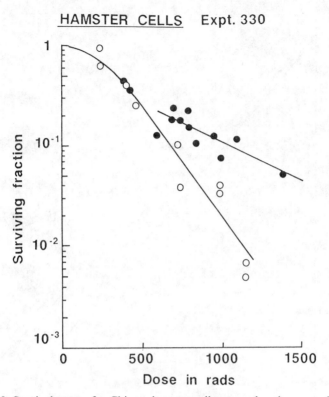

FIG. 12. Survival curves for Chinese hamster cells exposed under aerated conditions to Co-60 gamma-rays at a conventional dose-rate (100 rads/min) or to a pulse of electrons of 7 ns duration. (From Berry et al.[36]).

showed a break-point at around 500 rad, and before the break-point the survival curve coincided with that of gamma-irradiated cells. Beyond the break-point the slope of the survival curve for the X-irradiated cells was smaller than the gamma-irradiated cells. The break-point in the survival curve was also observed for HeLa cells irradiated at high dose rates as well.[37] The same phenomena for bacteria and mammalian cells were reported by Phillips and Worsnop.[38] Nias et al.,[39] however, reported that such a break-point in the survival curve of HeLa cells irradiated at a high dose rate was not observed, although there was a clear difference in the survival curve between irradiation under an aerobic condition and irradiation under an anaerobic condition. Todd et al.[40] did not find any

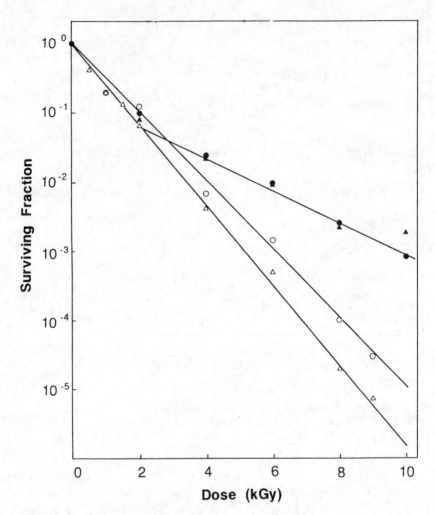

FIG. 13. Survival curves of *B. pumilus* spores irradiated in aqueous suspension. ○, electron beams, under aerobic conditions; ●, electron beams, under anaerobic conditions; △, gamma-rays, under aerobic conditions; ▲, gamma-rays, under anaerobic conditions. (From Takizawa *et al.*[31]).

difference in the inactivation of human kidney cells between high-dose-rate X-rays and low-dose-rate gamma-rays.

There is a report that such a break-point in a survival curve was observed both for gamma-irradiated and electron-irradiated spores under anaerobic conditions.[31] Takizawa *et al.*[31] prepared an aqueous suspension of bacterial spores by inactivating vegetative cells by treating bacterial suspension with heat, and then irradiated the aqueous suspension of spores without removing the dead vegetative cells. They observed the break-point of survival curves of spores of *Bacillus pumilus* at around 2 kGy (Fig. 13) but not for *Bacillus stearothermophilus*, when the spores were irradiated under anaerobic conditions. No break-point was observed for *Bacillus pumilus* and *Bacillus stearothermophilus* when they were irradiated under aerobic conditions. Under anaerobic conditions, *Bacillus pumilus* was more sensitive to gamma-rays than electron beams in the dose range of 0–2 kGy, while the sensitivity to gamma-rays and that to electron beams were the same at doses higher than 2 kGy (Fig. 13). They did not clearly define the cause of the break-point, but coexisting organic compounds from dead vegetative cells might have played a role in the occurrence of break-point. They presumed that the less efficient inactivation effect beyond the break-point was due to anoxia caused by reactions of organic compounds with residual oxygen in spores. Kigawa *et al.*[41]

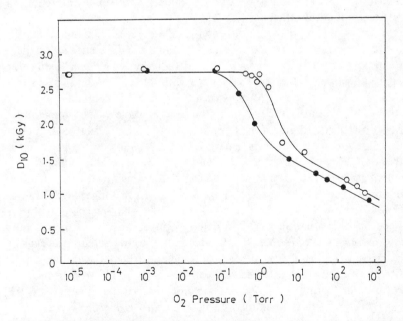

FIG. 14. Relationship between D_{10} values of *B. pumilus* spores and oxygen pressure for electron beams and gamma-rays. o, electron beams; •, gamma-rays. (From Watanabe *et al.*[42]).

reported that the presence of organic compounds such as ethanol and glycerin influenced the break-point of survival curves and oxygen effect in the inactivation of *Bacillus pumilus*. Watanabe *et al.*[42] reported that at extremely low oxygen concentrations in the atmosphere the difference in D_{10} of the spores of *Bacillus pumilus* was not observed between that for gamma irradiation and for electron irradiation, although the D_{10} for gamma-rays was lower than that for accelerated electrons at higher oxygen concentrations (Fig. 14). These results suggest that the depletion of oxygen in the spores caused by electron irradiation at high dose rates is responsible for the lower sensitivities of bacterial spores to electron beams than to gamma-rays.

4 EFFECT ON INSECTS

The difficulty in the study on the comparative effect of different types of radiation on insects is frequently brought about by the application of high doses of radiation (~ 100 krad) which masks small differences in relative biological efficiency.

Adem *et al.*[43] reported that gamma-rays are more effective in the inactivation of pupae than electron beams and that the difference in the inactivation effect between the two types of radiation was much smaller for eggs and larvae. They infested maize with several kinds of insects of *Sitophilus* spp. including *S. granarius*, *S. zeamais*, *S. oryzae* followed by storage for 3 months at 20°C, and then irradiated the maize with gamma-rays or accelerated electrons. Gamma-ray was more effective than electron beam in suppressing the emergence of *Sitophilus* spp., especially for the first 3 weeks after irradiation (Fig. 15). The difference in the emergence between gamma-rays and electron beams was less distinct during the storage of the infested maize for 4 weeks or longer after irradiation. Adem *et al.*[43] presumed that the adults that emerged within 3 weeks after irradiation were irradiated as pupae or pre-emergent adults and those emerged 4–6 weeks after irradiation were irradiated as eggs or larvae and reported that these results demonstrated the different sensitivity of each stage of insects to gamma-rays and electron beams.

Bull and Cornwell,[44] however, did not observe the difference in the emergence of adults of grain weevil (*Sitophilus granarius* (L)) from irradiated pupae between gamma-rays and electron beams. They compared the effects of the two types of radiation on the emergence of adults from pupae infested to grains with two experiments: for one experiment the

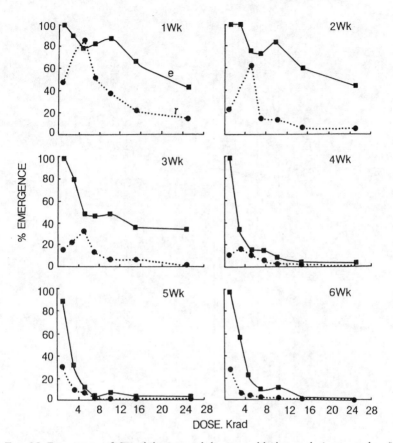

FIG. 15. Emergence of *Sitophilus* spp. adults at weekly intervals (expressed as %
of the controls) from maize irradiated with Co-60 gamma-rays and accelerated
electrons. (From Adem *et al.*[43]).

pupae were irradiated at 10, 15 or 20 krad and for the other experiment
the pupae were irradiated at 6, 10 or 16 krad. The results of the experiments showed no difference in adult emergence between the treatment with gamma-rays and that with electron beams irrespective of age
of pupae (Table 9).[44] The adults emerged from the pupae irradiated at
a dose of 10–20 krad produced no progeny irrespective of the type of
irradiation. However, the results on the adults emerged from the pupae
irradiated at lower doses of 6 and 10 krad indicated that accelerated elec-

TABLE 9

EMERGENCE OF ADULTS FROM IRRADIATED PUPAE. MEAN VALUES IN ANGLES, WITH STANDARD ERRORS. RETRANSFORMED PERCENTAGES IN PARENTHESIS

Treatment		Pupal experiment 1			Pupal experiment 2		
		Means		S.E.	Means		S.E.
		Gamma	Electron		Gamma	Electron	
Radiation ×	Control:	80·8 (97)	77·9 (96)		Control: 76·9 (95)	78·4 (96)	
dose (rads)	10 000:	38·3 (39)	38·0 (38)	±1·35	6 000: 42·4 (45)	37·0 (36)	±1·69
interaction	15 000:	27·7 (22)	29·3 (24)		10 000: 30·1 (25)	28·0 (22)	
	20 000:	20·9 (13)	24·9 (18)		16 000: 18·4 (10)	19·8 (12)	
Radiation ×	27–28:	27·0 (21)	29·4 (24)		27–28: 28·7 (23)	30·7 (26)	
pupal age	28–29:	32·1 (28)	33·1 (30)	±1·35	28–29: 32·8 (29)	33·9 (31)	±1·89
(days)	29–30:	52·5 (63)	51·8 (62)		29–30: 45·3 (51)	39·7 (41)	
interaction	30–31:	56·0 (69)	55·7 (68)		30–31: 46·9 (53)	45·4 (51)	
					31–32: 56·1 (69)	54·6 (66)	

(From Bull and Cornwell[44]).

trons were less effective than gamma-rays in inducing sterility (Table 10).[44] Although there was no difference in the survival of grain weevil between gamma-irradiated and electron-irradiated adults 12 days after the treatment, the difference became distinct during the next few days and the mortality of gamma-irradiated weevils was significantly higher than electron-irradiated weevils 28 days after irradiation (Fig. 16).[44]

TABLE 10

PROGENY PRODUCTION BY EMERGED ADULTS IRRADIATED AS PUPAE
(TOTALS FOR ALL DOSES)

Radiation	Wheat	Pupal age (days)	Progeny	
			Control	All doses
Gamma	English	30–31	13 330	0
		31–32	10 390	21
	Manitoba	30–31	7 210	3
		31–32	10 470	6
		Total	41 400	30
Electron	English	30–31	13 280	9
		31–32	11 920	46
	Manitoba	30–31	6 710	6
		31–32	9 170	134
		Total	41 080	195

(From Bull and Cornwell[44]).

Proctor et al.[45] also reported a higher sensitivity of the adults of Tribolium confusum to gamma-rays than to accelerated electrons or X-rays. The dose for 50% reduction in progeny was about 600 rad higher for electron-irradiated adults of the grain weevil than for gamma-irradiated ones and the dose for 99.9% reduction in progeny was 5–8 krad higher for electron beams than for gamma-rays.

Based on the results of the studies[43–45] on irradiated insects, it appears that the difference in the effect between gamma-rays and electron-beams is more distinct at later stage of growth of insects.

Bull and Cornwell[44] ascribed the difference in the effect on insects between gamma irradiation and electron irradiation to the difference in the dose rate which could lead to varied oxygen concentrations inside the insects. Cornwell et al.[46] observed the increase in the resistance of grain weevil to gamma-rays caused by the reduced oxygen during irradiation

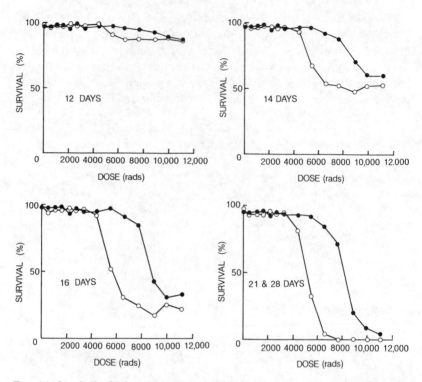

FIG. 16. Survival of adult *S. granarius* at intervals after treatment with gamma radiation or accelerated electrons. ○, gamma-rays; ●, electrons. (From Bull and Cornwell[44]).

(Table 11). Lower concentration of oxygen during irradiation resulted in the decreased sensitivity of insects to X-rays as well.[47]

In a low dose-rate region[48], however, the dose rate influenced the mortality of insects such as flour beetle (*Tribolium confusum*), grain beetle (*Oryzaephilus surinamensis*) and grain weevil (*Sitophilus granarius*) in a way different from in a high dose-rate region. The survival of the gamma-irradiated insects[48] increased with the decrease in dose rate in a range of 1·5–4·7 krad/h, which is inconsistent with the results[44] on the effect of gamma-rays and electron beams on the mortality and sterility of insects where the insecticidal effect of radiation was larger with the decrease in dose rate in a higher dose-rate range of $2·5 \times 10^2$ krad/h to $8·6 \times 10^3$ krad/h. The egg hatch ability of Mediterranean fruit fly (*Ceratitis capitata*) decreased as the dose rate of gamma-rays increased in the range

TABLE 11

AVERAGE PROGENCY PER PARENT (×10 000) FOR THE PERIOD 2–4 MONTHS AFTER IRADIATION OF 3 POPULATIONS OF ADULT GRAIN WEEVILS TREATED AT 10 000 AND 12 000 RADS UNDER ANOXIC AND ATMOSPHERIC CONDITIONS

Population	Irradiated under					
	Reduced oxygen			Atmospheric		
	Rads		Means	Rads		Means
	10 000	12 000		10 000	12 000	
7 500	4026	417	2222	51	6	28
75 000	2657	777	1717	31	18	24
750 000	394	366	380	28	3	15
Means	2359	520		37	9	

(From Cornwell et al.[46]).

of 7–97 rad/s.[49] The dose-rate effect in a lower region of dose rate was attributable to the biological recovery from radiation-induced damage during radiation treatment.

It should be emphasized that the dose rate influences the biological effects of radiation on both microorganisms and insects in different ways depending upon the dose-rate region. The dose-rate effect in a region of low dose rates can not be applied to the comparative effectiveness of gamma-rays and electron beams in food irradiation.

5 PHYSICAL EFFECT

Generally the comparative study on the physical effects of different types of radiation is not important in food irradiation, and only induced radioactivity should be investigated when new radiation is introduced to food irradiation. Even if new radioactive materials may be produced by accelerated electrons with high energies, the half-lives of the generated isotopes are so short that radioactivity is not present in measurable amounts if the energy of the electron is kept below 10 MeV. No change in the counts of gamma-ray and beta-ray of black pepper and white pepper was observed after the irradiation at 100 kGy with accelerated electrons at an energy of 10 MeV, and of course there was no difference in the radioactivity between gamma-irradiated and electron-irradiated peppers.[50-52] These results confirmed the conclusion of Joint FAO/IAEA/WHO Expert Committee on the Wholesomeness of Irradiated Food.[53]

6 CONCLUSION

Comparative studies on the effects of gamma-rays and electron beams are quite important for designing irradiation plants and establishing irradiation conditions for treating foods. Based on the results of the numerous studies on the comparative effects of gamma-rays and electron beams or of high dose rates of radiation and low dose rates of radiation, it can be concluded that the biological effects of electron beams on micro-organisms and insects are slightly lower than those of gamma-rays, although the difference in the effect between the two types of radiation is influenced by various parameters including oxygen concentration and water content. The difference in the effect on chemical reactions in foods between the two types of radiation is much smaller than that observed in the biological effects. It is difficult to judge whether the dose for treating foods with electron beams should be set at higher doses than that with gamma-rays, because the biological effect is influenced by various conditions. In some countries the dose for electron sterilization has been set at a slightly higher dose than gamma sterilization. For example, the minimum dose for sterilization of medical products with Co-60 gamma-rays is 3·2 Mrad and that with accelerated electrons is 3·5 Mrad in the Scandinavian requirements.[54] It is necessary to define the appropriate dose at each irradiation plant by taking account of the type of radiation, the conditions for irradiation and the storage following radiation treatment.

The comparison of biological and chemical effects between gamma-rays and electron beams is also important for the evaluation of the wholesomeness of the foods irradiated with accelerated electrons. Most of the studies on the wholesomeness of irradiated foods have been carried out with the foods irradiated with gamma-rays. It is frequently necessary to extrapolate the wholesomeness data on gamma-irradiated foods to electron-irradiated foods. The adequacy of this extrapolation can be determined based on the results on the comparative effects of gamma-rays and electron beams. Based on the data reported so far, it can be concluded that the biological and chemical effects of electron beams are the same or slightly smaller as compared with those of gamma-rays and that the foods irradiated with electron beams are wholesome at the same degree or higher as compared with gamma-irradiated foods.

REFERENCES

1. Gotoh, A., Yamazaki, K. and Oka, M., Some bacteriological observation on the radiation preservation of packed sliced ham, *Food Irrad. Jpn.*, 1972, **7**, 57–63.
2. Watanabe, Y., Ito, H. and Ishigaki, I., Sensitivity of *Bacillus pumilus* E601 spores to electron beams, *Food Irrad. Jpn.*, 1988, **23**, 88–92.
3. Mead, J.F., The irradiation-induced autoxidation of linoleic acid, *Science*, 1952, **115**, 470–472.
4. Brasch, A., Huber, W. and Waly, A., Radiation effects as a function of dose rate, *Arch. Biochem. Biophys.*, 1952, **39**, 245–247.
5. Azuma, K., Tsunoda, H., Hirata, T., Ishitani, T. and Tanaka, Y., Effects of electron beam irradiation on the amounts of volatiles from irradiated polyethylene film, *Agric. Biol. Chem.*, 1984, **48**, 2009–2015.
6. Taub, I.A., Kaprielian, R.A., Halliday, J.W., Walker, J.E., Angelini, P. and Merritt, Jr., C., Factors affecting radiolytic effects in food, *Radiat. Phys. Chem.*, 1979, **14**, 639–653.
7. Diehl, J.F., Thiamin in bestrahlten Lebensmitteln, *Z. Lebensm. Unters. Forsch.*, 1975, **157**, 317–321.
8. Diehl, J.F., Einfluss verschiedener Bestrahlungsbedlingungen und der Lagerung auf strahleninduzierte Vitamin E Verluste in Lebensmitteln, *Chem. Mikrobiol. Technol. Lebensm.*, 1979, **6**, 65–70.
9. Diehl, J.F., Ueber die Wirkung ionisierender Strahlen auf den Vitamin E Gehalt von Lebensmitteln und auf reines Tokopherol und Tokopherolacetat, *Z. Lebensm. Unters. Forsch.*, 1970, **142**, 1–7.
10. Gopal, N.G.S., Radiation sterilization of pharmaceuticals and polymers, *Radiat. Phys. Chem.*, 1978, **12**, 35–50.
11. Josephson, E.S., Thomas, M.H. and Calhoun, W.K., Nutritional aspects of food irradiation, *J. Food Proc. Pres.*, 1978, **2**, 299–313.
12. Johnson, B.C. and Moser, K., Amino acid destruction in beef by high energy electron beam irradiation. In: *Radiation Preservation of Foods*, Advances in Chemistry Series, Washington DC, 1967, **65**, 171–179.
13. Thomas, M.H., Atwood, B.M., Wierbicki, E. and Taub, I.A., Effect of radiation and conventional processing on the thiamine content of pork, *J. Food Sci.*, 1981, **46**, 824–828.
14. Merritt, Jr., C., Angelini, P. and Graham, R.A., Effect of radiation parameters on the formation of radiolysis products in meat, *J. Agric. Food Chem.*, 1978, **26**, 29–35.
15. Hayashi, T., Biagio, R., Saito, M., Todoroki, S. and Tajima, M., Effects of ionizing radiation on sterility and functional qualities of dehydrated blood plasma, *J. Food Sci.*, 1991, **56**, 168–171.
16. Josephson, E.S., Brynjolfsson, A., Wierbicki, E., Rowley, D.B., Merritt, Jr., C., Baker, R.W., Killoran, J.J. and Thomas, M.H., Radappertization of meat, meat products and poultry. In: *Radiation Preservation of Food*, STI/PUB/317, IAEA, Vienna, 1973, 471–490.
17. Heiligman, F. and Rice, L.J., Development of irradiation sterilized codfish cakes, *J. Food Sci*, 1972, **37**, 420–422.
18. Williams, P.C. and Hendry, J.H., The RBE of megavoltage photon and electron beams, *Br. J. Radiol.*, 1978, **51**, 220.

19. Ley, F.J., The influence of dose rate on the inactivation of micro-organisms, *Int. J. Radiat. Isotop.*, 1963, **14**, 38–41.
20. Beraha, L., Influence of gamma radiation dose rate on decay of citrus, pears, peaches, and on *Penicillium italicum* and *Botrytis cinerea* in vitro, *Phytopathology*, 1964, **54**, 755–759.
21. Beraha, L., Ramsey, G.B., Smith, M.A. and Wright, W.R., Effects of gamma radiation on some important potato tuber decays, *Am. Potato J.*, 1959, **36**, 333–338.
22. Dewey, D.L., An oxygen-dependent X-ray dose-rate effect in *Serratia marcescens*, *Radiat. Res.*, 1969, **38**, 467–474.
23. Tarpley, W., Ilavsky, J., Manawitz, B. and Horrigan, R.V., The effect of high energy gamma radiation from kilocurie radioactive sources on bacteria, *J. Bacteriol.*, 1953, **65**, 305–309.
24. Edwards, R.B., Peterson, L.J. and Cummings, D.G., The effect of cathode rays on bacteria, *Food Technol.*, 1954, **8**, 284–290.
25. Goldblith, S.A., Proctor, B.E., Davison, S., Kan, B., Bater, C.J., Oberle, E.M., Karel, M. and Lang, D.A., Relative bactericidal efficiencies of three types of high-energy ionizing radiations, *Food Res.*, 1953, **18**, 659–677.
26. Ito, H. and Tamura, N., Comparison of sensitivity of microbiological reference standards to gamma ray and electron beams for sterilization, *J. Antibact. Antifung. Agents*, 1985, **13**, 299–305.
27. Furuta, M., Katayama, T., Toratani, H. and Takeda, A., Radiation sterilization by 10 MeV electron beams, *Food Irrad. Jpn.*, 1987, **22**, 1–3.
28. Tabei, M., Sekiguchi, M., Minegishi, A., Sato, K., Katsumura, Y. and Tabata, Y., Electron-beam sterilization of surgical sutures, *J. Antibact. Antifung. Agents*, 1984, **12**, 611–618.
29. Emborg, C., The influence of preparation technique, humidity and irradiation conditions on radiation inactivation of *Streptococcus faecium*, Strain $A_2 1$, *Acta Pathol. Microbiol. Scand. Section B*, 1972, **80**, 367–372.
30. Ohki, Y., Ito, H., Watanabe, Y., Sunaga, H. and Ishigaki, I., Comparative sensitivity of endospores from some *Bacillus* species to gamma-rays, X-rays and electron beams for sterilization, *Food Irrad. Jpn.*, 1990, **25**, 71–74.
31. Takizawa, H., Hayashi, T., Suzuki, S., Suzuki, T., Takama, K. and Yasumoto, K., Comparative study on disinfection potency of spore forming bacteria by electron-beam irradiation and gamma-ray irradiation, *Food Irrad. Jpn.*, 1990, **25**, 89–93.
32. Sommer, N.F., Creasy, M., Romani, R.J. and Maxie, E.C., An oxygen-dependent postirradiation restoration of *Rhizopus stolonifer* sporangio-spores, *Radiat. Res.*, 1964, **22**, 21–28.
33. Dewey, D.L. and Boag, J.W., Modification of the oxygen effect when bacteria are given large pulses of radiation, *Nature*, 1959, **183**, 1450–1451.
34. Powers, E.L. and Boag, J.W., The role of dose rate and oxygen tension in the oxygen effect in dry bacterial spores, *Radiat. Res.*, 1959, **11**, 461.
35. Epp, E.R., Weiss, F. and Santomasso, A., The oxygen effect in bacterial cells irradiated with high-intensity pulsed electrons, *Radiat. Res.*, 1968, **34**, 320–325.
36. Berry, R.J., Hall, E.J., Forster, D.W., Storr, T.H. and Goodman, M.J., Survival of mammalian cells exposed to X rays at ultra high dose-rates, *Br. J. Radiol.*, 1969, **42**, 102–107.

37. Town, C.D., Effect of high dose rates on survival of mammalian cells, *Nature*, 1967, **215**, 847–848.

38. Phillips, T.L. and Worsnop, R.B., Oxygen depletion by ultra-high-dose electrons in bacteria and mammalian cells, *Radiat. Res.*, 1968, **35**, 545

39. Nias, A.H.W., Swallow, A.J., Keene, J.P. and Hodgson, B.W., Survival of Hela cells from 10 nanosecond pulses of electrons, *Int. J. Radiat. Biol.*, 1970, **17**, 595–598.

40. Todd, P.W., Winchell, H.S., Feola, J.M. and Jones, G.E., Inactivation by pulsed high-intensity X-rays of human cells cultured in vitro, *Radiat. Res.*, 1967, **31**, 644–645.

41. Kigawa, A., Tateishi, T., Iso, K., Kimura, T. and Mamuro, T., Effect of coexisting organic substances on radiation resistance of *Bacillus pumilus* spores suspended in water, *J. Antibact. Antifung. Agents*, 1987, **15**, 163–169

42. Watanabe, Y., Ohki, Y., Ito, H. and Ishigaki, I., Effect of dose rate on electron beam and gamma-ray sensitivities of *B. pumilus* E 601 spores, *Food Irrad. Jpn.*, 1990, **25**, in press.

43. Adem, E., Watters, F.L., Rendon, R.U. and De la Piedao, A., Comparison of Co-60 gamma radiation and accelerated electrons for suppressing emergence of *Sitophilus* spp. in stored maize, *J. Stored Prod. Res.*, 1978, **14**, 135–142.

44. Bull, J.O. and Cornwell, P.O., A comparison of the susceptibility of the grain weevil *Sitophilus granarius* (L) to accelerated electrons and Co-60 gamma radiation. In: *The Entomology of Radiation Disinfestation of Grain*, P.O. Cornwell (ed.), Pergamon Press, London, 1966, pp. 157–175.

45. Proctor, B.E., Lockhart, E.E., Goldblith, S.A., Grundy, A.V., Tripp, G.E., Karel, M. and Brogle, R., The use of ionizing radiations in the eradication of insects in packaged military rations, *Food Technol.*, 1954, **8**, 536–540.

46. Cornwell, P.B., Bull, J.O. and Pendlebury, J.B., Control of weevil population (*Sitophilus granarius* (L)) with sterilizing and substerilizing doses of gamma radiation. In: *The Entomology of Radiation Disinfestation of Grain*, P.O. Cornwell (ed.), Pergamon Press, London, 1966, pp. 71–95.

47. Slater, J.V., Yu, M.E. and Tobias, C.A., Oxygen dependence for radiation sensitivity during development in insects, *Radiat. Res.*, 1964, **22**, 236.

48. Jefferies, D.J. and Banham, E.J., The effect of dose rate on the response of *Tribolium confusum* DUv., *Oryzaephilus surinamensis* (L.) and *Sitophilus granarius* (L.) to Co-60 gamma radiation. In: *The Entomology of Radiation Disinfestation of Grain*, P.O. Cornwell (ed.), Pergamon Press, London, 1966, pp. 177–185.

49. Wakid, A.M., Amin, A.H., Shoukry, A. and Fadel, A., Factors influencing sterility and vitality of the Mediterranean fruit fly, *Ceratitis capitata* Wiedemann. In: *Sterile Insect Technique and Radiation in Insect Control*, IAEA, Vienna, 1982, p. 379–386

50. Furuta, M., Katayama, T., Toratani, H. and Takeda, A., Preliminary examination of induced radioactivity in peppers by 10 MeV electron irradiation, *Food Irrad. Jpn.*, 1988, **23**, 93–99.

51. Furuta, M., Katayama, T., Toratani, H. and Takeda, A., Preliminary examination of induced radioactivity in black peppers by 10 MeV electron irradiation, *Radioisotopes*, 1988, **37**, 390–393.

52. Furuta, M., Katayama, T., Ito, N., Mizohata, A., Matsunami, T., Toratani, H. and Takeda, A., Preliminary examination of induced radio activity in pepper by 10 MeV electron irradiation. Measurement of beta-ray, *Food Irrad. Jpn.,* 1989, **24**, 9–11.
53. WHO, Wholesomeness of irradiated food. Report of a Joint FAO/IAEA/WHO expert committee. Technical Report Series No. 659, WHO, Geneva, 1981.
54. Handlos, V., Sterilization by electron beam. *Radiat. Phys. Chem.,* 1981, **18**, 175–182.

Chapter 8

COMBINATION OF IRRADIATION AND THERMAL PROCESSING

B. Hozova & L. Sorman[†]

Department of Chemistry and Technology of Saccharides and Foods, Faculty of Chemical Technology, Slovak Technical University, Radlinskeho 9, 812 37 Bratislava, Czechoslovakia

1 INTRODUCTION

Recently, the technology and methods of food preservation has searched for ever better and less destructive methods and procedures which prolong the shelf-life of the primary products so that they do not undergo degradation processes. However, one can say that the way of improving the existing fundamental preservation methods is ever more difficult and expensive, and that possibilities for their improvement are gradually exhausted. Therefore, new non-traditional methods of preservation, as well as the procedures based on scientifically controlled combination of two or more preservation methods give increasingly higher chances for the improvement of the quality of canned products. Such combinations should reduce the intensity of the effects of adverse effects of separately applied preservation methods. The aim is to achieve synergic or additive effects of the decisive factors, which would ensure microbiological adequacy and storage stability of canned foods, including maximum retention of their nutritional and sensory characteristics. Such procedure should gradually improve and optimize preservation processes, thus obtaining foods of a higher quality.

However, if we evaluate in more detail the published scientific papers dealing with non-traditional methods of food preservation and application of combination of preservation methods, we see that there is no work which provides an analysis of this problem. Such an investigation would

[†]Deceased.

look for common features and try to generalize theoretically the knowledge of experimental research of established non-traditional preservation methods and their mutual combination. Therefore, it is necessary to pay particular attention to this question from the viewpoint of the study of theoretical fundamentals as well as from the viewpoint of practical aspects of binary, ternary or multiple combinations of preservation methods.

Application of ionizing radiation belongs to those modern methods of food preservation which reduce the number of microorganisms in foods. According to the conclusions of the Joint Commission of FAO, IAEA and WHO of 1980[1] irradiation of foods with a dose up to 10 kGy does not represent any toxicological risks. Although for more than 30 years attention has been paid to the problems of the preservation of foods, the existing results are not satisfactory[2-4] particularly because of interactions of ionizing radiation with the components of the irradiated foods, such as oxilabile vitamins, lipids, proteins, and the highest resistance is that of polysaccharides.[3] Decreasing the undesirable side-effects of radiosterilizing doses consists in reduction of doses and application of completing technological procedures is necessary. Special attention is paid particularly to the use of combinations of heat and irradiation, since this is suitable mainly for the preservation of meat products (ham, smoked meats[5,6]) but can also be applied to vegetables and other products.[7-10] The literature reveals that this combined method can ensure the necessary sterilization effect and provide products of better sensory quality than products which are only irradiated or only thermally treated, and in many cases it is more economical.[12,13]

1.1 Microbiological Aspects of Practical Application of Heat and Irradiation

Combination of sterilization and ionizing irradiation for preservation of foodstuffs offers the following possibilities: (a) use of large packages, (b) better choice of packing material, (c) conservation of fossil energy because the process is shorter, or it takes place at a lower temperature and (d) better product quality.[14]

The advantage of the combination of these two techniques is limited. Therefore, it is necessary to optimize process conditions. At higher temperatures, microorganisms are destroyed, but at the same time the level of labile components is unfavourably influenced and organoleptic characteristics are significantly worsened. This combination is of importance particularly from the viewpoint of enzyme inactivation. Joint effects of irradiation and heat can have synergic effects on inactivation of the cells of microorganisms [15,16] and irradiation resistant microorganisms are sensitivized. [17,18] Also some other factors, such as salt concentration, reduced a_w, temperature of freeze drying and the presence of antioxidants influence the radiation resistance of the vegetative forms of bacteria and of spores. [19]

Teufel [20,21] investigated the problem of highly radiation resistant microorganisms — Moraxella, Acinetobacter — and states that a much higher efficiency of reduction was achieved by using the combination of heat and ionizing radiation. The studies of these processes can be applied, for example, to the control of Vibrio parahaemolyticus contaminating fish and other products. [22]

The study by Gombas and Gomez [23] describes changes in the thermosensitivity of the spores of Clostridium perfringens, type A, gradually irradiated by 7 kGy and/or heated to 93—100°C. The spores were inactivated by heating and subsequent irradiation (40—94%). However, if the spores were irradiated before being heated, their sensitivity increased. The degree of thermosensitivity increased with the dose of ionizing radiation. Grecz et al. [24] found a higher sensitivity of spores of Bacillus subtilis 168 and Clostridium botulinum 62 A to heat treatment (90°C/10—30 min) after moderate irradiation (0·5—3 kGy). Combination of heat and ionizing radiation caused destruction of DNA and thus also inactivation of spores.

The effect of combined heat and irradiation procedures on germination and survival of the spores of Bacillus cereus was studied by Kamat and Lewis. [25] Spores of this microorganism isolated from shrimps were exposed to gamma rays (D_{10} = 4 kGy) and were compared with D_{10} values = 0·3 kGy for vegetative cells. Dipicolinic acid did not influence decisively the radiation resistance of these spores. Maximum germination of the irradiated spores was 60% in comparison to 80% in the case of non-irradiated spores. Radiation induced inhibition of germination did not depend on the radiation dose. Heat treatment (15 min at 80°C) activated germination; increase of the heating time (30 and 60 min) prolonged the lag phase of germination. The spores without DPA showed a lower thermal resistance than other spores and spores exposed to double phase inactivation.

Kiss and Farkas [26] have followed the kinetic of the growth of microflora in paprika and spices under the effect of radiation and heat. They compared the rate of growth in non-irradiated samples with that of the samples irradiated by doses of 1·6—4 kGy as the function of the time of heating and varying temperatures. Mesophilic aerobic bacteria of spices had a higher thermoresistance than the bacteria of paprika. Thermoresistance of the surviving microflora of irradiated spices decreased significantly. With increased dose of radiation also the thermal sensitivity increased. The thermal sensitivity quotient (ratio of D_{10} values of non-irradiated samples to irradiated ones) was about 4 for paprika and 9 for spices when irradiating by a dose of 4 kGy.

Anellis et al.[27] investigated the microbiological quality of packed beef inoculated with spores of Clostridium botulinum and irradiated by various doses after heat treatment. The limiting dose was 41·2 kGy at which the concentration dropped by 12 log cycles. The combination of heat and ionizing radiation (F_0 = 1·61 + 5, 10 kGy) also secured the microbiological quality of goose liver after 90 days of storage at 30°C.[28] Kampelmacher[29] reported that combinations of ionizing radiation and moderate thermal processes (pasteurization) increased inactivation processes of irradiation, particularly in the case of moulds. The synergic effect of heat and ionizing radiation at inactivation of moulds depends on the optimum combination of both processes. The most efficient process is irradiation after heat treatment.[30—32] The synergic effect lasts for 1—2 h, but it is gradually lost during 4—5 h. Radiation methods for yeast destruction are described by Petin and Berdnikova.[33]

1.2 Nutritional and Sensory Aspects of Practical Application of Heat and Irradiation

Many authors have dealt with the problems of reduction of losses of vitamin B during irradiation, both because of their essential nutritional role and from the theoretical point of view (mechanism of reaction of aqueous solutions, their constant rates in dependence on pH, protective effect of NO_2, oxygen and glucose in model systems[2,34]), but less from practical aspects. Some progress was recorded by the survey given at the international symposium in Colombo[11] in 1980. At this symposium Diehl[35] presented results of some authors in the field of combined preservation by sterilization and ionizing radiation. For example, Kennedy and Ley[36] describe effects of ionizing radiation (6 kGy), heating (4 min) and combinations of both methods and their effects on thiamine, riboflavin and niacin in cod fillet. The losses of thiamine due to irradiation were 47%, those by boiling 10%, and the losses caused by combination of both processes were 54%. Thus, the total losses of thiamine at combined preservation are comparable with the results corresponding to the total sum of losses at individual methods of preservation. The loss of riboflavin on irradiation was 6%, by heating 9%, and, in the case of combination of both methods, it was 16%.

Urbain[37] investigated the riboflavin content of pork irradiated by doses from 47 to 71 kGy and stored for 15 months. For comparison he presents retention of riboflavin during thermal preservation and freezing. The highest loss was at 15 months storage (from 2·80 to 1·69 mg kg^{-1}), whereas irradiation and thermal preservation did not significantly change the concentration of this vitamin. The losses of thiamine after 6—12 months long storage of beef irradiated by a dose of 0·6 kGy ranged from 5 to 10%.[38] The combination heat-cold-irradiation was applied to pork[3] to investigate thiamine retention. Vacuum packed product was thermally treated ($F_0 = 6$) and irradiated by doses of 15—75 kGy. The retention of thiamine decreased with the dose of irradiation (72—27%). The results have shown that the radiation sterilization depends on the phase, temperature and dose, and that irradiating processes lead to equivalent or higher retention of thiamine than do the thermal processes.

Nawar[40,41] has studied the mechanism of radiolysis of lipids and made a conclusion that the character of the products of degradation can be determined according to the composition and way of treatment. He followed degradation of model phospholipid by combination of heat and ionizing radiation. He found a difference in both quantitative and qualitative representation of particular n-alkanes, l-alkenes, 2,3,4-alkanones, γ-lactones, ethylesters and other components formed during thermal oxidation and radiolysis. The Schen study[42] deals with non-volatile products of unsaturated fatty acids after irradiation and thermal oxidation. The results were in agreement with the theory of lipid decomposition.

The problems of sensory changes in foods which occur as a consequence of reactions products of radiolysis in irradiated foods are usually less important. However, sensory characteristics of irradiated food change according to conditions at reduced doses of radiation, and their positive influence is rather problematic. All raw foods contain autochtonous enzymes which must be inactivated by thermal preservation. Radiation is not sufficient to inactivate enzymes; heating is the only practical and efficient method for obtaining adequate enzymic stability. It is sufficient to heat meat to 70—75°C before irradiation by gamma rays.[43,44] It was found that combination of radiation after previous inactivation of enzymes with vacuum packing and freezing permits reduction in the amount of nitrites in processed meat. When compared with products processed by conventional sterilization, better sensory quality is achieved (colour, consistency, flavour).[45]

Double and triple combinations of preservation methods aimed at improving the sensory characteristics of beef and pork products were applied in other experiments.[46] First, the semi-finished products were heated (3—4 min at 150—160°C) in vegetable oil with addition of spices, carrots and onion. The samples were packed into polyethylene films, irradiated by 0·6 kGy and stored either at 4—5°C or at room temperature. In addition to microbiological indices, sensory characteristics, peroxide thiobarbiturate numbers were followed. During one month storage, no changes in sensory or microbiological characteristics were observed when compared with fresh products. Further experiments of the same series were carried out with a dose of 0·8 kGy. The samples were vacuum packed and sensorially compared with products which were not irradiated and stored at −30°C. Based on the results obtained the authors recommended thermal treatment of semi-products from meat up to a centre temperature of 85°C, vacuum packing into impermeable foil and irradiation by a dose of 0·8 kGy. In this way stability of products during 6 months of storage at 4—5°C was guaranteed.

Boiled meat irradiated with gamma and beta rays in the absence of air losses its typical taste and flavour and off odours occur. It was found that these changes were smaller in pork because of the higher amount of fat. These changes probably cause a higher release of sulphane, mercaptan and carbonyl compounds during irradiation of heat treated meat. Nitrites play a protective function which can be explained by ability of NO and NO_2 to bind the hydroxy radical formed during irradiation of water with gamma rays particularly in the presence of oxygen. Cooking before irradiation or immediately after it has no marked influence on the change of colour. However, flavour was better in the samples irradiated after heat treatment. Irradiation of hams (cooked) was found to have an unfavourable effect on the colour of the product after 2—4 weeks. The interval between heat treatment and irradiation should be as short as possible.[47]

The study by Tarkowski and Beumer[48] presents the results obtained by irradiation (1 kGy) of a bi-component product inoculated by three species of pathogens. The chosen dose did not influence sensory characteristics of the product providing the meat was irradiated before the liquid was added. In general, however, opinion was in favour of non-irradiated samples (including colour); a panel of 20 members attributed them the following characteristics: more piquant, more tasty, more fresh, whereas the irradiated samples were evaluated as softer, acid, special, etc.

In the process of canned ham production (5—10 kGy + 70—80°C in the core) chemical changes were found which result in deterioration of the sensory properties of the product (so called 'off flavour', 'radiation flavour').[49] In the course of storage the difference between control and irradiated products was reduced which could be due to worsened quality of the control samples stored at room temperature. In the experiments attention was paid to the isolation of carbonyl compounds and amines from the samples treated by heat and irradiation. It was found that the content of carbonyl compounds separated as 2,4-dinitrophenyl hydrazones increases relatively with the radiation dose, whereas amines isolated in the form of hydrochlorides increase rapidly up to a relatively high dose of radiation (10 kGy). Thus, it is quite probable that particularly carbonyl compounds and thioaldehydes influence worsening of the sensory characteristics of canned meat products. The synergic effect of heat and ionizing radiation can be economically employed for keeping the quality of vegetable products. For example, in horse radish marked improvement of colour and texture was observed when 8 min heating was combined with 3 kGy irradiation instead of heating for 19·3 min at 115°C only; the texture of green beans also markedly improved when the heat treatment (17·5 min/121°C) was replaced by combination of heat treatment and irradiation (7·4 min/121°C + 3 kGy); so was the texture of green peas by substitution of heat treatment for 18 min at 121°C by combination of 6 min long heat treatment at 121°C and irradiation by 5 kGy. However, the taste of these vegetable products was not improved so markedly.[47] Better sensory characteristics in comparison with the traditional method were also recorded in the case of pickled carrot treated by blanching (7—10 min/90°C) and subsequently irradiated and stored for 3—18 months.[9]

2 THE FIRST SERIES OF EXPERIMENTS

Our previous experiments were focused on the study of the effects of combined preservation with heat and ionizing radiation on microbiological, nutritional and sensory quality indices of model mono- and bi-component products prepared from vegetable and meat raw materials. Concentrations of oxi- and thermolabile components were followed; they were chosen according to the type of the product and changed due to combination of a lower intensity of heat and subsequent ionizing radiation (^{60}Co) as well as by storage (20°C ± 2°C). Because of the limited number of pages only a brief survey of the experimental results is presented.[50]

2.1 Canned Beef in its Own juice

Combination: 35 min/121°C + 2, 4, 5 kGy
Analyses: thiamine, vitamin B$_6$, microbiological indices, SH groups, myoglobin, ammonia, fatty characteristics, sensory evaluation during 240 days long storage (1st, 21st, 114th, 240th).

With respect to the retention of thiamine and vitamin B$_6$,[51] as well as of other components (SH groups, myoglobin, ammonia)[52] the most suitable regime which we found was 35 min/121°C + 4 kGy. In comparison with the product prepared by the conventional method (60 min/121°C) the former has kept a proper nutritional and standard microbiological quality during the storage[53] (the dose of 2 kGy was insufficient from the aspects of required microbiological quality, the dose of 6 kGy negatively influenced the nutritional and sensory values). During combined preservation only slight changes in lipids were observed.[54] The changes in sensory characteristics evaluated by a point test and profiling of taste were non-significant during 240 days long storage, and neither was indication of 'off odour' and radiation 'off taste'.

2.2 Model Cauliflower in Salt Pickle

Combination: 3 min/90°C, 10 min/105°C + 2, 4, 6 kGy
Analyses: folacin, L-ascorbic acid, activity of peroxidase, microbiological parameters, SH groups, sensory evaluation during 240 days storage (1st, 25th, 42nd, 112th, 240th).

The losses of folacin by blanching were almost 40%, the control sample (20 min/115°C) had losses about 66%;[51] intensity of ionizing radiation and the length of storage did not significantly change the content of folacin (5—8%). On the other hand L-ascorbic acid is very sensitive to ionizing radiation which causes its oxidation into dehydroascorbic acid, as well as its destruction. In the sample with a regime of 10 min/105°C + 6 kGy retention of only 29·1% was recorded, and during storage further decrease was observed (by 7·9—12·8%). The SH groups are more sensitive to the intensity of sterilization, during combined preservation more total SH groups were retained than in the samples prepared by the conventional method; during storage 35·6—31·7% was retained, whereas in the sample treated only by sterilization it was 29·8%. The peroxidase activity was sufficiently eliminated by blanching (3 min/90°C) as well as by another heat treatment (10/105°C), and thus zero activity was recorded during the storage.[55]

The evaluation of the microbiological results has shown that all technological procedures which were used had lethal effects on the microorganisms.[53]

The sensory evaluation showed that the consistency was better, but the colour was worse when compared with the control samples (20 min/115°C). Even after 240 days of storage the samples of cauliflower in salt pickle sterilized for 10 min at 105°C + 2, 4, 6 kGy had a better profile of taste than the control.[55]

2.3 Model Canned Beef with Cauliflower

Combination: 30 min/121°C + 3, 5 kGy
For the purpose of calculation of the changes of the components studied also mono-component model products have been studied: canned beef in its own juice (30 min/121°C + 3, 5, 7, 9 kGy) and cauliflower in salt pickle (10 min/116°C + 3, 5 kGy) during 115 days long storage (1st, 24th, 50th, 115th).

Analyses: according to the type of product
Higher doses of ionizing radiation (9 kGy) unfavourably influenced thiamine (canned beef in its own juice — a loss of about 70%), in the meat-vegetable samples the retention was balanced (48·4 and 48·9%); retention did not decrease linearly with the increasing dose of ionizing radiation. During storage (115 days) as high as 85% losses of the thiamine concentration in a bi-component product (canned beef with cauliflower) were recorded.

As for the changes in the riboflavin concentration a relative stability to the heat process, doses of ionizing radiation and the time of storage was found (retention about 95%). Storage of bi-component products (115 days) revealed a rather low retention of vitamin B_6 due to the technological processes employed (about 25%).[56]

Microbiological examination of all components studied demonstrated good stability till the end of storage while the dose of ionizing radiation of 5 kGy was sufficient.[57]

It is evident from the results of evaluation of the changes in the SH groups in a bi-component product preserved in the regime of

30 min/121°C + 3, 5 kGy that the retention was in the interval 43·3—42·5%, i.e. it was higher than in the samples treated by sterilization only (60 min/121°C—31·5%). Storage resulted in further decrease of retention by 14·8 to 20%. The percentage of the retention of myoglobin in a bi-component product was higher in the case of combined preservation (about 50%) than in the case of conventionally prepared products (28·7%). The concentration of ammonia increased with regard to its initial value by 14—18%, and during storage the values were increasing even more.[58]

The values of fatty characteristics (acid and peroxide numbers) were increasing with the intensity of sterilization, dose of radiation and time of storage. On the contrary, the value of the thiobarbiturate number decreased due to the combined preservation and several months long storage.

The carbonyl number decreased with increasing intensity of heating, however, it was increasing with the increasing dose of ionizing radiation and storage.

From the sensory viewpoint the best sample was that which was sterilized by the regime of 30 min/121°C + 5 kGy. It was considered to be the best for the whole period of storage.

One can suggest from the results of this series of experiments that the applied combination of the methods is suitable for more extensive application in practice, particularly when using lower doses of ionizing radiation (4—5 kGy).

3 THE SECOND SERIES OF EXPERIMENTS — APPLICATION OF THE RESPONSE SURFACE METHOD

In the survey of experimental studies of various authors available till now, as well as in our experiments in the field of combined effects of heat and radiation sterilization a passive experiment was the criterion of evaluation which often does not cover the whole spectrum of possibilities of changes in the components followed. In the food technology a number of parameters should be considered (temperature, time of accomplishment, pressure, etc.) which frequently require higher material and energy costs, and can also be time consuming. Therefore, such methods were looked for which would provide sufficient information on the process with only minimum number of experimental measurements. One of them is the Response Surface Methodology (RSM) focused on optimization of technological processes.[59—61] At present, these methods find ever more extensive application possibilities, particularly in the research of such complicated systems like foodstuffs and food systems of experimental plans, the uniform central composite plans (ROUCEKOP) are often used. The number of experiments in such a plan for two factors — in our case the dwell time (t) and the dose of ionizing radiation (γ) — of which 5 experiments were carried out in the middle of the plan, and was used in our model experiments.

3.1 Model Canned Pork in Its Own Juice

For a model experiment of preservation with heat and ionizing radiation 6 heating regimens were used within the interval from 15 to 70 min and a constant temperature of 121°C. Control samples were exposed to the following regimens: 49 min/121°C, 21 min/121°C, 35 min/121°C, 55 min/121°C and 70 min/121°C without application of ionizing radiation. The doses of ionizing radiation determined by the design specification were within the interval from 0·5 to 9·5 kGy. These intervals were chosen so that maximum and minimum effects of varying factors on the studied nutritional and sensory characteristics are involved. A survey of model samples is shown in Table 1. The results obtained for the conditions defined by the design specification were processed using an EC 10 10 computer by means of the ROUCEKOP program, and were referred to the concentrations of substances in raw pork meat. By mathematical and statistical processing of the experimental values regression equations were formulated, and by their graphical representation contour diagrams were obtained which enable to follow the changes in characteristics within the real scope of the parameters of the combined process with respect to the concentration of the followed component in raw food material. The following substances were analysed: microbiological parameters, thiamine, riboflavin, vitamin B_6, SH groups, myoglobin, ammonia, creatine, creatinine, sulphane, lipids, amino acids in raw pork and in the model final product after preparation (1st day), 21st day and after 41 days long storage at room temperature (20°C ± 2°C). By the mathematical and statistical response surface method the optimum regime of application of a combined technological procedure was determined from contour diagrams and regression equations (because of limited scope of this contribution we present only examples for vitamin B and sensory analyses).

TABLE 1

SURVEY OF SAMPLES

Treatment	Heating regime (min/121°C)	F_0	E_{13}^{100}	Radiation dose (kGy)
Raw	—	—	—	—
Sterilized	18-49-15	2·08	129·4	—
Steriliz. + irrad.	18-49-15	2·08	129·4	8·1
Steriliz. + irrad.	18-49-15	2·08	129·4	1·9
Sterilized	10-21-10	—	2·03	—
Seriliz. + irrad.	10-21-10	—	2·03	8·1
Seriliz. + irrad.	14-21-10	—	2·03	1·9
Seriliz. + irrad.	18-35-17	1·05	33·8	9·5
Sterilized	18-35-17	1·05	33·8	—
Seriliz. + irrad.	18-35-17	1·05	32·6	0·5
Sterilized	13-55-15	6·6	167·7	—
Steriliz. + irrad.	13-55-15	6·6	167·7	5·0
Steriliz. + irrad.	10-15-10	—	0·53	5·0
Steriliz. + irrad.	15-35-17	1·05	33·8	5·0
Steriliz. + irrad.	27-35-16	1·05	33·8	5·0
Steriliz. + irrad.	27-35-16	1·1	34·6	5·0
Steriliz. + irrad.	20-35-17	1·08	34·2	5·0
Steriliz. + irrad.	20-35-17	1·08	34·2	5·0
Sterilized	20-70-15	10·31	632	—

3.1.1 Changes in Microbiological, Nutritional and Sensory Characteristics due to Combination of Sterilization and Irradiation

3.1.1.1 Microbiological analyses (CSN 56 0100[62]). The quantitative determination of microorganisms in raw pork provided the following picture: mesophilic aerobic microorganisms $5·4 \times 10^5$, coliform microorganisms $1·2 \times 10^4$, aerobic spore-forming microorganisms $1·8 \times 10^2$, yeast $2·5 \times 10^5$ and fungi $3·3 \times 10^1$. After preparation of the model product the microbiological control showed that all followed groups of microorganisms were inactivated by the combined procedure. After storage for 41 days only the spores of aerobic microorganisms survived (5×10^1) in the sample with the regime of 15 min/121°C + 5 kGy; the sample treated by the regime 21 min/121°C + 1·9 kGy contained $1·5 \times 10^1$ of spores, confirming thus the assumption of its insufficient effects. Sporadically, spores occurred at the regime 21 min/121°C + 8·1 kGy and 35 min/121°C + 0·5 kGy. Other results were negative during the followed time of storage.

3.1.1.2 Thiamine (CSN 56 0052[63]). By mathematical and statistical processing of the obtained results of changes in the thiamine concentration in dependence on the change of factors (intensity of sterilization — dose of ionizing radiation) regression equation 1 was obtained (1st day):

$$y = 103 \cdot 735 - 5 \cdot 596.\gamma - 0 \cdot 350.t + 0 \cdot 580.\gamma^2 - 0 \cdot 055.t.\gamma \qquad (1)$$

The coefficients of the regression equation are calculated for real variables — time of sterilization (t) and dose of ionizing radiation (γ). This equation is valid within the followed scope of changes in the factors. By graphical representation of the regression equation the contour diagram was obtained — Fig. 1.

Regression Eqn. 2 describes changes in the concentration of thiamine in dependence on the presented factors and period of storage (21st day):

$$y = 221 \cdot 794 - 21 \cdot 177.\gamma - 3 \cdot 730.t + 1 \cdot 966.\gamma^2$$
$$+ 0 \cdot 045.t^2 - 0 \cdot 085.\gamma.t \qquad (2)$$

what is graphically illustrated in Fig. 2.

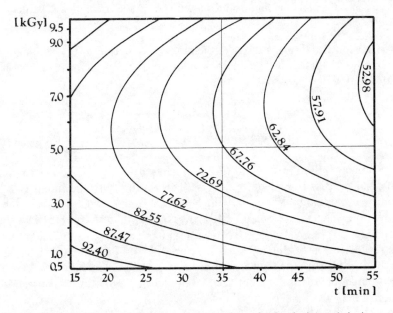

FIG. 1. Contour diagram of changes in retention of thiamin in pork in its own juice canned by combination of thermosterilization and irradiation (1st day).

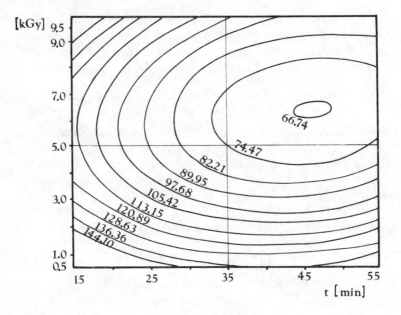

FIG. 2. Contour diagram of changes in retention of thiamin in pork in its own
juice canned by combination of thermosterilization and irradiation (21st day).

After the 41st day of storage of the model product the effect of changes
in individual factors on the thiamine concentration is given by regression
Eqn. 3 and Fig. 3:

$$y = 190 \cdot 223 - 20 \cdot 681.\gamma - 3 \cdot 713.t + 2 \cdot 111.\gamma^2$$
$$- 0 \cdot 049.t^2 - 0 \cdot 098.\gamma.t \tag{3}$$

From regression Eqn. 1 and Fig. 1 known sensitivity of thiamine to the
thermal process is evident[64—66] as well as sensitivity to the intensity of
ionizing radiation. When applying regime 49 min/121°C + 4 kGy the
retention of thiamine was 62·8%, analogous to that at combination 40
min/121°C + 7 kGy. In the control without irradiation (70 min/121°C) the
lowest retention of thiamine was measured — 42%. On the other hand,
in the sample with the regime 55 min/121°C + 9·5 kGy retention of 54·6%
was achieved. It results from the mentioned that radiation resistance is
higher than thermal.

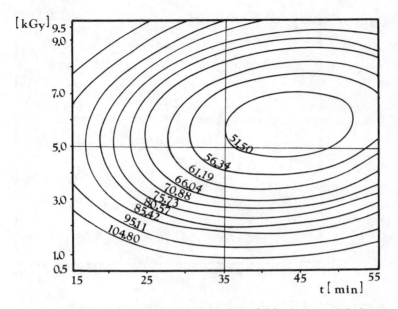

FIG. 3. Contour diagram of changes in retention of thiamin in pork in its own juice canned by combination of thermosterilization and irradiation (41st day).

Figure 2 demonstrates a slight increase of thiamine after the 21st day of storage if compared to the non-stored samples, which according to some authors[66—68] is due to thiamine release from the bonded forms during heat process and storage. Maximum losses were recorded in the case of the regime 47 min/121°C + 6·1 kGy − 41% (s = ±0·035 − 0·055).

One can see from regression Eqn. 3 and Fig. 3 that the concentration of thiamine after 41 days of storage has a decreasing tendency. Maximum losses — 52·2% were found at the regime 44 min/121°C + 6 kGy. A more objective picture on the rate of the loss of thiamine would certainly be provided after longer storage (at least several months).

3.1.1.3 Riboflavin (CSN 56 0054[69]). By mathematical and statistical processing of the results of changes in the concentration of riboflavin in dependence on changing factors (1st day) regression Eqn. 4 was obtained:

$$y = 112·499 − 4·044.\gamma − 0·347.t + 0·404.\gamma^2 \qquad (4)$$

Its graphical representation is in Fig. 4.

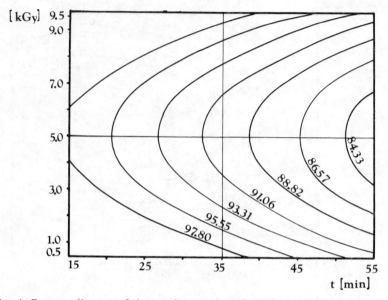

FIG. 4. Contour diagram of changes in retention of riboflavin in pork in its own juice canned by combination of thermosterilization and irradiation (1st day).

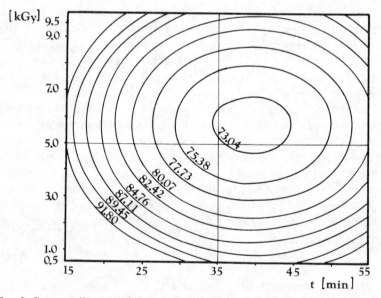

FIG. 5. Contour diagram of changes in retention of riboflavin in pork in its own juice canned by combination of thermosterilization and irradiation (21st day).

The changes in the concentration of riboflavin in dependence on the presented factors and storage (21 days) are expressed by regression Eqn. 5 and Fig. 5.

$$y = 135 \cdot 758 - 6 \cdot 648.\gamma - 2 \cdot 2246.t + 0 \cdot 588.\gamma^2 + 0 \cdot 0282.t^2 \qquad (5)$$

Regression Eqn. 6 and Fig. 6 represent the effect of followed factors on riboflavin after 41 days of storage:

$$y = 132 \cdot 483 - 7 \cdot 556.\gamma - 1 \cdot 924.t + 0 \cdot 756.\gamma^2 + 0 \cdot 023.t^2 \qquad (6)$$

Retention of riboflavin during storage ranged from 70·3% to 97·8% when referred to the initial value of 1·82 mg kg^{-1}. The achieved results confirmed higher resistance of riboflavin (in comparison with that of thiamine) to technological processes and storage.[64,65] It is evident from regression Eqn. 4 and Fig. 4 that the retention of riboflavin was higher in the samples treated by combination of sterilization and ionizing radiation (about 90%) than that in the control samples without irradiation (80·2%).

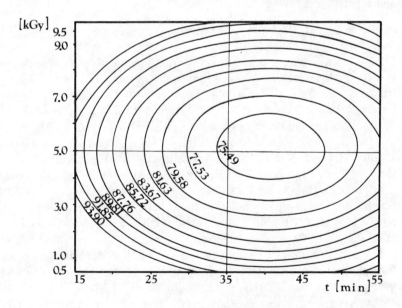

FIG. 6. Contour diagram of changes in retention of riboflavin in pork in its own juice canned by combination of thermosterilization and irradiation (41st day).

During 21 day long storage (Fig. 5) the concentration of riboflavin decreased in all samples. The course of contours in the diagram shows that the highest losses were recorded at the regime 40 min/121°C + 5·7 kGy (about 30%).

Generally, one can state that during the chosen storage interval the retention of riboflavin was relatively balanced within the limits of the applied doses of ionizing radiation from 4·5 to 5·5 kGy and at the heating regime up to 35 min/121°C.

3.1.1.4 Vitamin B₆ (CSN 56 0056[70]). By mathematical and statistical processing of the achieved results of changes in the retention of vitamin B_6 in dependence on the change of the studied factors (1st day) regression Eqn. 7 was obtained:

$$y = 96·676 - 0·124.\gamma - 0·723.t \ (0.\gamma) + 0·0062.t^2 \tag{7}$$

Regression Eqn. 8 describes changes in the concentration of vitamin B_6 in dependence on the presented factors and time of storage (21 days):

$$y = 101·404 - 3·084.\gamma - 1·065.t + 0·308.\gamma^2 + 0·013.\gamma.t \tag{8}$$

and regression Eqn. 9 after 41 day long storage:

$$y = 96·665 - 0·025.\gamma - 1·074.t + 0·229.\gamma^2 + 0·011.t^2 \tag{9}$$

It is evident from the achieved results that vitamin B_6 is more sensitive to a thermal process,[71–74] less to ionizing radiation.[75] Retention during storage varied from 52·7 to 87·7%, particularly in dependence on the intensity of the heat process. It is evident from regression Eqn. 7 — at identical doses of radiation (5 kGy) and at different regimes the differences in retention are as follows: 55 min/121°C – 76·4%; 35 min/121°C – 77·4%; 15 min/121°C – 86·7%. In the control without irradiation (70 min/121°C) the lowest retention of vitamin B_6 was recorded — 66·2%. On the other hand, at the regime 35 min/121°C + 9·5 kGy the retention was about 77% which also proves higher radiation than thermoresistance. At the same regime (35 min/121°C) but with different applied doses (9·5 kGy and 0·5 kGy) was the difference in retention 1% only.

The concentration of vitamin B_6 was changing in dependence on the time interval of storage (Eqn. 8). The highest storage stability was found in the samples with the regime 21 min/121°C + 1·5 kGy (8·1 kGy), i.e. 75·5%. Maximum losses were found at the regime 43 min/121°C + 5 kGy — 30%.

From regression Eqn. 9 a more marked decrease of vitamin B_6 after 41 days of storage is evident. Maximum losses were at the combination 47 min/121°C + 5·5 kGy, i.e. 28·8%

3.1.1.5 Sensory analyses. The samples were evaluated by a committee of 10 members on the first day after preparation and after 41 days of storage at a temperature of 20°C ± 2°C. The model product was evaluated cold (unheated) and after 20 min boiling and cooling to the temperature of consumability.

(a) Point-scale evaluation. This involved evaluation of colour, appearance, odour, taste and consistency within a scale of 0—5 points for particular sensory characteristics. Numbers 5 and 4 refer to very good and good quality, numbers 1 and 2 refer to bad or insufficient quality.[76]

By mathematical and statistic processing of results regression equations for sensory evaluation of product were obtained.

Unheated (1st day)

$$y = 7·56072 + 2·01107.\gamma + 0·7989.t - 0·229489.\gamma^2$$
$$- 0·01074176.t^2.0·014981483.\gamma.t \qquad (10)$$

and after heating:

$$y = 6·34863 + 1·95758.\gamma + 0·855404.t - 0·231206, \gamma^2$$
$$- 0·01149435.t^2 - 0·0132528.\gamma.t \qquad (11)$$

The effect of changes in the parameter of the process on the sensory value of model products after 41 day storage is expressed by regression Eqn. 12.

After heating:

$$y = -19·672836 + 1·0949628.\gamma + 1·889984.t$$
$$- 0·1160513.\gamma^2 - 0·0220166.t^2 \qquad (12)$$

By mathematical and statistical processing of the results optimum conditions for combination of doses of ionizing radiation and intensity of sterilization were obtained (Table 2).

(b) Profiling of tastiness. Profiles of odour and taste were evaluated by intensity of particular descriptors in general perception according to a scale with from 0—5 in comparison to a control sample prepared by the

TABLE 2

OPTIMUM PARAMETERS OF COMBINATION OF THERMOSTERILIZATION AND
IONIZING RADIATION FOR TOTAL SENSORY EVALUATION

Maximum sensory value (points)	Dwell time of steriliza- tion (min)	Radiation dose (kGy)	Storage (days)	Conditions of evaluation
24·775	34·99	3·2	1	Unheated
24·615	35·00	3·2	1	Heated
23·470	44·92	4·7	41	Heated

conventional method (70 min 121°C ($F_o = 10·3$). The evaluated samples
were ascribed the following tastes: meaty, salty, fatty, bitter, off flavour,
and others; odours were marked as follows: meaty, roasty, metallic, off
odour and irradiation.

These were evaluated by mathematical and statistical method: partial
taste — meaty, and partial odour — meaty. Other taste and odour
descriptors could be evaluated in this way because of unexpressive
representation, eventually equal representation in the total perception at
more combinations of irradiation doses and time of sterilization.

For better illustration, differential graphs were made which
demonstrate differences of the individual descriptors of taste and odour
with respect to the conventionally sterilized product (70 min 121°C). An
example of a different profile of taste on the 1st day and after storage is
in Figs. 7 and 8. Of the descriptors of taste most points were ascribed to
the meaty taste. All unheated samples were evaluated by more than 4
points except the sample with higher dose of radiation (35 min/121°C +

TABLE 3

OPTIMUM PARAMETERS OF COMBINATION OF THERMOSTERILIZATION AND
IONIZING RADIATION FOR PROFILING OF TASTINESS

Maximum sensory value (points)	Dwell time of sterilization (min)	Dose of ionizing radiation (kGy)	Storage (days)	"Meaty" description
4·696	33·864	4·221	1	Odour
4·714	41·306	4·543	41	Odour
4·897	34·509	2·909	1	Taste
4·709	45·499	3·413	41	Taste

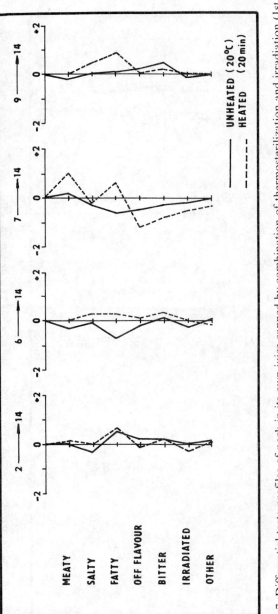

FIG. 7. Differential taste profiles of pork in its own juice canned by combination of thermosterilization and irradiation (1st day). Sample: (2) 49 min/121°C + 1·9 kGy; (6) 35 min/122°C + 0·5 kGy; (7) 55 min/121°C + 5·0 kGy; (9) 35 min/121°C + 5·0 kGy; (14) 70 min/121°C.

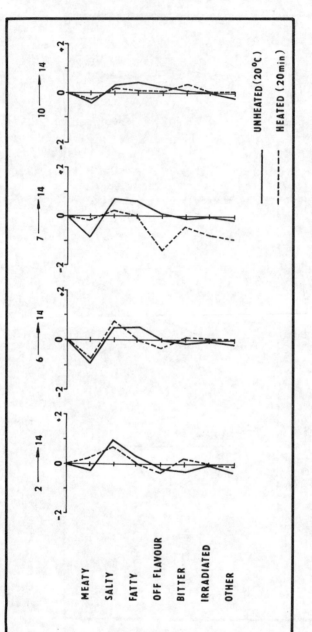
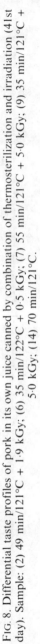

FIG. 8. Differential taste profiles of pork in its own juice canned by combination of thermosterilization and irradiation (41st day). Sample: (2) 49 min/121°C + 1·9 kGy; (6) 35 min/122°C + 0·5 kGy; (7) 55 min/121°C + 5·0 kGy; (9) 35 min/121°C + 5·0 kGy; (14) 70 min/121°C.

9·5 kGy) which was evaluated by 3·7 points. Less than 4 points were ascribed to samples with either higher heating regime (55 min/121°C) or a higher dose of radiation (8·1—9·5 kGy) in which tastes characterized as off flavour, bitter, radiation, were recorded (they were more marked in the heated products).

After 41 days storage only slight changes were observed in comparison to the evaluation on the 1st day after preparation.

When evaluating the descriptors of odour the highest number of points was allotted to the meaty one (3·2—4·8) in both unheated and heated products. As for the odour, the roasty was given more points in the case of unheated products than after heating, and was more marked in the samples with shorter heating time in combination with higher doses of radiation. Metallic odour was recorded when evaluating both unheated samples and the samples after 20 min heating in all samples at approximately the same level. Off flavours were recorded in the samples with higher doses of radiation (occurrence of the off flavours is due to the presence of benzene, phenol and sulphur compounds produced by degradation of phenylalanine, tyrosine and methionine).[77]

From the evaluation of sensory analyses (total point scoring, profiling of taste and odour) optimum time of sterilization result (for short-term storage) 35—45 min/121°C and a radiation dose of 3·4—4·7 kGy.

3.1.1.6 Other nutritionally and sensorically important components. From the results of changes of other components studied which are important from the aspects of nutrition and sensory characteristics (SH groups, myoglobin, ammonia, creatine, creatinine, sulphane, lipids, amino acids) expressed by regression equations and contour diagrams in the chosen food material during short term storage optimum time of sterilization has resulted which is 35 min/121°C in combination with doses of ionizing radiation being 5 kGy.

From a comprehensive survey of the achieved results it is possible to make a conclusion that the irradiation processes (especially lower doses of 4—5 kGy) in combination with a lower intensity of sterilization ($F_o = 1·1$) will find in future extensive application possibilities in the foodstuff industry. Until then it is necessary to enrich the gamut of the studied food material by other types and suitable combinations of bi- and multi-component canned products.

4 SUMMARY AND CONCLUSIONS

In the last decade the interest of scientific workers in ionizing radiation has significantly increased, and it is at a higher level in combination with other preservation methods, for example, with thermal sterilization, which — in addition to economic advantages — can provide products with prolonged shelf life without undesirable nutritional and sensory changes.

In the effort to contribute to objectivization of knowledge in this sphere of research the effect of binary combination of preservation methods — sterilization and subsequent ionizing radiation (^{60}Co up to 10 kGy) on changes in some microbiological, nutritional and sensory characteristics of the quality of model vegetable and meat products during storage has been studied. Both a passive experiment (the first series of the experiments) based on empiric knowledge presented in the literature, and planned experiment (the second series of the experiments) — RSM (Response Surface Methodology) were used.

By mathematical and statistical evaluation an optimum applied regime of combination heat + ionizing radiation was determined on the basis of regression equations and contour diagrams — 35 min/121°C ($F_0 = 1.1$) +5 kGy, which was satisfactory from the microbiological viewpoint as well as for retention of nutritionally and sensory important components. Optimum time of sterilization 35—45 min/121°C ($F_0 = 1.1 - 1.3$) with a dose of ionizing radiation from 3·4 to 4·7 kGy resulted from the evaluation of sensory tests for chosen food material.

One can deduce from our results that the chosen combination of methods as a progressive preservation method is suitable for more extensive application in industry and that it creates real prerequisites for proposal of optimum and less destructive preservation procedures.

ACKNOWLEDGEMENTS

The authors would like to thank Dr. S. Dudasova, Slovak Technical University, Bratislava for her assistance at the evaluation of the experimental results from the computer EC 10 10, to Dr. Z. Salkova, Research Institute of Foodstuffs, Bratislava, for help at irradiation of model samples and to A. Horvathova, Slovak Technical University, Bratislava, for her excellent technical assistance.

REFERENCES

1. FAO/IAEA/WHO Expert Committee, Wholesomeness of Irradiated Food. Report of Meeting 27 October-3 November 1980, Geneva, Wld Hlth Org. Tech. Rep. Ser. No 659, 1981.

2. Basson, R.A., Recent advances in radiation chemistry of vitamins. In: *Recent Advances in Food Irradiation*, P.S. Elias and A.J. Cohen (eds.), Elsevier Science Publish. Co. Inc., Amsterdam, The Netherlands, 1983, 361.
3. Rauch, P., Kyzlink, V. and Rusz, J., Present state of problems dealing with the conservation of food by irradiation. *Food Ind.*, 1983, **34**, 570—573.
4. Grünewald, Th., Lebensmittelbestrahlung. *Z. Lebensm. Unters. Forsch.*, 1985, **180**, 513.
5. Diehl, J.F. Hitzersterilisierung und Westerhaltung von Lebensmitteln. Darmstadt. 1971, 121—130.
6. Rusz, J. and Bystrá, K., Application of irradiation with the purpose of prolonging the retention of semicans. *Meat Ind. Rep.*, 32—39.
7. Rao, V.S. and Vakil, U.K., Effects of gamma-radiation on cooking quality and sensory attributes of four legumes. *J. Food Sci.*, 1985, **50**, 372—375.
8. Urbain, W.M., Combined methods. In: *Food Irradiation. Advances in Food Research*, New York, Academic Press, 1978, **24**, 155—216.
9. Kudryaševa, A.A., Dejstvije slabogo nagrevanija i gama-oblučenija ^{60}Co na kačestvo i mikrofloru natural'nych konservov iz morkovi. *Voprosy Pitanija*, 1974, **33**, 83—84.
10. Bertolini, P. and Restaino, F., Effect of gamma-irradiation on microbial and physiological changes in carrot. *Frutticoltura*, 1978, **40**, 49.
11. Combination processes in food irradiation. Proc. Int. Symp. Colombo, 24-28 November, 1980, Int. Atomic Energy, Vienna, 1981, 467.
12. McCormick, R.D., Processed prepared food, 1982, **151**, 97—98, 100—101.
13. Niemand, J.G., Van Der Linde, H.J. and Holzapfel, W.H., Shelf life extension of minced beef through combined treatments involving radurization. *J. Food Protection*, 1983, **46**, 791—796.
14. Harrigan, W.F. and McCance, M.E., Laboratory methods in food and dairy microbiology. London Academic Press, 1976, 452.
15. Okazawa, Y. and Matsuyama, A., Variation of combined heat: irradiation effects on cell inactivation in different types of vegetative bacteria. In: *Food Preservation by Irradiation*, I, 1978, 251—262.
16. Langerak, D.I., Irradiation of Foods: technological aspects and possibilities. In: *Food Irradiation Now*, 1983, 40—59.
17. Kiyoshi, T., A scientific status summary by the Institute of Food Technologists' Expert. Panel on Food Safety and Nutrition: Radiation Preservation of Food. *Food Technol.*, 1983, **37**, 48—53.
18. Stegeman, H. and Leveling, H.B., Studies concerning complementary effects of irradiation and other preservation techniques on microorganisms for improvement of food hygiene. In: *Application of Atomic Energy in Agriculture*. Annual Report 1979 Assoc. Euratom ITAL, Wageningen, The Netherlands, 1980, 105—108.
19. Ma, K. and Maxcy, R.B. Factors influencing radiation resistance of vegetative bacteria and spores associated with radappertization of meat. J. Food Sci., 1981, **46**, 612.
20. Teufel, P., Microbiological implications of the food irradiation process. *Rev. Food Irrad. Inform.*, 1981, **3**, 3—47.
21. Teufel, P., Microbiological aspects of food irradiation. In: *Recent Advances*

In Food Irradiation, P.S. Elias and A.J. Cohen, (eds.), Amsterdam, The Netherlands, 1983, 361.

22. Kim, J.H., Stegeman, H. and Farkas, J., Preliminary on radiation resistance of thermophilic anaerobic spores and the effect of gamma radiation on their heat resistance. *Int. J. Food Microbiol.,* 1987, **5**, 129—136.

23. Gombas, D.E. and Gomez, R.F. Sensitization of Cl. perfringens spores to heat by gamma radiation. *Appl. Environ. Microbiol.,* 1978, **36**, 403—407.

24. Grecz, N., Bruszer, G. and Amin, I., Effect of radiation and heat on bacterial spores DNA. In: *Combination processes in food irradiation.* IAEA, Vienna, 1981, 3—20.

25. Kamat, A.S. and Lewis, N.F., Influence of heat and radiation on the germinability of B. cereus BIS-59 spores. *Ind. J. Microbiol.,* 1982, **23**, 198—202.

26. Kiss, I. and Farkas, J., Combined effect of gamma irradiation and heat treatment on microflora of spices. In: *Combination Processes in Food Irradiation.* IAEA, Vienna, 1981, 107—115.

27. Anellis, A., Rowley, D.B. and Ross, E.W., Microbial safety of radappertized beef. *J. Food Prot.,* 1979, **42**, 927—935.

28. Farkas, J. and Andrassy, E., Combined effect of reduced water activity, heat and irradiation on microbial stability of canned goose-liver. *Lecture,* 1981, 131—139.

29. Kampelmacher, E.M., Irradiation of food: a new technology for preserving and ensuring the hygiene of foods. *Fleischwirtschaft,* 1984, **64**, 322—327.

30. Odamtten, G.T., Appiah, V. and Langerak, D.I., In vitro studies on the effect of the combination treatment of heat and irradiation on spores of Aspergillus flavus LINK NRRL 5906. *Acta Alimentaria,* 1985, **14**, 139—150.

31. Padwal-Desai, S.R., Bongiwar, D.R. and Sreenivasan, A., Control of foodborne moulds by combination of heat and radiation. *Indian Food Packer,* 1979, **33**, 15—63.

32. Langerak, D.I., Combined heat and irradiation treatment to control mould contamination in fruit and vegetables. In: *Application of Atomic Energy in Agriculture.* Annual Report 1979, Assoc. Euratom — ITAL, Wageningen, The Netherlands, 1980, 113—129.

33. Petin, V.G. and Berdnikova, I.P., Responses of yeast cells to heat applied alone or combined with gamma-rays. *Int J. Radiat. Biol.,* 1981, **39**, 281.

34. Murray, T.K., Nutritional aspects of food irradiation. In: *Recent Advances in Food Irradiation.* P.S. Elias and A.J. Cohen, Elsevier Science Publish. Co. Inc., Amsterdam, The Netherlands, 1983, 203—215.

35. Diehl, J.F., Effects of combination processes on the nutritive value of food. *Com. Proc. Food Irrad. Proc. Int. Symp.* Colombo, 24-28 November, 1980, 349—360.

36. Kennedy, T.S. and Ley, F.J., *J. Sci., Fd Agric.,* 1971, **22**, 146.

37. Urbain, W., Wholesomeness of irradiated foods. In: *Food Irradiation. Advances in Food Research,* New York, Academic Press, 1978, **24**, 209—213.

38. Siunyakova, Z.N. and Karpova, I.N., *Voprosy Pitanija,* 1969, **25**, 52.

39. Thomas, M.H., Atwood, B.M., Wierbicki, E. and Taub, I.A., Effect of radiation and conventional processing on the thiamin content of pork. *J. Food Sci.,* 1981, **46**, 824—828.

40. Nawar, W.W., Thermal and radiolytic decomposition of lipids. Food Technol., 1984, **38**, 49.

41. Nawar, W.W., Comparison of chemical consequences of heat and irradiation treatment of lipids. In: *Recent Advances in Food Irradiation.* P.S. Elias and A.J. Cohen, Elsevier Science Publish. Co. Inc., Amsterdam, The Netherlands, 1983, 115—127.

42. Shen, S.S., A comparative study of the nonvolatile products in irradiated and thermally oxidized unsaturated fatty acids. Dissertation Abstracts Int., B.45/10.31: order DA 8500136, 1985, 206.

43. Dempster, J.F., Hawrysh, Z.J., Shand, P., Lahola-Chomiak, L. and Corletto, L., Effect of low-dose irradiation (radurization) on the shelf life of beefburgers stored at 3°C. *J. Food Technol.,* 1985, **20**, 145—154.

44. Shults, G.W. and Wierbicki, E., Changes in non-protein nitrogen content and the sensory characteristics of beef steaks as affected by the heat treatment. *Tech. Rep. 75-8-Fel.* US Army Natick Lab., Massachusetts, 1974.

45. Wierbicki, E., Technology of irradiation preserved meats. *Lecture Proc. Eur. Meeting of Meat Research Workers,* 1980, **26**, 194—197.

46. Gelifand, S.Yu. and Nomerotskaya, N.F., Combined thermal and irradiation treatment of meat. *Lecture Proc. European Meeting of Meat Research Workers,* 1980, **26**, 210—212.

47. Anon. Emballages traitements ionisants des denrées. *Ind. Aliment. et Agricol.,* 1983, **100**, 717—721.

48. Tarkowski, J.A. and Beumer, R.R., Low-dose gamma irradiation of raw meat. *Int. J. Food Microbiol.,* 1984, **1**, 13—23.

49. Rusz, J., Effect of ionizing radiation on the sensory characteristics of canned meat products. Acta Alimentaria Polonica, 1976, **27**, 175—183.

50. Šorman, L., Hozová, B., Rajniaková, A. and Salková, Z., Study of the influence of chosen combinations of preservation methods on the retention of food products. *Food Ind.,* 1988, **39**, 177—180.

51. Hozová, B. and Šorman, L., Effect of irradiation and heat on the retention of some B-group vitamins in canned foods. *Proc. Euro Food Chem. III,* March 26-29, 1985, Antwerp, Belgium, 107—112.

52. Šorman, L., Rajniaková, A. and Hozová, B., Influence of irradiation and thermosterilization on the change of sulphhydryl groups, myoglobin and ammonia in beef. *Bulletin PV,* 1987, **26**, 55—62.

53. Hozová, B. and Šorman, L., Influence of combined heat treatment and irradiation on the microflora of canned foods. In: *Collection of lectures from the seminar in Liblice,* 6-8 May, 1985, 29—33.

54. Šorman, L., Rajniaková, A. and Salková, Z., Influence of combined irradiation and thermosterilization on the change of lipids of beef. *Bulletin PV,* 1987, **26**, 71—76.

55. Rajniaková, A., Šorman, L. and Kiss, R., Influence of combined preservation on the retention of L-ascorbic acid, SH groups and changes in sensoric properties in cauliflower in salt pickle. *Bulletin PV,* 1988, **27**, 473–483.

56. Hozová, B., Šorman, L., Salková, Z. and Richter, P., Combined effect of heat and irradiation on thiamin and riboflavin in modelled types of canned products (beef in natural juice and beef with cauliflower without culinary treatment). *Collection ÚVTIZ, Food Sciences,* 1986, **4**, 197—204.

57. Hozová, B., Šorman, L., Salková, Z. and Fazekašová, H., Combined effect of thermosterilization and irradiation on the retention of canned foods. *Bulletin PV,* 1986, **25,** 263—273.
58. Rajniaková, A., Šorman, L. and Homolová, B., Influence of combined preservation (thermosterilization-irradiation) on the content of thiol groups, myoglobin and ammonia in bi-component canned product. *Collection ÚV-TIZ, Food Sciences,* 1987, **5,** 287—293.
59. Giovanni, M., Response surface methodology and product optimization. *Food Technol.,* 1983, **37,** 41—44, 83.
60. Fishken, D., Consumer-oriented product optimization. *Food Technol.,* 1983, **37,** 49—52.
61. Schutz, H.G., Multiple regression approach to optimization. *Food Technol.,* 1983, **37,** 46—48, 62.
62. CSN 56 0100. Microbiological testing of foods, materials of current use and environment of food plants. *ÚNM,* Praha, 1969, 62—76.
63. Czecchoslovak standard 56 0052. Thiamin determination. *ÚNM,* Prague, 1969, 5.
64. Davídek, J., Janíček, J. and Pokorný, J., *Food Chemistry,* Prague, SNTL, Alfa, 1983, 595.
65. Príbela, A., *Analysis of Foods,* Bratislava, Ed. Slovak, Tech. Univ., 1987, 394.
66. Wilska, J. and Zajac, J., Effect of sterilization and storage of canned vegetable-meat products on the stability of different forms of thiamin. *Przemysl Spozywczy,* 1983, **37,** 477—478, 500—503.
67. Mikkelsen, K., Rasmussen, E.L. and Zinek, O., Retention of vitamin B_1 and B_6 in frozen meats. Manuscript, 1984, **251,** 778—781.
68. Rogowski, B., Popp, J. and Kaiser, A., Influence of storage conditions, time and temperature on some compounds determining the nutritious value of sterilized meat dishes. *Fleischwirtschaft,* 1980, **60,** 1226—1229.
69. Czechoslovak standard 56 0054. *Riboflavin determination.* Prague, 1971, 6.
70. Czechoslovak standard 56 0056. *Vitamin B_6 determination.* Prague, 1981, 9.
71. Nadolna, I., Rakowska, M., Zielinska, Z. and Werner, J., Sterilization and thiamin and vitamin B_6 content, as exemplified by canned meat with vegetables. *Przemysl Spozywczy,* 1976, **29,** 442—445.
72. Jonsson, L. and Danielsson, K.K., Vitamin retention in foods handled in food service systems. *Lebensmitt.-Wise and Technol.,* 1981, **14,** 94—96.
73. Engler, P.P. and Bowers, J.A., Vitamin B_6 in reheated hold and freshly cooked turkey breast. *J. Am. Diet. Assoc.,* 1975a, **67,** 42—44.
74. Engler, P.P. and Bowers, J.A., Vitamin B_6 of turkey cooked from frozen, partially and thawed states. *J. Food Sci.,* 1975b, **40,** 615—617.
75. Kösters, W.W. and Kirchgessner, M., Effect of UV and Gamma irradiation on vitamin B_6 content and protein constituents of foods. *Landwirtschaftlische Forschung,* 1979, **29,** 194—203.
76. Pokorný, J., Marcín, A. and Pavliš, J., Elaboration of descriptors lists for the evaluation of sensoric profile. In: *Science for Assistance in Practice,* 1981, 131—135.
77. Graham, H.D., *The safety of foods.* Second Edition AV Publish. Co., Westport, Connecticut, 1980, 591.

Chapter 9

FOOD IRRADIATION IN THE UNITED STATES

G.H. PAULI

Division of Food and Color Additives, Food and Drug Administration, 200 C St S.W., Washington, D.C. 20204, USA

1 INTRODUCTION

Since 1963, some irradiated foods have been permitted for sale in the United States. Yet, at the time of this writing, commercial application has been limited to irradiation of a relatively small fraction of the spices and seasonings used as ingredients in other foods. It should not be surprising that adoption of this technology has been so slow, however, for many factors go into acceptance of new processing and handling steps to produce food. A process will not be adopted unless there is a substantial consensus among government, food industry officials, and consumers that the process is in some respects advantageous compared to other practical options.

The government acts as a representative of all the people through an orderly, deliberative process, establishing policy by law and applying that policy through data analysis, evaluation and regulatory procedures. In the United States, broad policy concerning food irradiation is established by the Congress through legislation and appropriation of funds, and is implemented by a variety of agencies whose responsibilities may lie predominantly in food safety, for environmental protection (including protection of agricultural areas from pests or protection of workers and the general population from radiation hazards), for industrial development, or for improving the food supply for either a targeted group or the population in general. Thus, government can act both as a gatekeeper by regulating safety and environmental issues and as a promoter by funding research and development and by purchasing food.

While the government acts on behalf of the consumer on technical issues such as safety, consumers themselves play a major role through their own purchasing decisions. No rational food producer will invest time and money to produce a food that consumers will not buy. On the other hand, it is very difficult to determine whether consumers will accept a food without market testing a real product.

In addition to the need for acceptance by government and consumers, the food producing industry must decide to develop products that satisfy consumer needs and desires. Irradiated foods are often described as if they were a class of foods to be offered to consumers, but it is as individual products that they must succeed in the marketplace. Industry must judge the likelihood of commercial success of individual products, considering the costs and demand for the same or competing products produced by alternative means.

Considering the large number of favorable individual decisions necessary to adopt a new food technology, one should be cautious about prematurely projecting rapid developments in application. This consideration has often been neglected in the many articles written about food irradiation, a number far disproportionate to the amount of irradiation that has taken place. Unlike other processing methods, which are not usually of popular interest before commercialization, the public premarket safety evaluation process has stimulated many articles in the news media. These articles have often ignored factors that are important for determining the likelihood of commercial application. Thus, as long as the demonstration of safety has been the primary goal to be accomplished before marketing irradiated foods, there has been a tendency among some writers to treat this as the only factor crucial to commercial application.

In this chapter, the author will discuss the current situation regarding irradiated foods in the United States and how it developed. The author writes from experience as a government regulator concerned primarily with ensuring the safety of food and, therefore, this chapter will emphasize that aspect, although it will also consider the crucial role played by consumers and industry.

2 REGULATORY FACTORS AFFECTING APPLICATION

The food industry is a thoroughly regulated industry in the United States, with Federal legislation concerning food safety dating from the beginning of the century and state and local legislation even earlier. Responsibilities for enforcing the legislation are shared among several agencies, depending on the major mission of the agency and the food issue in question. In addition, the use of nuclear technologies introduces additional regulatory needs that are generally met by government agencies other than those regulating foods. For a clear understanding, it is essential to distinguish between regulatory concerns generally applicable to food and those applicable to radiation safety or to the handling of radioactive materials.

2.1 Food and Drug Administration (FDA)

The FDA is the major agency for determining whether foods may be irradiated because the Federal Food, Drug, and Cosmetic Act, the major law enforced by FDA, requires a premarket demonstration of safety before an irradiated food may be marketed.

2.1.1 Applicability of the Food Additives Amendment of 1958

Prior to 1958, there was no requirement to show that use of a food substance was safe prior to marketing, although it was illegal to sell a food containing an added poisonous or deleterious substance. The government was responsible for proving whether an added substance was poisonous or deleterious. Increasing urbanization, growing advances in food technology, and increasing consumption of processed foods brought concern that the safety of a new method of manufacturing a food should be determined before the food is marketed, rather than afterwards. Thus, in 1958, Congress passed the Food Additives Amendment which prohibited (a) any food that contained a food additive unless its use conformed to a regulation prescribing safe conditions of use and (b) any food that had been intentionally irradiated unless the irradiation conformed to a regulation prescribing safe conditions.

It is clear from the legislative history that Congress believed that the majority of the food industry consisted of responsible companies that had always tested their products for safety before introducing them to the market. It is also clear that Congress was attempting both to promote responsible developments in food technology and to restrain irresponsible elements that might introduce new technologies without proper safety considerations. Thus, this legislation was intended both to assure safety and to provide fair rules that would require all segments of industry to meet the same standard.

The Food Additives Amendment is broad and applies to various categories of products that may reasonably be expected to add new components to food or affect the characteristics of a food. It applies to ingredients, processing aids, packaging materials and processing equipment (generally because components of packaging materials or processing equipment can migrate to food). It also defines sources of radiation to be food additives, such as radioactive isotopes, particle accelerators, and X-ray machines cited in the legislative history, presumably because their use can affect food characteristics.

This linking of *sources of radiation* in the same amendment that applies to food ingredients, however, has caused considerable confusion among those unfamiliar with the statute and its legislative history. It is of utmost importance for anyone trying to understand the requirements of this legislation to read the specific wording of the statute itself (and for greater understanding, the legislative history) rather than to speculate on presumed requirements.

In context, it was reasonable for Congress to combine all new applications of food processing that were considered to require premarket safety evaluation into one piece of legislation. In that context, it is also reasonable that one term would be used to describe all the substances falling under the control of this legislation. However, the term used to describe these substances, 'food additive', often conveys a concept of a food ingredient that is quite different from the legal meaning given to it by Congress.

The term 'food additive', as defined in the Food, Drug, and Cosmetic Act, does not apply to all food ingredients. For example, to avoid a burdensome regulatory approval process, Congress exempted, *from the definition of the term food additive*, food ingredients generally recognized as safe (GRAS) by food safety experts or subject to a prior sanction or approval by the Federal government under the then relevant food safety laws. Likewise, pesticides that are subject to pesticide approval requirements are exempted from the requirements of food additives. Other food ingredients, however, as well as packaging materials, processing aids such as boiler water additives or ion exchange columns, and equipment such as filters or X-ray machines are considered to be food additives within the meaning of law. Thus, the term 'food additive' has a specific legal meaning and any discussion of legal requirements outside of this meaning is likely to be misleading.

Because of a lack of clear understanding that the term 'food additive' is applied to the sources of radiation (i.e., the radiation equipment) as it is to other food processing equipment, such as filters, many writers have incorrectly inferred that the statute requires a 'process' to be considered an 'additive' and have made false assumptions about what the statute requires. For example, well-respected food scientists have written reports contending that food irradiation should be considered a process, not an additive,[1,2] unaware that there is no legal dispute that irradiation is a process. Similarly, considerable time and effort have been spent considering nonsubstantive aspects of proposed legislation. In 1985, the U.S. Congress considered a proposed bill (The Federal Food Irradiation Development and Control Act of 1985) that never went to a vote. Most attention seemed to be focussed on those provisions of the bill that would have removed sources of radiation from the food additive definition and added 'or food irradiation process' after the term 'food additive' throughout the Federal Food, Drug, and Cosmetic Act. Such a change in wording would have had no substantive effect on any requirement for permitting or regulating irradiated foods because all requirements applicable to food additives would still apply to the 'food irradiation process'. Yet, Congressional testimony by several interested persons showed evidence of widespread belief that the terminology used, whether 'process' or 'additive', somehow would affect the standards applied for determining safety or labeling. Similarly, many people thought that such an editorial change would change the labeling requirements, although labeling provisions of the Federal Food, Drug, and Cosmetic Act do not refer specifically to 'additives'.

2.1.2 Standard for approving use of food additives
The legal standard in the United States for permitting irradiation of foods or the use of any food additive is simple in concept, although sometimes difficult to apply in practice. Namely, adequate evidence must be presented such that a fair evaluation of the entire record will lead FDA scientists to conclude, with a reasonable certainty, that no harm will result from consuming foods produced under the proposed conditions. The specific data necessary to achieve that reasonable certainty are not prescribed by law. For example, there is no requirement for animal feeding studies, although the legislative history shows an underlying presumption that scientists will ordinarily need such studies for making judgments on the safety of a substance in food.

Safety can be considered to be the absence of harm. Because only harm, not its absence, can be demonstrated conclusively, one can always speculate about situations where harm could occur but for which no data exist. Thus, one can never be absolutely certain that any safety requirement can protect everyone in all circumstances. One can do all that is reasonable, however, to eliminate any foreseeable source of harm.

Because of the complex biochemistry of the human response to ingestion of a substance, it is difficult to predict possible toxic consequences solely in biochemical terms. Therefore, feeding studies conducted with mammalian models provide a comprehensive approach to an analysis of potential toxicity. Individual variations among humans and among animals and variations between humans and any animal species provide significant potential errors for any judgment concerning safety. Confidence in a safety decision can be increased by studies that provide a margin of safety, i.e. studies that show no adverse effect in animals consuming a substance in significantly higher proportions of their diet than would occur in human diets. Confidence is also increased by chemical characterization of the substance being tested, allowing scientists to estimate its potential hazard based on studies with similar substances.

In the 30 years since the 1958 amendment became law, FDA scientists have attempted to use the best science of the time to judge whether the use of a substance has been demonstrated to be safe. On one hand, reasonable certainty is needed and certainty generally increases as the amount and quality of data increase. On the other hand, once all reasonable questions have been addressed adequately, the likelihood decreases that increasing amounts of data will have a significant effect on a safety decision. Thus, FDA scientists have sought to balance the objective of obtaining enough information to reach a conclusion with the objective of not requiring unnecessary data after sufficient assurance has already been attained.

To determine that use of a substance is safe one must estimate the likelihood that something that is not known could prove harmful. This is a difficult concept to communicate to the general public because it requires technical knowledge and experienced judgment to distinguish between safety questions of sufficient importance to require experimental data from those of trivial significance. Traditionally, these judgments have relied heavily on Paracelsus' observation that 'the right dose differentiates a poison and a remedy'.[3] Thus, even with no knowledge of a substance's toxicity, one can say that, generally, the less of a substance that is consumed, the less potential exists for it to harm the public. A knowledge of probable consumption alone is not enough, of course, because the toxic potency of substances can vary widely, but such knowledge is critical for assessing the need for toxicity data.

In the 1980s, FDA discussed a concept called a 'concern level' which attempts to quantify the potential for harm even in the absence of toxicological data.[4] FDA toxicologists have always used such concepts, but increasingly there has been an effort to articulate this judgmental process. For example, FDA toxicology requirements for indirect food additives, such as components of packaging material or processing equipment, could be as minimal as the results from routine acute toxicity studies that are performed on all industrial chemicals, provided that consumption is sufficiently small ($<5 \times 10^{-8}$ of the diet) and that available evidence does not indicate a high toxicity. On the other hand, food ingredients directly added to food are consumed in much larger amounts and are typically tested with chronic studies in three mammalian species, in addition to testing for effects on reproduction or other possible adverse effects.

Equally important to the safety standard is the fact that the standard for approval does not authorize consideration of benefits to counterbalance any potential for harm. Apparently, Congress believed that the food supply was so safe and plentiful that any benefit from a food additive could not justify any potential for harm. This contrasts with the pesticide provisions of the Act, enacted four years earlier, requiring that appropriate consideration be given to the necessity for the production of an adequate, wholesome, and economical food supply. (21 U.S.C. 346).

The decision of Congress on the standard for approval also was based on the premise that the benefits accrued would be technical benefits to be judged in the marketplace. It did not want the government to interfere in that marketplace decision, once use of the additive had been shown to be safe.

2.1.3 Application of approval standard to irradiated foods

Studies necessary to demonstrate the safety of irradiated foods also depend on the specific situation and an analysis of the 'concern level' raised by any food use. In the 1960s, safety decisions on irradiated food relied almost completely on animal feeding studies because the chemical differences between irradiated and non-irradiated food were not well-characterized. This nearly complete reliance on animal studies accentuated the importance of obtaining a safety factor by feeding exaggerated amounts of test substance. Significant exaggeration is always difficult and sometimes impossible when testing a macronutrient such as a whole food. Indeed, attempts to feed exaggerated amounts of an irradiated food can compromise a study by imposing a dietary stress that produces confounding effects. Many of the early studies did show adverse effects of unknown origin, leading authorities to require additional animal feeding studies of greater sophistication and power.

In spite of these difficulties, during this time FDA found the data to be sufficient to authorize the low-dose irradiation of white potatoes to inhibit sprouting, wheat and wheat flour to control insects, and high-dose radiation sterilization of canned bacon to produce a shelf stable product. FDA later revoked the radiation-sterilized bacon regulation, however, because reevaluation of raw data from the animal feeding studies showed that the data could not be interpreted unambiguously. (A more complete discussion of regulatory decisions on irradiated food in the United States prior to 1987 can be found in the review by Pauli and Takeguchi.[5]) No additional approvals were given by FDA for other irradiated foods until it obtained a greater knowledge of the chemical changes occurring during irradiation.

The rapid increase in knowledge during the 1970s concerning the chemical effects caused by irradiating foods allowed FDA scientists to rethink how to test such foods for safety. In 1979, FDA's Bureau of Foods (now the Center for Food Safety and Applied Nutrition) established an *ad hoc* committee of six scientists with expertise and experience in toxicology, nutrition, chemistry and regulatory issues. This committee, the Bureau of Foods Irradiated Food Committee (BFIFC), considered the amount of chemical change that would occur during irradiation as a function of dose, the types of radiolytic products formed, the likelihood of unexpected chemicals being formed, and the resulting toxicological consequences; these factors were used to estimate the level of concern that should be applied to any outstanding questions related to formation of toxic substances. BFIFC also considered the potential for further testing to resolve any toxicological concerns. It concluded that, because of the types and amounts of radiolytic products that would be formed during irradiation, there was no need for further toxicological testing of foods irradiated at a dose below 1 kGy or for ingredients constituting $<0.01\%$ of the diet irradiated at a dose up to 50 kGy.[6] (Throughout this chapter, reference to a FDA dose limit is to the maximum dose, not the average dose.) BFIFC recommended toxicological testing for other irradiated foods consisting of a battery of four short-term mutagenicity tests and two 90-day feeding studies (one rodent and one non-rodent mammalian species). BFIFC did not evaluate those animal feeding studies already conducted with irradiated food to determine the extent to which these proposed criteria had already been met, however.

2.1.4 FDA Rulemaking in the 1980s

FDA requested comments on the BFIFC report and announced that it was considering proposing a regulation to permit irradiation of any food at a dose not exceeding 1 kGy or, alternatively, requiring limited petitions on a case-by-case basis that would demonstrate the efficacy of the irradiation but that would not require additional safety data.[7] FDA also announced that it was considering a policy that would permit irradiation (up to 50 kGy) of a food class comprising only a minor portion of the diet based upon minimal biological testing. Such testing was not intended to consider toxicity questions but, rather, would demonstrate the efficacy of irradiation for killing microorganisms.

2.1.4.1 Regulations based on petitions. While FDA was developing a regulation to implement the BFIFC recommendations, it also received petitions for specific actions that relied on those recommendations. In 1983, in response to a petition, FDA authorized irradiation of several spices and seasonings at a dose up to 10 kGy.[8] To define the scope of this regulation, the spices and seasonings were listed by name. FDA also issued regulations in response to petitions for irradiation of pork to control trichina, irradiation of dry enzymes to control microorganisms, and irradiation of additional spices, herbs and seasonings.[9-11] These petitioned uses of radiation were all within conditions where the BFIFC report recommended no requirement for toxicology data.

2.1.4.2 Regulations initiated by FDA. Based on the findings of BFIFC that the effect of radiation on some foods is so small that additional safety data were not needed, FDA decided that it would be more efficient to promulgate regulations permitting such uses than to respond to many petitions. The record supporting a regulation must be complete to withstand a challenge from anyone opposed to the regulation. A petition would ordinarily provide data relevant to toxicity, microbiology, and nutrition; efficacy of the process, considering dose; potential effects on the environment; and methods for enforcing the regulation. FDA concluded that it would be more expedient to develop a regulation that could be supported by the data it had already reviewed than to provide guidance to petitioners on what would be needed for specific petitions. FDA initiated rulemaking could also be designed to solicit maximum input from the public, to ensure that no significant issues would be overlooked.

In 1984, FDA proposed to permit irradiation of food at a dose up to 1 kGy to accomplish certain technical effects; namely, maturation (e.g., sprouting, ripening), inhibition of fresh fruits and vegetables and insect disinfestation of foods.[12] FDA also proposed to establish a new dose limit for the list of dry spices and seasonings of 30 kGy, rather than 50 kGy as recommended by BFIFC, because it could find no evidence that a higher dose was necessary to control microorganisms.

The 1984 proposal was limited to situations where a petition would not be required to provide additional data to show that the irradiated food is safe. BFIFC had already concluded that no toxicological data were needed to conclude that irradiation of foods at a dose below 1 kGy, or of minor food ingredients at a higher dose, is safe. Also at doses below 1 kGy, microbiological effects of safety concern are negligible, as are nutritional losses. The agency concluded that insect control and maturation inhibition could be accomplished at a dose below 1 kGy, obviating the need for efficacy data although, as discussed below, efficacy data are needed to establish a minimum dose for meeting quarantine requirements. There was no need for concern that radiation-resistant pathogenic organisms would grow without signs of spoilage on dry foods irradiated at higher doses or that the higher doses to be permitted would adversely affect nutritional value because dry spices and seasonings do not support microbiological growth and are not sources of nutrients. Potential environmental effects appear to be the same regardless of dose. In sum, none of the proposed uses of radiation required a review of data beyond that already evaluated by the agency.

FDA also designed its proposal to enhance regulatory control of the process by eliminating incentives for violating requirements. For example, because of the difficulties in determining by laboratory analysis that a food has been irradiated, regulatory control is maintained by inspection of records. Post-harvest treatment of foods for insect control is often done to meet quarantine requirements, requiring documentation of proper treatment to move the product in trade. Thus, there is no incentive to hide the fact that the food was treated. Also, 1 kGy is high enough to control any problem due to insects or maturation and 30 kGy is high enough to control any microorganism in dry spices or seasonings. Moreover, the major commodities for insect control are grains or fresh fruits and vegetables. These, however, often cannot tolerate a dosage significantly above 1 kGy. Thus, there is no incentive for a manufacturer to exceed those dose limits.

The proposed rule also considered what labeling requirements would be appropriate for irradiated food. When FDA issued its first regulation authorizing irradiation of a food in 1963, no specific requirements pertained to labeling such food, other than the general, statutory requirement that the labeling not be false or misleading in any particular. In 1966, however, FDA issued a regulation requiring the wording 'treated with ionizing radiation' or 'processed by ionizing radiation' on foods irradiated at a low dose and a high dose, respectively. This requirement applied only to foods that were irradiated, not to foods containing ir-radiated ingredients.

In its 1984 proposal, FDA reopened the question of labeling irradiated foods and discussed different options, including whether special labeling was needed. At that time, FDA tentatively concluded that a labeling re-quirement might be interpreted as a 'warning' even though a warning was neither necessary nor appropriate. FDA proposed not to require special labeling for foods sold directly to the consumer but solicited comments on any future labeling policy, especially requesting comments on the various options discussed.

FDA received over 5000 comments on that proposal. Very few of the comments raised specific, substantive issues, but an overwhelming number of those coming from consumers asked for some type of special labeling requirement.

2.1.4.3 Current FDA regulations. After reviewing all the comments and all the toxicological data in its files, including a series of new studies on radiation-sterilized chicken received in 1984, on April 18, 1986, FDA issued a regulation[13] that was generally consistent with its 1984 proposal but with some modifications as follows:

(a) The intended use for 'insect' control was changed to 'arthropod pest' control, to avoid an unnecessarily narrow restriction.

(b) The restriction of maturation inhibition to 'fresh fruits and vegetables' was changed to 'fresh foods' to avoid confusion over what constitutes a fruit or vegetable.

(c) The listing of specific spices, herbs and seasonings was replaced by a reference to aromatic vegetable substances of certain kinds. This reference has been further clarified to specify aromatic vegetable substances that are used as ingredients in small amounts solely for flavor-ing or aroma but that do not appear to be vegetable substances eaten for their own sake.[14]

(d) The proposal to require no special labeling on retail packages of foods was changed to require the international 'radura' symbol and the phrase 'Treated with radiation' or 'Treated by irradiation'. The require-ment for the specific wording was set to expire April 18, 1988, but was extended to April 18, 1990, because the 'radura' symbol was not yet well recognized by US consumers. The labeling requirement for retail packages applies to foods that have been irradiated but not to foods that merely contain an irradiated ingredient. A food to be shipped to a food manufacturer or processor for further processing, labeling, or packaging must be labeled with one of the statements listed above along with the words 'do not irradiate again' if any portion of the food has been irradiated, as FDA had proposed.

Under US law, the government must set forth the facts on which it relies in reaching a decision and opponents to a decision are given an opportunity to rebut such facts. FDA received several objections to its 1986 decision, some of which requested an opportunity to present evidence at a hearing. After a prolonged evaluation, which included giving one of the objectors an opportunity to clarify its position, FDA concluded on December 30, 1988, that none of the objections provided any factual basis for challenging FDA's decision.[14] Rather, the objections were simply unsupported opinions and allegations.

2.1.4.4 FDA consideration of Codex Alimentarius Commission Recommended Standard. Concurrent with this activity, FDA was considering the 1980 recommendations of the Joint FAO/IAEA/WHO Expert Committee on the Wholesomeness of Irradiated Food (JECFI). FDA's earlier committee, BFIFC, had issued recommendations pertaining only to what toxicology data should be required before reaching a decision on the safety of irradiated food. It had not reviewed the available toxicology data to determine whether those requirements had been met and it had not considered microbiological or nutritional data relevant to safety.

Therefore, in late 1981, FDA established a second *ad hoc* committee, the Irradiated Foods Data Task Group, to review all available toxicology data on irradiated foods and to identify any consistencies and/or adverse findings, looking for patterns and trends among the studies. Because of the enormous volume of data to consider, the Task Group developed a workplan that concentrated on (a) those studies that were unassailable in design, conduct and reporting and (b) those studies that, even if faulty in part, reported data that might suggest evidence of an adverse effect related to eating an irradiated food. Studies were rejected without a detailed evaluation if the type of food, the irradiation dose or the animal group size was not reported; if the group size or length of experiment was too small to be useful; if the diet was clearly inadequate; if the dose was <0.1 kGy or >100 kGy; if there were no controls; if the food was administered other than orally; and if the studies were conducted by Industrial Biotest Laboratories, a US firm that was once a major testing laboratory but that had later been shown to have become untrustworthy. Approximately one third of the studies were rejected for these reasons. The Task Group made no effort to recover additional details, data or explanations not present in reports of the studies because that could have significantly delayed completing their report.

They did, however, give thorough consideration to any reported adverse effects to determine whether such effects could be attributed to toxic substances produced by irradiating the food. Comparisons were made between studies reporting similar effects to see if any patterns emerged that would not have been considered definitive when reviewing one study alone. The Task Group concluded that studies with irradiated foods do not appear to show adverse toxicological effects. The Task Group also recommended that regulations permitting irradiation of foods at high dose not be adopted at that time because the data it reviewed were not of sufficiently high quality and diversity to evaluate the safety of all foods irradiated at medium to high doses and constituting major contributions to the diet.

Although FDA's rule is not the same as the recommendation of the Codex Alimentarius Commission (Codex), FDA has not concluded that the Codex recommendations are unsound. Rather, FDA chose its regulatory approach based on the quality and quantity of safety data available to it and on the requirements of the particular regulatory system under which it controls the safety and quality of food. When FDA issues a regulation permitting irradiation of food, the regulation itself must prescribe any conditions necessary for safe use. Any food processor in the United States or abroad may sell food irradiated under the terms of the regulation in the United States without any further consideration by FDA. Thus, any particular regulatory concerns of the agency must be resolved before issue of the regulation.

FDA did not issue a regulation prescribing safe conditions of use for all foods irradiated at a dose up to 10 kGy, as recommended by the Codex, for three reasons:

(a) The majority of animal feeding studies available to FDA were either reported incompletely or used experimental designs that do not meet current standards. FDA found no evidence that any irradiated food caused a toxic effect but concluded that it needed more thorough documentation of safety before reaching a safety conclusion applicable to all foods.

(b) Some foods may undergo significant nutritional loss when irradiated at doses between 1 and 10 kGy. To properly assess whether this would have any nutritional effect on the diet, FDA must consider the quantity of expected nutrient loss in a food in the context of the significance of that food to the diet and the different foods likely to be irradiated. FDA has not yet assessed the possible or probable nutritional impact of unrestricted irradiation at these doses. In those instances when the agency permitted irradiation at levels > 1 kGy, the agency concluded that the foods involved are not significant sources of nutrients.

(c) There may be a need for specific regulatory requirements for the irradiation and subsequent handling of foods that have been irradiated at doses that extend the shelf life of food but that do not destroy all pathogens. Although this concern is equally valid for other processing methods, FDA's lack of experience with irradiated food and its responsibility to require premarket approval for this process convinced it to proceed on a case-by-case basis.

Thus, FDA's decision allows for authorization of additional irradiated foods in the future on a case-by-case analysis of data. FDA does not plan to commit resources to change these regulations, however, unless the industry submits petitions with relevant data, sufficient to document the safety of the food irradiated under the conditions it requests.

2.1.4.5 Enforcement of irradiated food regulations. As discussed above, once a regulation is issued, anyone may irradiate a food in compliance with the regulation without further permission from FDA; that is, the regulation itself is the full authorization. There are no FDA licensing procedures but FDA monitors food processing by periodic unannounced inspections of facilities handling products under FDA jurisdiction. In planning inspections, FDA attempts to balance the need for coverage of an entire industry with the need to target certain activities or companies where violations are more likely to be found or where protection of public health requires particular attention. Inspectors seek to determine whether facilities are being operated according to current good manufacturing practices and whether all regulations applicable to food are being followed. If violations are found, the company may be issued a warning intended to correct the violation, a product may be seized, or an injunction may be sought to halt operation until corrections are made. The enforcement remedy used is determined by the seriousness of the violation and the company's history of compliance with the law. For flagrant violations, criminal prosecution may be sought.

Presently, enforcement of the irradiated food regulations is primarily through on-site or records inspection. Research is being conducted to develop sensitive methods that can determine unequivocally whether a food has been irradiated. Such research based on detection of long-lived free radicals, protein derived radiolytic products, and lipid derived reaction products, has been conducted at the National Institute of Standards and Technology (formerly the National Bureau of Standards) and the University of Massachusetts, under contract with the US Department of Agriculture, as well as at the FDA.

Under the Federal Food, Drug, and Cosmetic Act irradiated food products imported into the United States must meet the same standard as those produced domestically. Foods presented for import that appear to be in violation of US law, however, may be held pending a decision on admission. With a few exceptions, irradiated food may be exported from the United States to another country if the food is in compliance with the law of the importing country, regardless of the regulations pertaining to sale in the United States, provided that such food was produced solely for export.

2.2 Department of Agriculture (USDA)

Two components of USDA, the Food Safety and Inspection Service (FSIS) and the Animal and Plant Health Inspection Service (APHIS) have a significant role to play in regulation of certain irradiated foods. Another component, the Agricultural Research Service (ARS) has been active in research.

2.2.1 Food Safety and Inspection Service (FSIS)

FSIS has the responsibility to assure that meat and poultry products sold in interstate commerce in the United States and foreign trade are safe, wholesome, and accurately labelled.[15] This responsibility derives from the Federal Meat Inspection Act and the Poultry Products Inspection Act. These statutes provide for a much more intensive regulation of meat and poultry products than the Federal Food, Drug, and Cosmetic Act does for other food products. In addition to observing the FDA regulations for assuring the safe use of radiation sources for treating food, FSIS must establish rules for the use of ionizing radiation on meat and poultry products.

On January 15, 1986, in response to an industry petition, FSIS amended its regulation concerning substances for use in the preparation of products (9 CFR 318.7) to permit irradiation of pork products to control *Trichinella spiralis*.[16] To irradiate pork, a processor must obtain a grant of inspection and have an approved partial quality control program for each facility where pork would be irradiated. In addition, each irradiated product must have a prior approved label. Although irradiation of pork is permitted, at this time no company has obtained approval of a partial quality control program. Thus, there has been no commercial irradiation of pork in the United States.

FSIS has also taken the initiative to petition FDA for a regulation permitting irradiation of poultry at doses from 1·5 to 3 kGy to control pathogenic microorganisms. This petition is in addition to a related petition submitted by a privately owned radiation processing company, Radiation Technology, Inc., in 1979. At the time of this writing, FDA's decision on these two petitions is imminent.

FSIS has also played an important role in irradiated food labeling. When FDA issued its first regulations permitting the sale of irradiated foods in the 1960s no special labeling was required. One of these regulations, later revoked because of inadequate data to demonstrate safety, permitted radiation sterilization of bacon. When FSIS amended its regulations to permit irradiated bacon in conformance with FDA's regulation, in 1965, it required the labels of such foods to carry an approved term such as 'Processed by Ionizing Radiation'.[17] It was only after this requirement was issued by FSIS that FDA decided to impose a similar requirement on other irradiated foods.

2.2.2 Animal and Plant Health Inspection Service (APHIS)

APHIS conducts regulatory and control programs to protect and improve animal and plant health for the benefit of humans and their environment. In cooperation with State governments, APHIS administers Federal laws and regulations pertaining to animal and plant health and quarantine and the control and eradication of pests and diseases. Quarantine inspection officials administer Federal regulations that prohibit or restrict the entry of foreign pests or plants, plant products, animal products and byproducts, and other materials that may harbor pests. APHIS also inspects and certifies domestic commodities for export.

The role of APHIS in regulating irradiated food concerns the efficacy of the process as it pertains to problems requiring quarantine treatment. While FDA's regulatory concerns are primarily with ensuring that the food is safe and honestly represented, APHIS must ensure that foods do not carry unwanted pests that could harm this country's agricultural sector. Thus, it must have data demonstrating that applied processing conditions are effective for controlling the specific pest of concern when harbored by the food being irradiated. Considering the variety of combinations of pests and foods that may be subject to quarantine requirements, much needs to be done for irradiation processing to be adopted widely to meet quarantine needs. In 1987, APHIS proposed to permit irradiation as a quarantine treatment for papaya fruit from the State of Hawaii to prevent the introduction of the oriental fruit fly (*Dacus dorsalis* [Handel]), the Mediterranean fruit fly (*Ceratitis capitata* [Wiedemann]) and the melon fly *Dacus cucurbitae* [Coquillett]) into the continental United States and into uninfected offshore areas.[18] In 1989, APHIS issued a final rule that established the conditions necessary for certification of papayas for movement from Hawaii. Among other things, APHIS required a minimum absorbed dose of 150 Gy (15 krad).[19] This regulation is the only authorization for radiation processing as a US quarantine treatment method. Anyone interested in the use of radiation processing to meet US quarantine requirements should contact APHIS.

2.3 Other Federal and State Requirements

Essentially all industrial activity today must conform to laws designed to protect worker safety and the environment. The food industry is no exception, although the use of radiation processing would impose additional requirements not normally applicable to food processing.

For example, manufacturers of machine sources of radiation must submit a report to FDA's Center for Devices and Radiological Health under the Radiation Control for Health and Safety Act.[20] While this imposes no requirement on food processors, a processor should ensure that any equipment used has been reported by the manufacturer. Under this legislation, the Secretary of Health and Human Services prescribes performance standards for electronic products to control the emission of radiation if he determines that such standards are necessary to protect public health and safety. At this time, however, standards have not been deemed necessary for industrial processing equipment. Standards have been issued only for television receivers, cold-cathode gas discharge tubes, diagnostic X-ray systems, radiographic equipment, fluoroscopic equipment, computed tomography equipment, and cabinet X-ray systems. Some states, however, require registration or licensing for facilities using machine-generated radiation.[21]

The use of radioactive source material for radiation processing always requires a license, although the licensing agency varies with locale. The US Nuclear Regulatory Commission regulates the use of radioactive substances in approximately half of the states and maintains agreements with the other half whereby a state agency is responsible for regulatory control.[21] In addition, laws administered by the Occupational Safety and Health Administration for worker safety and by the Department of Transportation for transportation of hazardous materials apply also to conditions and radioactive materials that would be used in food processing plants.

The discussion above attempts to explain the normal workings of the Federal regulatory system as it applies particularly to irradiated foods. The Federal control system concentrates on food producers, processors, manufacturers and distributors. Retail stores and restaurants are normally monitored by state and local authorities in coordination with Federal agencies. In recent years, groups not satisfied with Federal regulations have attempted to negate or supplement Federal rules and policies through state and local legislation. Although state or local legislation has been initiated on a variety of issues normally addressed by the Federal government, this has been particularly widespread with irradiated foods. It may be worth noting that several local governing bodies have passed resolutions in opposition to irradiated foods. In addition, in 1987, nine states (Alaska, Hawaii, Illinois, Maine, Massachusetts, New Jersey, New York, Oregon, and Pennsylvania) considered legislation related to irradiated food. Maine passed legislation prohibiting the knowing sale of irradiated food. This law did not prohibit sales of foods containing irradiated foods as ingredients. Alaska passed a resolution requesting FDA to adopt certain labeling requirements. All other proposed legislation failed to be enacted. The proposed legislation in only one of the remaining seven states, Hawaii, would promote irradiation.

All other legislation would have prohibited irradiated foods or would have enacted labeling requirements other than those issued by FDA. In 1988, bills that would have prohibited irradiated foods or enacted labeling requirements were submitted in seven states (Alaska, Massachusetts, Minnesota, New Hampshire, New Jersey, New York and Vermont). None were enacted, although a few have been carried over to 1989. New York passed a 2-year moratorium on the sale of irradiated foods in the summer of 1989.

Federal law can be used to preempt local legislation, and sometimes preempt ion may be an explicit part of Federal legislation, when the local laws either interfere with interstate commerce or defeat the purpose of Federal law. In general, however, Federal agencies such as FDA prefer to cooperate with state and local officials to the extent possible rather than to use such preemptive authority. Similarly, state and local officials generally strive to ensure that any legislation they enact is consistent with

Federal law. Nevertheless, the relatively small voting districts for state and local legislators make such officials 'inviting targets' for pressures from lobbying groups that can affect voting patterns. This can be seen by the number of legislatures that have taken up consideration of food irradiation, a subject that they would not normally address.

In summary, responsibility for assuring safe processing of food using sources of radiation is shared by many government agencies, based on the particular operation or material involved. Government agencies cooperate, of course, to minimize redundancy of requirements and to maximize efficiency and effectiveness. Nevertheless, the multiplicity of agencies involved must be bewildering to the food manufacturer considering adoption of radiation processing and especially to those, whether from the food industry or other governments, who would like to find one person knowledgeable in all areas.

3 COMMERCIAL FACTORS AFFECTING APPLICATION OF FOOD IRRADIATION

As stated above, commercial application of food irradiation has been minimal to date. While the regulatory burden discussed above is by no means inconsiderable, that alone does not explain the situation. After all, similar burdens have been taken on numerous times by food processors or ingredient manufacturers developing successful non-irradiated products. Before initiating a major venture, however, there needs to be a reasonable probability that one's investment will be returned through future profits. In general, food manufacturers in the United States have not invested heavily in radiation technology but have limited their investments to contingency planning to be sure that a major opportunity is not missed. (For an excellent analysis of the requirements for success of radiation processing see Masefield and Dietz.[22])

For a new technology to be commercially successful, it must show clear advantages in cost (or in quality, justifying a higher price) compared to alternative approaches for accomplishing similar objectives. This has been a formidable barrier to food irradiation in the United States because food production and technology is very highly developed, resulting in a relatively inexpensive and bountiful food supply. Thus, food irradiation technology has had to compete against solidly established technologies such as chemical treatment, thermal processing and refrigeration.

Moreover, because development of the technology has not been perceived to benefit individual companies in the competitive arena, most of the research and development work has been carried out by the government. Such an effort is nearly always product-oriented rather than profit-oriented, developing products that are acceptable although not necessarily profitable. The major research effort was divided between the former Atomic Energy Commission, for products treated at a relatively low dose, and the US Army, primarily for high-dose radiation-sterilized meats. (It is interesting to note that the first commercial application with dry spices and seasonings falls into neither of these categories.) Government-sponsored research, just as industry-sponsored research, must be justified to the source of funding (the taxpayer). Since the taxpayers or their representatives are not going to make a profit in any case, government-sponsored research is less likely to emphasize minor projects that might make a profit, as opposed to ambitious projects that show a vision for future change. Thus, government-sponsored research has often emphasized new developments that would make a significant difference in food handling practices, but that are not necessarily ready to compete commercially against established technologies.

Until the 1980s, regulatory approvals were issued only to delay sprouting of white potatoes or to control insects in wheat and wheat flour. These objectives have been attained cheaply for years by chemical treatment with a much lower capital investment in facilities. A food manufacturer is unlikely to abandon an accepted procedure for one untested in the marketplace, unless there is a sizeable economic advantage or no other alternative.

Spices and seasonings became good candidates for radiation processing because (1) their sale for use as ingredients in other foods typically requires processing for antimicrobial control and (2) the future of the major alternative procedure (ethylene oxide fumigation) was under question because of its toxicity. Moreover, irradiation appears to reduce microbial contamination to a greater extent than fumigation. Because large amounts of spices and seasonings are sold for use in other foods, irradiation processing could be marketed to manufacturers based on its technical advantages. This contrasts with typical marketing strategies used for sale to the ultimate consumer. Thus, there was a natural incentive for food manufacturers to consider change, even if they were satisfied with the alternative procedure.

4 CONSUMER FACTORS AFFECTING APPLICATION OF FOOD IRRADIATION

Consumer acceptance can be measured accurately only by determining whether sufficiently large numbers of consumers purchase a commercially available product. Premarket surveys can help estimate how consumers will act but such estimates may be proven wrong in the marketplace. Thus, there has been no totally valid measure of acceptance of irradiated foods to date. Consumer surveys do show, however, that most people are either neutral or willing to accept foods produced by the irradiation process if the products are of a good quality and sold at an acceptable price.[23-25] This is shown particularly in the one survey where irradiated papayas were available.[26]

The acceptance of irradiated foods by consumers, however, faces unusual barriers. Marketing of most consumer products does not need to contend with groups organized to prevent their acceptance. Groups knowledgeable in mass communication, able to capture the attention of the news media and of legislators, have been organized to oppose food irradiation, however. They have reviewed the irradiated food safety literature sufficiently to find examples of questions that have been raised at one time and that would still appear to raise issues of real concern, if one were not familiar with the work done to answer such questions. Frequently, they have quoted experts out of context to make it appear that a scientist has raised serious reservations about the safety of an irradiated food when, in fact, the opposite is true. In other cases they have interpreted a laboratory finding to have a significance for public health far beyond, or contrary to, that which any expert in the field would conclude. Many of these misrepresentations or misinterpretations are illustrated in objections to FDA's 1986 rule.[14]

Allegations arising from such errors can be dealt with fairly and properly in the context of a formal regulatory process but their effect on public opinion is harder to determine. A person who hears such allegations for the first time without any knowledge of the entire context is likely to be hesitant about any change. Most consumers will not have the time, interest, or expertise to check the credibility of those who misrepresent scientific findings. Likewise, legislators and food producers may have little incentive to expend resources necessary to resist pressures raised by protesters. Whether or not most consumers find the allegations of such opposition groups credible, any rumor about the safety of food poses one

more barrier for introduction of a new process. Even a small barrier may
be significant to the food producer because profits will not necessarily increase appreciably by selling irradiated food.

For example, the Coalition for Food Irradiation, a group composed
primarily of food commodity processing, packaging and retail industries,
was established to promote consumer acceptance of irradiated foods. Its
membership list was circulated among opposition groups, however,
which used it as a focus for protests. Major companies dropped out of
the Coalition[27] and in December 1988, the Coalition closed its
Washington office, effectively ending its activity. Thus, industry leader-
ship was neutralized by protest groups. It has often been said that many
food companies would like to be the second company to market an irradiated food.[27] Being first, however, poses public relations risks.

5 FUTURE CONSIDERATIONS

The future use of food irradiation in the United States will depend primarily on the interest and commitment of the food industry in implemen-
ting the technology and on the public mood for accepting products.
Other than spices and seasonings, the greatest interest to date has come
from the Hawaiian papaya industry. Its products are expected to be the
first to be marketed directly to the consumer with full labeling for consumer choice. Hawaiian papaya must undergo an approved quarantine
treatment before it may be shipped to the continental United States and
irradiation conditions to meet quarantine requirements have been
established.[19] Appropriate conditions have also been established for
Florida grapefruit although the Federal government does not have
quarantine requirements for this product. One would also expect growth
in the amount of spices and seasonings being irradiated but reports to
date do not support such a judgment. Although the amount of spices and
seasonings being irradiated grew rapidly after such processing was first
permitted, it has been reported that growth quickly reached a plateau and
the amount being irradiated even declined slightly.[27] The lack of continued growth has been attributed to fears by food processors that use
of irradiated ingredients could lead to opposition groups leading
boycotts of their products. Thus, even if irradiation is considered to be
a superior processing method, marketing considerations based on public
relations dominate decision making.

The most likely candidates for new approvals in the United States are
foods where public health benefits may be obtained by control of food-

borne pathogens. FDA expects to reach a decision on irradiation of poultry for pathogen control soon. Decisions will not be made on other products until industry commits the time and money to provide reliable data sufficiently complete to establish safe conditions for irradiation of each product. This may require supplementing available reports with additional raw data to remove any ambiguity in interpretation. It might also require establishing adequate quality control conditions that would preclude any likely safety problems that would be anticipated in routine commercial practice.

Whether any of this will occur in the near future depends on the public willingness to accept a new processing technique. Concerns of the public that are based on fact can endure a long time but concerns based solely on emotion, while often more strongly held, can fade quite quickly. Commercial adoption of the process for a specific product may require a simultaneous recognition by the industry and the public that irradiation presents a feasible solution to a real problem.

Meanwhile, the regulatory need for a reliable and efficient method for monitoring what is done to food provides a stimulus for developing analytical procedures for determining that a food has been irradiated. Recent research in the United States has concentrated on three approaches: (1) measuring reaction products from radiation-stimulated hydroxylation of protein fragments;[28] (2) measuring degradation products from irradiated fats;[29] (3) measuring long-lived free radicals in dry solid matter such as bone, shell, or dry foods.[30] While considerable progress has been made, no method has been validated sufficiently to be used in regulatory actions.

Finally, development of the technology also requires development of standard radiation measuring techniques to be used by the processor to control and monitor the process. The American Society for Testing Materials (ASTM) has established committees to develop dosimetry standards for food processing.[31] Undoubtedly, such work will continue as the need requires.

REFERENCES

1. Coon, J.M. and Josephson, E.S., *Is Radiation a Food Additive/Comments from CAST,* Council for Agricultural Science and Technology, Ames, Iowa, May, 1984.
2. Elias, P.S., Task force on irradiation process — wholesomeness studies in food protection technology. In: *Food Protection Technology,* C.W. Felix (ed.), Lewis Publishers, Inc., Chelsea, Michigan, 1987.

3. Doull, J. and Bruce, M., Origin and scope of toxicology. In: *Toxicology,* J. Doull and L.J. Cassarett (eds.), Macmillan Publishing Company, New York, 198-, p. 3.

4. Kokoski. C.J., Regulatory food additive toxicology. In: *Chemical Safety Regulation and Compliance,* F. Homburger and J.K. Marquis (eds.), S. Karger, New York, 1985, pp. 24–33.

5. Pauli, G.H. and Takeguchi, C.A., Irradiation of foods — an FDA perspective, *Food Rev. Int.,* 1986, **2**, 79–108.

6. Brunetti, A.P., Frattali, V., Greear, W.B., Hattan, D.G., Takeguchi, C.A. and Valcovic, L.R., In: *Recommendations for Evaluating the Safety of Irradiated Foods,* US Food and Drug Administration, Washington, D.C., 1980.

7. Federal Register, **46**, 18992, US Government Printing Office, Washington D.C. (March 27, 1981).

8. Federal Register, **48**, 30613, US Government Printing Office, Washington D.C. (July 5, 1983).

9. Federal Register, **50**, 15415, US Government Printing Office, Washington D.C. (April 18, 1985).

10. Federal Register, **50**, 24190, US Government Printing Office, Washington D.C. (June 10, 1985).

11. Federal Register, **50**, 29658, US Government Printing Office, Washington D.C. (July 22, 1985).

12. Federal Register, **49**, 5714, US Government Printing Office, Washington D.C. (February 14, 1984).

13. Federal Register, **51**, 13376, US Government Printing Office, Washington D.C. (April 18, 1986).

14. Federal Register, **53**, 53176, US Government Printing Office, Washington D.C. (December 30, 1986).

15. Engel, R.E., The implementation of ionizing radiation for the control of trichinosis in fresh pork cuts in the United States. In *2nd World Congress Foodborne Infections and Intoxications Final Congress Documents Institute of Veterinary Medicine,* Robert von Ostertag-Institute, W. Berlin, 1986.

16. Federal Register, **51**, 1769, US Government Printing Office, Washington D.C. (January 15, 1986).

17. Federal Register, **30**, 5702, US Government Printing Office, Washington D.C. (April 22, 1965).

18. Federal Register, **52**, 292, US Government Printing Office, Washington D.C. (January 5, 1987).

19. Federal Register, **54**, 387, US Government Printing Office, Washington D.C. (January 6, 1989).

20. US Code of Federal Regulations, Part 1002, US Government Printing Office, Washington D.C., 1989.

21. Ballantine, D.S., Radiation processing and the regulatory process, *Rad. Phys. Chem.,* 1979, **14**, 245.

22. Masefield, J. and Dietz, G.R., Food irradiation: the evaluation of commercial opportunities, *CRC Crit. Rev. Food Sci. Nutr.,* 1983, **19**(3), 259–272.

23. Wiese Research Associates, Inc., Consumer Reaction to the Irradiation Concept, *US Department of Energy Contract No. DE-SC04-84 AC 24460,* March, 1984.
24. Consumer Research Department, Women's Attitudes toward New Food Technologies, *A Good Housekeeping Institute Report,* New York, 1985.
25. Brand Group, Inc., *Market Survey for Seafood,* US Department of Commerce, National Marine Fisheries Service, Chicago, 1986.
26. Bruhn, C.M. and Noel, J.W., Consumer in-store response to irradiated papayas, *Food Technol.,* 1987, **41**, 83–85,
27. Lochhead, C., The high-tech food process foes find hard to swallow, *Insight.* The Washington Times, Washington, D.C. (November 7, 1988), pp. 42–45.
28. Karam, L.R. and Simic, M.G., Detecting irradiated foods: use of hydroxyl radical biomarkers, *Analyt. Chem.,* 1988, **60**, 1117A–1119A.
29. Weiss, R., The gamma-ray gourmet: scientists cook up tests for irradiating food, *Science News,* 1987, **132**, 398–399.
30. Yang, G.C., Mossoba, M.M., Merin, U. and Rosenthal, I., An EPR study of free radicals generated by gamma-radiation of dried spices and spray-dried fruit powders, *J. Food Qual.,* 1987, **10**, 287–294.
31. Farrar IV, H., International effort to establish dosimetry standardization for radiation processing, Proceedings of the 7th International Meeting on Radiation Processing, Noordwijkerhout, The Netherlands. April 23–28, 1989. In: *Radiation Chemistry and Physics* (in press).

Chapter 10

THE IMPACT OF IRRADIATED FOOD ON DEVELOPING COUNTRIES

P. POTHISIRI & P. KIATSURAYANONT
Food and Drug Administration, Ministry of Public Health, Devaves Palace, Bangkok 10200, Thailand

&

C. BANDITSING
Office of Atomic Energy for Peace, Ministry of Science and Technology, Bangkok, Thailand

1 INTRODUCTION

Many approaches to the problem of meeting the increasing worldwide demand for food have received attention. A study by the U.N. Food and Agricultural Organization (FAO) on the future world food situation concluded that 'by the end of the century, shortage of land will have become a critical constraint for about two-thirds of the population of developing countries. Developing countries demand for food and agriculture products is expected to double during this time.'[1] One of the most important approaches of those which have been accepted is a reduction in food losses. Most fruits and vegetables are perishable and have a limited marketable life especially fruits and vegetables grown in tropical countries which have high ambient temperatures and humidity, because of post-harvest diseases, insect infestation, physiological changes and other pathological breakdowns. There are a number of technologies which could be used to extend the marketable life of these commodities, e.g. chemical preservatives, drying, freezing, canning and irradiation. Use of chemicals may prove to be quite efficient and practicable in terms of industrial application. However, such practices may introduce potential health hazards. Most conventional physical processes, while not directly involved with health hazard problems, do not always retain the original

261

properties and nutrients of the commodities. For this reason, much work have been done on the use of irradiation in the preservation of food. Microorganisms are not only involved in food losses but are often also of concern in food safety. The incidence of foodborne infection and diarrhoea as a consequence of pathogenic microorganisms is not only one of the major public health problems but also gives rise to economic disadvantages such as loss of wages and medical expenses. Precise statistical information on diarrhoeal diseases and the consequent economic losses in the Asian and Pacific region is not available. However, the Economic Research Service of the United States Department of Agriculture has recently estimated that the annual costs in the United States of America may exceed US$ 1000 million in terms of medical expenditures and loss of wages because of outbreaks of salmonellosis, toxoplasmosis and campylobacteriosis caused by contaminated pork and chicken. The same study estimated that irradiation of chickens at 2·5 kGy could achieve public health benefits of US$480 million to US$805 million annually, and that irradiation of pork in the USA at the same dose could bring about human health benefits of US$200 million to US$300 million annually.[2]

The potential for food irradiation in the control of foodborne diseases is enormous and of particular importance in the developing countries of the Asian and Pacific region where considerable human resources are lost because of these diseases.

2 TRADITIONAL PRESERVATION TECHNOLOGIES

Food preservation varies from comparatively simple methods, such as sun-drying, to highly sophisticated processes requiring complex equipment and specially trained personnel. The ability to preserve food for relatively long periods had been discovered a long time ago by primitive people.

The use of fire for food preservation can be traced to the pre-Neolithic period. Other methods; salting, smoking, drying, fermentation, and freezing are known to have been used by Neolithic people 10 000 or more years ago.

Even early peoples did not understand, however, why drying, smoking, freezing, and other methods prevented food from going bad. The role of microbes in food spoilage was not discovered until the time of Pasteur; human ingenuity produced sophisticated food-processing techniques.

The first successful process for preserving food was heating process. Later dehydration of food was stimulated and recently freeze-drying and irradiation technique have been discovered.

The traditional methods of preserving food can be divided into five major groups: drying, heat treatment, freezing, fermentation and chemical treatment.

2.1 Drying

Drying is one of man's oldest methods of food preservation and seems to be important in developing countries where dried fish is an important protein source. There are chemical and biological forces acting upon the food supply that man desires. Drying is biological force action which controls food quality by reducing the free water content by heating and thus inhibits bacteria by dessication.

Dehydration can also make foods suitable for subsequent processing that may, in turn, facilitate handling, packaging, shipping, and consumption. Both physical and chemical changes take place during food drying, but not all of them are desirable. Drying may significantly alter food texture, pigments, flavor, form, and rheological properties. In addition to changes in bulk density, foods may undergo unwelcome colour changes, such as browning; they may also lose nutritional value, flavour, and even the capacity to reabsorb water.

There are many types of driers used in the dehydration of foods. Forced-air drying is used largely with grains, fruits, and vegetables. Other methods that can produce powder from liquid are drum-drying, spray-drying and freeze-drying.

2.2 Heat Treatment

Various forms of heat treatment, baking, broiling, roasting, boiling, frying and stewing, are among the most widely used techniques in the food industry as well as in the home. Heat not only produces desirable changes in food, but can also lengthen safe storage times. Heating reduces the number of organisms and destroys some life-threatening microbial toxins. It inactivates enzymes that contribute to spoilage, makes foods more digestible, alters texture, and enhances flavour. But heating can also produce unwanted results, including loss of nutrients and adverse changes in flavour and aroma.

As a method for reducing the number of microorganisms, heat treatment of food consists primarily of blanching, pasteurization, and sterilization. Blanching, exposing food briefly to hot water or steam, is normally used before foods are further processed by freezing, drying, or canning. In addition to cleansing the raw food product, blanching reduces the microbial load, removes accumulated gases, and inactivates enzymes. Pasteurization is a treatment which can destroy susceptible non-spore-producing organisms especially pathogens by mild heat.

2.3 Freezing

Freezing has long been recognized as the best method for the long-term preservation of food, especially for perservation of meat and fish. Frozen food retains most of its original flavour, colour, and nutritive value. Despite its superiority, however, freezing often produces detrimental effects on food texture as a result of ice formation. Fast freezing minimizes this problem.

Preservation by freezing is achieved by lowering the temperature of the food to at least $-18°C$, which crystallizes all the water in the product to ice. At these low temperatures, microbial growth ceases and destructive enzyme activity, while not completely stopped, is reduced to an acceptable level. With some foods, such as vegetables, where enzyme activity during storage or thawing is critical, heat treatment, or some other means of destroying enzymes, is carried out prior to freezing. Food can be frozen before or after packaging. Unpackaged foods freeze faster but are subject to considerable water loss unless they are frozen very rapidly.

Initially, the practice was to freeze food by placing it in a cold room ($-18°C$ to $-40°C$) and allowing air to circulate slowly over the food, a technique known as sharp-freezing. Later, air blast freezers were developed for both batch and continuous processing. Their use has significantly reduced processing time and improved the quality of frozen products.

Obviously, frozen food must be maintained at or below freezing temperatures at all times if this method of preservation is to be effective. In addition, frozen food must be packed in containers that prevent moisture loss and oxidation, i. e., freezer burn. While the overall costs of thermal treatment and freezing are similar up to the completion of the processing operation, the need for an unbroken chain of transportation and storage at freezing temperatures places serious economic constraints on the use of freezing for the preservation of food.

2.4 Fermentation

While food preservation systems in general inhibit the growth of microorganisms, all such organisms are not detrimental, in fact, some are commonly utilized in food preservation. The production of substantial amounts of acid by certain organisms creates unfavorable conditions for others.

Fermentation technique is based on this principle started by the selective removal of the fermentable substrate and the consequent development of an unfavourable environment for spoilage organisms. Microorganisms are used to ferment sugars to alcohol or acids. A number of factors determine what kind of product is obtained by fermentation; the kind of organism used, the material being processed, the temperature, and the amount of available oxygen determine whether the end-product of fermentation will be beer, wine, leavened bread, or cheese.[3]

2.5 Chemical Treatment

Preservation of food by the addition of chemicals is sometimes used when (a) the product cannot be given a suitable terminal treatment or (b) as a supplement to another method of preservation to reduce the intensity of treatment with a resultant improvement in quality.

The substances employed in food preservation are of two general kinds: (a) common food ingredients, such as sugar and salt and (b) specific substances that prevent or retard food deterioration. In the latter category are the so-called food additives and certain other chemicals of value in lengthening the shelf-life of fresh foods or preventing infestation of grains and other food during bulk storage.

The preservative effect of sugars (in concentrations of 65% or more) results because (a) lowering the water activity of food and hence inhibiting the growth of microorganisms and (b) the osmotic pressure of the solution is raised, thus causing plasmolysis of microbial cells. Products such as fruit preserves, jams, and syrups are commonly processed with sugar. Although, this process can inhibit growth of microorganisms, some of their spores can survive for long periods in high concentrations of sugar, such as yeasts, or molds can grow on the surfaces of highly sugared foods, especially when temperature cycling results in accumulation of surface moisture which may have a lower sugar content than the product.

Salt can act in several ways to preserve foods. It keeps spoilage organisms under control and acts as a drying agent, again by reducing the water activity. Furthermore, the increased osmotic pressure produced by salt causes plasmolysis of cells. Other detrimental effects of salt on microorganisms include reduction of the solubility of oxygen in water, sensitization cells to carbon dioxide and interference with the action of proteolytic enzymes. However, some food-poisoning microorganisms, e.g. *Staphylococcus aureus* are most resistant to the effects of salt and can grow at brine concentrations of 14–15% when the pH value is between 5 and 7. In addition, in recent years, changes in taste, combined with growing concern about health hazards associated with a high intake of salt, have led to a significant lessening in the use of salt as a food preservative.

Today, modern chemical additives are big business in many developing countries. Even figures of additives intake in developing countries are not available now, it is estimated that processed foods which now exceed 50% of the total food market contain one or more additives.

Among the food additives approved for use as chemical preservatives in many countries are propionic acid, benzoic acid, sorbic acid, and their salts and derivatives. Sulfur dioxide and sulfites have a long history as important preservative agents but due to the hazard concern, recently their use has been severely limited in several countries, namely Singapore, Malaysia and Thailand. All these substances are most effective in foods which are dry or fairly acidic; they are of limited or no value in watery low acid foods, such as mushrooms and certain green vegetables. Several other chemicals, notably methyl bromide, ethylene dibromide, and ethylene oxide, have been widely used as antimicrobial agents and as fumigants to destroy insects in various foods, such as spices, copra, and walnuts. Evidence that ethylene dibromide and ethylene oxide are harmful to man has led to their being banned by some national regulatory authorities. The use of ethylene bisdithiocarbamates (EBCDs) namely mancozeb, maneb, zineb and metiram are currently also proposed to revoke or reduce tolerances by US.EPA because of the potential dangers to human beings.

Each method used to control spoilage and deterioration of food and to protect the consumer against foodborne disease has both advantages and disadvantages. Research is being undertaken, however, in many countries to make all these methods more effective and efficient.

3 FOOD IRRADIATION IN PRACTICE

Food irradiation is a simple and effective way to process food, one whose potential has not been recognized fully. It is a physical treatment of food similar to heating, freezing, or subjecting the food to microwaves or ultraviolet light. The process involves exposing the food, either packaged or in bulk, to gamma rays, X-rays, or electrons in a special chilled room for a specified period of time, to achieve the technical purpose. Cobolt-60 and Caesium-137 are used as a source of gamma rays. Electrons and X-rays are produced by machines and involve no radioactive source.

4 HEALTH IMPACT

Despite great advances in modern technology, keeping food safe remains a worldwide public health problem. Illness caused by contaminated food is a leading cause of suffering and death in the developing world, and affects millions in all countries.

Food and waterborne disease is one of the major causes of diarrhoea with up to 1000 million cases every year in children under 5 and 5 million deaths.[4] Other estimates are even higher. Table 1 indicates the type of diseases reported in some southeast Asian countries. An estimate of the economic impact of diarrhoea was made in Bangladesh where 238 000 person days of diarrhoea were experienced by 53 000 adults. This is equivalent to a loss of 4·5 work days per person each year. In Indonesia there are an estimated 300 000–500 000 deaths per year because of severe diarrhoea. In Malaysia this type of mortality ranks 10th (0·6/100 000). For Indonesians diarrhoea is the second cause of death after lower respiratory tract infections.

Traditionally, it was thought that contaminated water supplies were the main source of the infectious agents causing diarrhoea, but food is now recognized as a vehicle contributing to a high proportion of the cases, especially in children fed improperly prepared breast-milk substitutes. Problems have occurred with these substitutes, for instance, in the Caribbean area because of poor quality, inadequately labelled product or incorrect administration to infants. Temperature abuse of the prepared drink may lead to *Bacillus cereus* intoxication. *B. cereus* grown in cooked rice, widely eaten in developing countries, has caused illness in India and in China and probably many other countries but the disease tends to be under-reported. The organism was found at $\geq 10^4$/g in 81 of 211 samples surveyed in Nanjian, China (foods included milk powder, moon cakes, pickled vegetables and overnight stored cooked rice). Whereas the main foodborne disease organisms of industrialized nations (*Salmonella*, *S. aureus*, *Clostridium*) are also important in developing countries, greater numbers of cases are probably caused by *Vibrio parahaemolyticus*, *V. cholerae*, *E. coli*, parasites and viruses.

TABLE 1

CASES OF GASTROENTERITIS IN SOUTHEAST ASIAN COUNTRIES[a]

Country	Population	Cases of gastrointestinal illnesses where pathogens isolated							Cases of general gastroenteritis reporting to government health facilities or estimated	
		Cholera	Typhoid/ paratyphoid fever	Amoebiasis/ bacterial dysentery	Food poisoning (bacterial)	Infectious	Total	No./ 100 000	Total	No./ 100 000
Indonesia	146·8	47 056	19 038	—	—	—	66 094	45·0	60 000 000	40 872
Japan	120·0	55	338	997	31 125	—	32 515	27·1	—	—
Philippines	48·1	1 898	3 222	27 378	719	9 578	42 795	89·0	250 000	520
Thailand	44·8	645	6 584	12 984	36 041	14 200	70 454	157·3	170 000	380
Malaysia	13·8	67	2 000	1 545	1 700	2 223	7 535	54·6	150 000	1 087
Singapore	2·4	40	188	—	619	601	1 378	57·4	82 552	3 440

[a]From D. Mossel.[17]

Although the consumption of certain meats are proscribed for adherents of several religions, there is a growing trend away from a vegetarian type diet (cereals and pulses) to one that includes meat, poultry and fish. This will probably result in a changing foodborne diseases pattern. Table 2 shows the prevalence of meatborne diseases in Nigeria. Tuberculosis (20–45%) and brucellosis (18%) are much higher than salmonellosis (2·1%) or trichinosis (no record of cases).[5] In addition WHO estimated that in developing countries, the ratio between actual and reported cases may be as high as 100:1.

TABLE 2
PREVALENCE OF MEATBORNE DISEASES IN NIGERIA[a]

Disease	Location	Prevalence
Tuberculosis	Nigeria	20–45%[b]
Brucellosis	Nigeria	18%[c]
Taeniasis	Northern Nigeria	0·9–4·5%
	Western Nigeria	2·4–5·5%
Q fever	Nigeria	4·5%
Salmonellosis	Nigeria	2·1%
Onchocerciasis	Western Nigeria	2·0%
Paragonimiasis	Eastern Nigeria	1–4%
Trichinosis	Nigeria	No record

[a]From E· Todd.[5]
[b]Serological survey.
[c]10% of cases due to bovine strain.

The economic impact of foodborne disease in developing countries has never been estimated but is probably much greater than industrialized countries, not perhaps in absolute dollars but in the impact on family, local and national economies. The social cost has to be considered in light of high infant mortality, malnutrition, chronic diarrhoea, lost work and child care. If an estimate of US$50 per case is considered (to include value of deaths) then the cost of the 1000 million children alone suffering from diarrhoea would be US$50 000 million.[5]

The potential for food irradiation in the control of food-borne diseases is enormous and of significant importance in the developing countries where considerable human resources are lost due to these diseases. The use of ionizing radiation for the elimination of bacterial and parasitic enteropathogens from raw foods of animal origin has been studied widely since the middle 1950s as shown in Tables 3 and 4.

TABLE 3

EFFECT OF IRRADIATION ON THE MICROBIAL COMMUNITY STRUCTURE OF FROZEN CHICKEN[a]

Organisms	log_{10} CFU/g after irradiation at (kGy):				
	0	1	2	3	4
Mesophilic colony count	6·8	5·8	4·6	4·1	3·6
Psychrotrophic colony count	5·8	5·7	4·0	<2·8	<1·8
Enterobacteriaceae	5·5	<2·8	1·0	0·4	−0·4
Lactabacillus	6·0	4·1	4·2	3·1	<2·8
Lancefield D streptococci	5·1	3·7	3·9	3·2	<2·0
S· aureus	4·6	2·2	<−0·5	<−0·5	<−0·5

[a]From D. Mossel.[17]

TABLE 4

EFFECT OF IRRADIATION ON THE MICROBIAL COMMUNITY STRUCTURE OF FROZEN MALAYSIAN SHRIMPS[a]

Organisms	log_{10} CFU/g after irradiation at (kGy):			
	0	2	4	6
Mesophilic colony count	6·8	4·8	3·2	2·8
Psychrotrophic colony count	6·2	4·2	<2·8	<2·8
Enterobacteriaceae	3·2	<−0·5	<−0·5	<−0·5
Lactobacillus	5·2	<2·8	<2·8	<2·8
Lancefield D streptococci	4·9	1·0	<−0·5	<−0·5
Staphylococcus aureus	3·5	<−0·5	<−0·5	<−0·5

[a]From D. Mossel.[17]

5 ECONOMIC IMPACT (BENEFIT OF FOOD IRRADIATION)

Despite the availability of established technologies for food preservation, canning, freezing, chemical preservative, aseptic packaging and traditional technologies such as sun-drying, fermentation, salting and smoking, the world is still experiencing high post-harvest losses of food. A conservative estimate shows that ≈25% of the world's food production is lost after harvesting. The losses are especially important in developing countries which are situated mainly in tropical zones having high ambient temperature and humidity and where the need for food is greatest. Losses of agricultural produce are due to hot climate accelerating the ripening of fruits and sprouting of vegetables, spoilage microorganisms, pathogenic microorganisms, and insect infestation. As early as 1976, the US National Academy of Sciences projected that the minimum post-harvest food loss in developing countries would amount to over 100 million tons at a value of over US$1 billion in 1985.

In ASEAN countries, the post harvest losses of grains are estimated at 30% in perishable crops such as fruits, vegetables at 20–40% and up to 50% for fish. This loss occur in various stages of distribution starting from the farmer to the market place.[6]

In relation to Latin American countries, large problems exist, such as the huge post-harvest losses, which sometimes represent more than 30–50%, especially where the climate conditions are adverse and/or the commercial practices are not adequate. In addition, there are also other problems which need to be solved such as public health and/or the quality of the products for export. It is necessary to take into account that Latin American countries export enormous quantities of agricultural and sea products, so their economies depend heavily on the currency these products can generate.[7]

Developing countries in Europe and Middle East region are characterized by a number of both developed and developing countries in which a large variety of temperate and semi-tropical foods are produced. Major food losses are rare in this region. However the interest in this region is to apply technology on sprout inhibition root crops, insect disinfestation, decontamination of spices and seasoning, shelf-life extension of fresh fruits and vegetables particularly strawberries and mushrooms, fish and meat and decontamination of poultry products.[6]

In Africa, it was reported that the shortage of food supply in that continent ranges from chronic to acute. Post harvest food losses are one of the fundamental causes of the shortage. For cost estimation purposes, minimum overall losses of 10–12% for durables, 20% for roots and tubers and 30% for fruits and vegetables can be assumed.[8] An extrapolation in monetary terms of these minimum losses estimates for 1980 is that 1 million tons of cereals are lost, while almost the same quantity of maize and rice is imported. The average food import bill per country in 1982 stood at about US$100 million and it is forecasted may exceed US$125 million by 1990.

For solving many of the above problems, food irradiation technology can benefit almost every food. Fruit can be preserved, freed of insects, their ripening delayed and senescence inhibited. Similar benefits occur with vegetables, cereal grains and legumes can be freed of insects. Meats and poultry can be radappitized to match thermal sterilization, but with superior results, radurization extends the shelf-life of fresh meats and animal feeds can be freed of *Salmonellae*. The list of potential uses in treating foods with radiation seems endless.

The introduction of food irradiation technology depends not only on the technological studies, but also on the economic feasibility. Both aspects must be very closely linked in food irradiation plans. Besides, the potential application should elicit positive responses to questions concerning volume-logistics, and competitive technologies.

Techno-economic feasibility studies proved that food irradiation is an economic, efficient and convenient method for processing of foods in many countries. Studies on economic feasibility of radiation preservation of dried fish and fishery products, onions, potatoes, garlic, showed economically viable in Thailand,[9] Bangladesh,[10] the Philippines,[11,12] Indonesia[13] and Pakistan.[14] In addition, other irradiated agricultural commodities, such as mungbean, poultry, sausage, are also found economically viable in Thailand.[9] Furthermore cost/feasibility studies in Latin America (Brazil, Chile, Cuba, and Uruguay) were found to benefit calculation in some of agricultural commodities in these countries (see Table 6).[7]

Economic feasibility studies of food irradiation plants have been carried out in some developing countries and have proven to be feasible. The multi-purpose agricultural pilot plant demonstration facility (100 kG) in Thailand operated by the staff of the Office of Atomic Energy for Peace (OAEP) was estimated economically viable and would obtain an adequate return from the capital invested (the internal rate of return is 16·59). In addition, the benefit-cost ratio is 1·12 at an interest of 12%.[9] It was anticipated to give service for irradiation of eight selected food items, initially for 6084 operating hours.

An economic feasibility study was also conducted on radiation disinfestation of Mexico-grown mangoes and citrus (oranges and tangerines) (see Table 5).[15] Irradiation was being evaluated as an alternative quarantine treatment method to chemical fumigation. Two levels of throughput were assumed for each of the two irradiator plants: 50 and 100 million kg of mangoes irradiated per year, and 100 and 150 million kg of citrus irradiated per year. Results of the study show that the radiation disinfestation process at an assumed minimum absorbed dose of 0·30 kGy for controlling the emergence of Mexican fruit flies from mangoes and citrus is economically feasible in terms of plant size and all cost factors, and will provide the opportunity for the packers to realize a profit

TABLE 5

UNIT PROCESSING COST FOR PACKING AND IRRADIATING MANGOES AND CITRUS FOR DISINFESTATION AT 0·30 kGy AT TWO LEVELS OF THROUGHPUT PER YEAR (IN MILLION DOLLARS FOR TOTAL COST; IN DOLLAR FOR UNIT COST)[a]

| | Mango | | | | Citrus | | | |
| | 50 million kg/year | | 100 million kg/year | | 100 million kg/year | | 150 million kg/year | |
	Packing	Irradiation	Packing	Irradiation	Packing	Irradiation	Packing	Irradiation
Total fixed cost (US$)	0·6973	0·7152	1·0088	1·0027	1·0088	0·9897	1·4246	1·3506
Total variable cost (US$)	28·2115	1·3465	53·558	1·650	23·530	1·5938	35·309	1·9880
Total fixed and variable cost (US$)	28·909	2·062	54·567	2·653	24·539	2·504	36·734	3·339
Unit cost/kg (US$)	0·578	0·041	0·546	0·027	0·245	0·026	0·245	0·022
Total unit cost/kg (US$)	0·62		0·573		0·271		0·267	

[a]From J. Moy.[15]

TABLE 6

RESEARCH ON THE COST/FEASIBILITY[a]

| Product | Country | | | | | | | | | |
	Argentina	Brazil	Columbia	Cuba	Chile	Ecuador	Mexico	Peru	Uruguay	Venezuela
Beans		+[b]								
Corn		+[b]								
Fish meal					+[b]				+	
Garlic										
Onions		+[b]			+[b]				+	
Potatoes	+	+[b]			+[b]				+	
Rice/rice products		+[b]								
Spices				+[b]						
Sugar cane					+[b]					
Wheat/wheat products		+[b]								

[a]From T. Rubio.[7]
[b]Benefit calculation.

from the sales of their packed and irradiated fruits. The irradiation costs were US$0·026–0·041 per kg of mangoes, and US$0·022–0·026 per kg of citrus. Still lower cost can be realized if the cost of the irradiator facility is less than indicated, or if the plant size is increased because of the availability of scale of economies in all cases.

6 FOOD IRRADIATION POTENTIAL (FOOD AND AGRICULTURE IN THE PROMOTION OF FOOD TRADE)

Up to one-third of the fresh produce harvested is discarded because of spoilage resulting from handling, storage and transportation. The potential for food irradiation in the promotion of the food trade is increasing since it is a process which is technically and economically feasible. As shown in Table 7, substantial increases have been registered in food production in Bangladesh, Burma, China, India, Indonesia, Malaysia, Pakistan, the Philippines, Sri Lanka and Thailand.

TABLE 7
PRODUCTION INCREASES, 1978–1988

	Cereals (%)	Potatoes (%)	Groundnuts (%)	Fruits (%)
Bangladesh	15·04	39·60	−17·86	7·94
Burma	37·59	146·30	20·47	13·72
China	28·98	3·38	137·14	145·11
India	22·85	73·79	17·59	29·13
Indonesia	62·44	87·55	109·26	54·42
Malaysia	12·58	—	−77·27	29·7
Pakistan	27·67	91·50	66·66	81·34
Philippines	26·22	165	−7·89	51·94
Sri Lanka	30·16	175·86	25	−20·25
Thailand	27·95	83·33	32·81	5·01

Table 8 shows the percentage increases in the production of pigs, chickens and hen eggs from 1978 to 1988 in the same countries. Similarly, the increases in fresh water and marine fish production from 1978 to 1988 are shown in Table 9. These two tables show that, over the decade, increases in these easily spoiled food commodities have been large enough to warrant safe storage measures. Low to medium dose radiation can

TABLE 8
MEAT PRODUCTION INCREASES, 1978–1988

	Pigs (%)	Chickens (%)	Hen eggs (%)
Bangladesh	—	—	25·58
Burma	17·57	115·91	104
China	133·87	87·43	156·43
India	48·28	141·94	68·08
Indonesia	96·43	286·61	371·74
Malaysia	23·73	112·10	51·64
Pakistan	—	366·67	174·65
Philippines	79·25	83·59	1·32
Sri Lanka	0	25	166·67
Thailand	36·33	70·71	16·84

reduce the microbial load and thus prevent losses of these valuable commodities.

In addition, the trade figures in thousands of tonnes of fish and fish products for the developing countries of the Asian and Pacific region for the triennium ending 1974 and 1987 are 539·8 and 724·2 thousand tonnes, respectively, showing that the export of fishery products increased by 34·16%. Further increases in trade can be expected, provided the number

TABLE 9
FISH PRODUCTION INCREASES, 1978–1988

	Inland (%)	Marine (%)
Bangladesh	9·16	11·73
Burma	4·39	42·42
China	265·94	59·67
India	40·38	16·09
Indonesia	54·95	70·59
Malaysia	246·15	−2·87
Pakistan	176·74	43·14
Philippines	102·08	15·93
Sri Lanka	178·63	22·12
Thailand	34·64	−3·2

of storage facilities is increased and the quality of food products improved and maintained to meet the prescribed standards for import and export.

Developing countries account for 30% of world imports of agricultural products; the share is higher (50–90%) for typical tropical commodities such as coffee, cocoa, tea and spices. They also account for 64% of the world's supply of shrimp and almost 35% of the supply of fresh horticultural products.[16]

Trade in food products is a major factor of regional or international commerce, and markets are growing. Today, the inability of countries to satisfy each other's quarantine and public health regulations is a major barrier in food trade. Some countries such as the United States and Japan have banned the use of certain fumigants identified as health hazards, thus causing acute problems for developing countries whose economies are still largely agricultural based. Radiation technique suggests an alternative choice to aid their global food trade.

Practical application of food irradiation as a process of food preservation and decontamination has been proposed by the Joint FAO/IAEA/WHO Expert Committee on the Wholesomeness of Irradiated Foods concluding that irradiation of any food commodity up to an overall average dose of 10 kGy causes no toxicological hazard and introduces no special nutritional or microbiological problems. This conclusion was endorsed by the Codex Alimentarius Commission in July 1983.[3]

7 TREND OF FOOD IRRADIATION IN DEVELOPING COUNTRIES

Alongside traditional methods of processing and preserving food, the technology of food irradiation is gaining more and more attention around the world. In 37 countries, of which 15 countries are developing countries, health and safety authorities collectively have approved irradiation of 40 different foods altogether, ranging from spices to grains to fruits and vegetables (Table 10). Some of these developing countries are actually applying the process, some have plans to do so, in six countries no facilities have been built or planned (Table 11).

Nearly 70% of these approvals have come during this decade alone, as food irradiation's safety and usefulness have become more widely

TABLE 10

LIST OF CLEARANCES AMONG DEVELOPING COUNTRIES (AS OF MARCH 31, 1991) (GROUP ACCORDING TO COUNTRY)[a]

Product	Country[b]														
	Arg.	Ban.	Bra.	Chil.	Cuba	India	Indo.	Korea	Mexico	Pak.	Phil.	Syria	Thai.	Urug.	Viet.
Fruit															
Strawberry	√	√	√	√								√	√		
Papaya		√	√	√								√	√		
Mango				√								√	√		
Date				√								√	√		
Potatoes and onions	√	√	√		√	√	√	√			√¹	√	√	√	√¹
Garlic	√	√						√	c		√³				√¹
Rice		√		√				√		√		√	√		√²
Maize			√							√					
Wheat and its product			√												√²
Pulses and cereals		√	√	√			√	√				√	√		
Cocoa bean				√	√		√						√		
Spices	√	√	√	√		√		√		√	√	√	√		
Chicken		√	√	√				√				√	√		
Fish		√		√		√						√	√		√²
Shrimps and Froglegs		√													
Poultry			√												

[a]Source: IAEA, 1990.
[b]1, provisional; 2, experimental batches; 3, test marketing.
[c]A number of products have been approved by Ministry of Health in 1988 but no detailed information available.

TABLE 11

APPLICATIONS OF FOOD IRRADIATION AMONG DEVELOPING COUNTRIES[a]

Country	Location (starting date)	Food item	Facilities being built or planned
Argentina	Buenos Aires (1986)	Spices, spinach, cocoa powder	
Bangladesh			Dhaka, Chittagong
Brazil	Sao Paulo (1985)	Spices, dehydrated vegetables	
Chile	Santiago (1983)	Spices, dehydrated vegetables, onions, potatoes, chicken	
Cuba	Havana	Potatoes, onions	Habana
Korea,	Seoul (1985)	Garlic powder	Seoul (2)
Malaysia			Kuala Lumpur
Pakistan			Lahore
Thailand	Bangkok	Onions, fermented sausages	Bangkok

[a]From IAEA.[16]

acknowledged after years of scientific research and testing (Figs. 1 and 2). Further impetus has come from the adoption, in 1983, of a worldwide standard covering irradiated foods by the Codex Alimentarius Commission, a joint body of the Food and Agriculture Organization (FAO) and World Health Organization (WHO) representing 130 countries that sets global food standards. In addition, the decision of five developing country governments (Greece, Vietnam, German Democratic Republic, Peru, and Bulgaria) to join the International Consultative Group of Food Irradiation in 1990 indicated the positive trend of food irradiation technology and its application.

Food irradiation technique seems to be of interest in developing countries for many reasons:

(1) Increasing awareness of potential and real problems of toxicity of chemical residues in food by the consumer and importers. Irradiation offers strong potential to replace chemical treatment of food for the same purpose.

(2) Irradiation offers a cold process which can decontaminate frozen food without causing undesirable changes in their physicochemical and organoleptic properties.

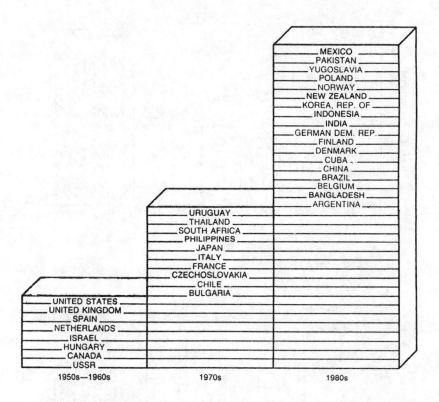

FIG. 1. Trends in National Approval of Irradiated Foods. Thirty-five countries have granted regulatory approval of irradiated foods. In 21 of these countries, food irradiation is in actual use. (From IAEA[16])

(3) Irradiation is a broad spectrum process, so it can be used to treat a variety of foods in a wide range of shapes and sizes, with little or no adoption of the handling and processing practices.

(4) Irradiation would facilitate the distribution and sale of fresh fruits, vegetables, and meat by increasing the shelf-life of these commodities, without changing the food quality which serves consumer demand while other preservation methods cannot.

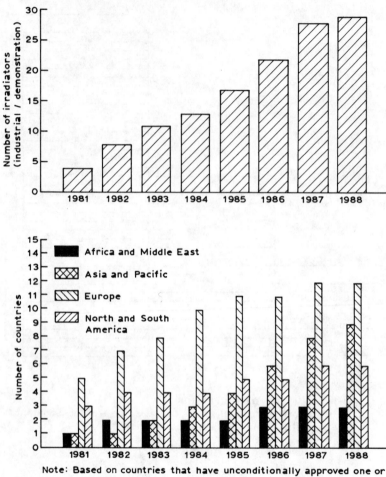

FIG. 2. *Top:* Trends in the use of irradiators for food processing on a commercial scale. *Bottom:* Regional trends in approvals of irradiated food items. (From IAEA[16])

8 SOME CONSTRAINTS IN TRADE PROMOTION OF IRRADIATED FOOD

In spite of all the problems with food preservation and the apparent solutions offered by irradiation, widespread use of irradiation is not a reality.

Some of the constraints in the international trade promotion of irradiated foods can be identified as follows:

(1) As a consequence of the association of radiation with the emotive subjects of nuclear power and radioactivity, consumers demand assurances of adequately controlled and regulated processes of food irradiation. Unless that is provided, consumer acceptance of the irradiated food will remain a problem.

(2) Since legislation in individual countries for the control of irradiated food and clearances is not harmonized, exporting countries find it difficult to permit food irradiation if their main trading partners do not accept irradiated food. Although some countries are treating several food commodities on a commercial scale, international trade in irradiated foods is still almost non-existent. However, each year about half a million tons of food products and ingredients are irradiated worldwide. This amount is small in comparison to total volumes of processed foods.

REFERENCES

1. International Atomic Energy Agency, Nuclear energy and development, *IAEA News Features,* 1988, **4**, 1–12.
2. Pothisiris, P., *Regulatory control of food irradiation process for consumer protection,* Proc. Acceptance, Control of and Trade in Irradiated Food, Geneva, 1988, International Atomic Energy Agency, Vienna, 1989, 87–102
3. World Health Organization, *Food Irradiation: A Technique for Preserving and Improving the Safety of Food,* WHO, 1988.
4. World Health Organization, *The role of food safety in health and development,* Technical Report Series No. 705, WHO, Geneva, 1984, 9–10.
5. Todd, E., *Impact of spoilage and food borne diseases on national and international economics,* Paper presented at the International Consultative Group on Food Irradiation Workshop on the Use of Irradiation to Ensure Hygienic Quality of Food, Wageningen, The Netherlands, 14–25 March, 1988.
6. Loaharanu, P., *Worldwide status of food irradiation and the role of IAEA and other international organizations,* Proc. 18th Japan Conf. on Radiation and Radioisotopes, Tokyo, 1987, pp. 491–508.
7. Rubio, T., Food Irradiation activities in Latin American countries. In: *Food Irradiation Processing,* Proc. IAEA/FAO Symposium, Washington D.C., 4–8 March, 1985, 203–213.
8. Chinsman, B., Food irradiation development in Africa. In: *Food Irradiation Processing,* Proc. IAEA/FAO Symposium, Phase I, Washington D.C., 1985, 185–203.
9. Banditsing, C., Prinksulka, V., Sutantawong, M., Noochpramool, K., Prachasitisakdi, Y., Piadang, S. and Songprasertchai, S., The Thai Multipurpose food irradiator and experience in food irradiation. In: *ASEAN Workshop on Food Irradiation,* Proc. ASEAN Food Handling Bureau, Bangkok, Thailand, 26–28 November, 1985.

10. Martin, Commercialization storage and transportation studies of irradiated dried fish and fishery products and onions. In: *Final meeting of project committee of RPFI phase II.* Office of Atomic Energy for Peace, Bangkok, 1988.
11. Lustre, A.O., Ang, L., Dianco, A., Cabalfin, E.G. and Navarro, Q.O., *Philippines' experience in marketing irradiated foods,* Proc. ASEAN Workshop on Food Irradiation, Bangkok, 1985, Association of South East Asian Nations Food Handling Bureau, Kualalumper, 1985, 52–59.
12. Singson, Pilot plant studies on the techno-economic feasibility of food irradiation in the Philippines. In: *Final meeting of project committee of RPFI phase II.* Office of Atomic Energy for Peace, Bangkok, 1988.
13. Maha Technology transfer of Irradiation of frozen shrimps, dried fish and spices. In: *Final meeting of project committee of RPFI phase II.* Office of Atomic Energy for Peace, Bangkok, 1988.
14. Khan, T., Commercial trial on radiation preservation of onions under tropical conditions. In: *Final meeting of project committee of RPFI phase II.* Office of Atomic Energy for Peace, Bangkok, 1988.
15. Moy, J., In: *Economic feasibility study. Irradiation of Mexican Fruits,* IAEA Project MEX/5/011.03, IAEA, Salazar, Mexico, 1984.
16. International Atomic Energy Agency, Food processing by irradiation: world facts and trends, *IAEA News Features,* 1988, **5**, 1–12.
17. Mossel, D.A.A., Processing for safety of meat and poultry by radicidation: progress, penury, prospects. In: *Elimination of Pathogen Organisms from Meat and Poultry,* Elsevier Science Publisher, B.V., 1987, 305–310.

FURTHER RECOMMENDED READING

Anonymous, In: *Regional Cooperative Arrangements for the Promotion of Nuclear Science and Technology in Latin America, 1985–1990.*

Banditsing, C., Prinksulka, V., Piadang, S., Sutantawang, M., Noochapramool, R. and Prachasitisakdi, Y., The multi-purpose food irradiation plant in Thailand. In *Food Irradiation Processing,* Proc. IAEA/FAO Symposium, Washington D.C., 4–8 March, pp. 365–377

Banditsing, C., Food irradiation for national development. *J. Nutr. Assoc. Thailand,* 1988, **4**, 435–443.

Desrosier, N., In: *The Technology of Food Preservation,* The AV1 Publishing company, INC., Westport, Connecticut, 1970.

Food and Agriculture Organization of the United Nations, In: *Selected indicators of Food and Agriculture Development in Asia-Pacific Region, 1978–1988,* RAPA, FAO, Bangkok, 1989.

Haigh, M.R., Regulatory aspects of irradiated food in the EEC, In: *Trade Promotion of Irradiated Food,* IAEA-TECDOC-391, IAEA, Vienna, 1986, 49–58.

International Consultative Group on Food Irradiation, In: *Report of Workshop on Food Irradiation for Food Control Officials,* Budapest, May, 1985.

International Atomic Energy Agency, In: *Trade Promotion of Irradiated Food,* IAEA-TECDOC-391, IAEA, Vienna, 1986.

Johnson, A. and Peterson, M., In: *Encyclopedia of Food Technology,* The AVI Publishing Company, Inc., Westport, Connecticut, 1974.

Kaferstein, F.K., In: *The Global Nature and Extent of Food Safety Problems,* ICGFI Workshop on the Use of Irradiation to Ensure Hygienic Quality of Food, Wageningen, Netherlands, March, 1988.

Kurian. T., In: *Encyclopedia of the Third World,* 3rd Edition, Vols. 1–3, Facts on File Inc., New York, 1987.

Oosterkamp, J., In: *Measures of Quality Assurance in the Food Processing Role of Irradiation in the Food Processing Industry,* ICGFI Workshop on the Use of Irradiation to Ensure Hygienic Quality of Food , Wageningen, Netherlands, March, 1988.

Qureshi, R.U., In: *Food irradiation: its role in food safety,* Proc. ASEAN Workshop on Food Irradiation, Bangkok, 1985, Association of South East Asian Nations Food Handling Bureau Kuala Lumpur, 1985, 118–124.

Urbain, J., In: *Promotion of nuclear science and technology in Latin America.* Phase 1, 1985–1990.

Chapter 11

EFFECTS OF IONIZING RADIATION ON VITAMINS

D.W. Thayer, J.B. Fox, Jr. & L. Lakritz
Eastern Regional Research Center, USDA, ARS 600 East Mermaid Lane
Philadelphia, PA 19118, USA

1 INTRODUCTION

Vitamins have been known to be sensitive to the effects of ionizing radiation almost since the discovery of radioactivity. Sugiura and Benedict[1] fed young albino rats an artificial casein diet containing either gamma-treated (radium) or untreated yeast and measured growth responses for 34 days and concluded that the growth promoting factors in yeast were partially inactivated. Sugiura[2] found that an X-ray dose of 120 000 roentgens destroyed at least 50% of the vitamin B content in dry wheat seeds. Wet germinated seeds exposed to a dose of 60 000 roentgens lost almost all of the vitamin B content.

Since most foods contain a large proportion of water, the most probable reaction of the ionizing radiation will be with water; and as vitamins are present in very small amounts compared with other substances in the food, they will be affected indirectly by the radiation. Goldblith[3] summarized the evidence for the indirect nature of these reactions as (a) the dilution effect (on a percentage basis the greater the dilution of a given solute the greater the destruction that will take place), (b) the temperature effect (freezing the solutions should decrease the rate of diffusion of the radicals and decrease the amount of destruction) and (c) the protection effect (irradiation of mixtures of solutes will produce different results from those obtained by irradiation of pure solutes). The overall validity of Goldblith's conclusions about the effects of ionizing radiation on vitamins in foods and model systems will be apparent in the discussion which follows.

The reader should see the following reviews for details of the interaction of ionizing radiation with water: Simic,[4] Swallow,[5] and von

285

Sonntag.[6] When water is irradiated at 25°C the following overall reaction takes place: $H_2O \rightarrow e_{aq}^-$ (2·8), H_3O^+ (2·8), OH (2·8), H (0·5), H_2 (0·4), H_2O_2 (0·8).[4] The relative amounts are expressed as G-values (number of species per 100 eV absorbed). The process thus produces highly reactive free hydroxyl radicals, hydrated electrons, and hydrogen atoms. The OH radical is a strong oxidising agent, and the hydrated electron e_{aq}^- is a strong reducing agent. The fact that hydrated electrons can be quantitatively converted to OH radicals by reaction with N_2O allows the reaction of the OH radical to be studied in the absence of competing reactions of e_{aq}^-.

2 WATER SOLUBLE VITAMINS

2.1 Ascorbic Acid
2.1.1 Function
Ascorbic acid, vitamin C, is essential to man for the maintenance of the mesothelial tissue and the structures derived from it, such as collagen. Ascorbic acid plays a major role in electron transport by reducing metals facilitating the transportation of molecular oxygen by enzymes.[7] These include the enzymes that synthesize hydroxyproline for collagen and noradrenaline synthesis. Ascorbic acid has an important biological function in the destruction of free radicals derived from oxygen, such as the hydroxyl, singlet oxygen, and superoxide.[8] A lack of sufficient vitamin C in the diet leads to fatigue, anorexia, muscular pain, greater susceptibility to infection and stress, and increased tissue fragility. At first the patient simply bruises easily, but this can progress into scurvy when degeneration of the muscles and hemorrhages occur especially in the gums and joints.

2.1.2 Source
Fruits and vegetables are the major sources of vitamin C; however, animal tissues contain small amounts and the liver, adrenal gland, and thymus contain significant amounts.

2.1.3 Chemistry
The enediol group of ascorbic acid possesses semiaromatic acidic properties and the reducing properties for the molecule.[7] In solution the vitamin is easily oxidized to a quinone-type triketo compound (dehydroascorbic acid), and the lability to oxidation increases with increasing pH.

Under alkaline conditions dehydroascorbic acid can be converted to 2,3-diketo-L-gulonic acid by opening of the lactone ring, which is not biologically active.

There are several possible stereoisomers and simple homologs of ascorbic acid, but activity requires a D-configuration and a side chain attached at carbon 4 with a hydroxyl group in the 5 position. Further, all hydroxyl groups must be free.[9] Thus, D-ascorbic acid is inactive and D-araboascorbic acid and isoascorbic acid possess only about 1/20 of the activity of ascorbic acid. The biological activity of dehydroascorbic acid is equal to that of ascorbic acid.[10]

2.1.4 Radiation Chemistry

Barr and King[11] and Proctor and O'Meara[12] found that the rate of gamma-ray induced oxidation of ascorbic acid was inversely proportional to the concentration of the solute. Barr and King[11] found the stoichiometry of irradiation-induced oxidation of ascorbic acid to be $AH_2 + O_2 \rightarrow A + H_2O_2$. They postulated that the results were consistent with a reaction scheme involving hydrogen and hydroxyl radicals.

Ogura et al.[13] found evidence for reactivity of the H atom with ascorbic acid in gamma-irradiated deaerated solutions. Both hydrogen and dehydroascorbic acid were formed. The authors suggested the formation of gulono-gamma-lactone. Bielski et al.[14] suggested that ascorbic acid radicals could exist in protonated, neutral, and anion forms and suggested that they could be generated by oxidation of ascorbic acid by hydroxyl radicals ($H_2O \rightarrow H$, e_{aq}^-, OH, H_2O_2, $H_2 \cdot AH_2 + OH \rightarrow AH\cdot + H_2O$), by reduction of dehydroascorbic acid by hydrated electrons ($A + e_{aq}^- \rightarrow AH\cdot$), or by atomic hydrogen ($A + H \rightarrow AH\cdot$).

Laroff et al.[15] studied the radicals produced by the pulse radiolysis of ascorbic, araboascorbic, reductic, and α-hydroxytetronic acids directly within the ESR cavity and found the electron to be spread over a highly conjugated tricarbonyl system. The radicals resulted mainly from the reaction of OH radicals and were identical to those found in other oxidation processes. The radical produced from ascorbic acid under acidic conditions persisted in the anionic form down to a pH of 0. However, at pH values < 6 a second radical was also present, which apparently was formed by the reaction of OH radicals with the neutral form of ascorbic acid. Fessenden and Verma[16] found with time-resolved ESR spectroscopy that in basic solution the ascorbate mono-anion radical was formed by oxidation by OH radicals. In acid solutions (pH 3–4·5) saturated with N_2O the ascorbate mono-anion radical and two OH adducts were formed. Fessenden and Verma[16] speculated that the OH adduct at C2, but not the other adduct, loses water rapidly to form the ascorbate radical, approximately doubling its concentration.

Sadat-Shafai et al.[17] found that ascorbic acid was oxidized by gamma-radiation to dehydroascorbic acid in a linear manner in the presence of oxygen. Bielski et al.[18] suggested that the ascorbate radical ion was in equilibrium with a dimer, which with the hydrogen ion or other proton donors, reacted to form the ascorbate ion and dehydroascorbic acid because the radical decay constant k_{obsd} rose rapidly with increasing acidity from pH 8 to 4. The rate of change, however, decreased and appeared to be independent of acid concentration at lower pH values. Above pH 8, the rate also decreased presumably indicating that the dimer also reacted with the much weaker proton donor, water. Bielski et al.[18] postulated that the ascorbate radical $A^{\bar{}}$ forms a dimer which then reacts with a hydrogen ion in acidic solution or with water under alkaline conditions, the overall reaction being: $2A^{\bar{}} + H^+ = HA^- + A$. The individual reactions postulated were:

$$2A^{\bar{}} == A_2^{2-}$$
$$A_2^{2-} + H^+ \rightarrow HA^- + A$$
$$A_2^{2-} + H_2O \rightarrow HA^- + A + OH^-$$

2.1.5 Food Irradiation

The use of ionizing radiation (0·05–0·15 kGy) treatments for the inhibition of sprouting of white potatoes was approved in the United States in 1964 and the advantages and disadvantages were reviewed by Matsuyama and Umeda.[19]

Salkova[20] investigated the effects of gamma radiation from ^{60}Co on sprouting and spoilage of the potato exposed to 10 000 to 20 000 roentgens. The concentration of ascorbic acid, but not that of dehydroascorbic acid, decreased as a function of radiation dose. This was an immediate effect because during storage the differences between irradiated and non-irradiated potatoes tended to disappear.

Schreiber and Highlands[21] irradiated (0·14–0·41 kGy) 17 500 pounds of Russet Burbank and Katahdin potatoes. An initial loss of ascorbic acid occurred in both varieties of potatoes during the first few months of storage followed by a buildup of ascorbic acid with the onset of warmer temperatures. In contrast, Panalaks and Pelletier[22] reported that irradiated Katahdin potatoes had higher initial ascorbic acid contents but contained significantly less ascorbic acid after storage. In both varieties, the greatest drop in ascorbic acid content occurred during the first 4·5 months storage. After 9 months storage differences in ascorbic acid content between the irradiated (maximum dose 0·14 kGy) and non-irradiated potatoes disappeared. Similar results were reported by Schwimmer et al.[23]

Cotter and Sawyer[24] studied the effect of gamma irradiation on the incidence of black spot (a non-enzymatic darkening of the cooked tuber) and the ascorbic acid content of potato tubers, confirming that the incidence of black spot was directly related to the radiation dose. The irradiation treatment resulted in an immediate decrease of the ascorbic acid content and an increase in glutathione content, but no specific cause and effect could be demonstrated. However, no analysis was performed for dehydroascorbic acid.

Romani et al.[25] studied the effects of doses of gamma-radiation up to 16 kGy on ascorbic acid and dehydroascorbic acid content of the juice of Valencia oranges and Eureka lemons. The ascorbic acid content of orange juice decreased immediately after irradiation but recovered somewhat after 24 h. Ascorbate losses were accompanied by nearly equivalent increases in dehydroascorbic acid. Maxie et al.[26] observed a dramatic dose-related loss in the ascorbic and citric acid content of Eureka lemon fruit exposed to 2·0–4·0 kGy of gamma radiation from ^{60}Co and stored at 15°C for 40 days. Lee and Salunkhe[27] reported losses of ascorbic acid in irradiated (0–18 kGy) freeze-dried apples; a dose of 3 kGy resulted in a loss of 16% of the ascorbic acid. The method of analysis for ascorbic acid was not cited; thus, it is uncertain if total ascorbic acids were measured. Maxie and Sommer[28] reported that mature green Earlypak No. 7 tomatoes irradiated to 4·0 kGy and then ripened showed an 8·6% loss in ascorbic acid while irradiated table-ripe fruits lost 20·4%, causing them to conclude that the physiological state of the fruit prior to irradiation was very important. Beyers et al.[29] found no detrimental changes in the vitamin contents (ascorbic acid, riboflavin, niacin, and thiamine) of four mango, two papaya, one litchi, and two strawberry cultivars exposed to 2·0 kGy of gamma radiation (^{60}Co). Beyers and Thomas[30] compared the ascorbic acid contents of canned, frozen, and gamma-irradiated mangoes, papayas, and litchis and concluded that irradiated mangoes and litchis lost not more than 17% of their ascorbic acid. The differences between the irradiated fruits and the controls were not significant. Thomas and Beyers[31] found that the natural variation in the ascorbic acid content of two mango and two papaya cultivars as the fruits ripened from mature green to the edible ripe stage was greater than that due to gamma radiation doses of up to 2·0 kGy. In 1982 Moshonas and Shaw[32] reported small (<6%) decreases in the ascorbic acid content of the juice from Marsh seedless grapefruit exposed to 0·60 kGy from X-ray, or gamma radiation from either ^{60}Co or ^{137}Cs. Moshonas and Shaw[33] did not find significant differences in the ascorbic acid content of the juice of Florida March seedless white and Ruby Red grapefruit exposed to gamma radiation up to 0·90 kGy when compared with those of un-treated fruit. Nagai and Moy[34] found that Valencia oranges could withstand a radiation dose of 0·75 kGy without changes in ascorbic acid content providing that they were stored at low temperature (i.e., 7°C).

Only limited conclusions can be drawn from studies of irradiated fruits and vegetables because of the variations in irradiation technique and doses used; however, there is an immediate reduction in ascorbic acid in most products, some of which appears to have been converted to dehydroascorbic acid. Since tubers and fruits are living tissues, many of the apparent losses disappear when they are stored following irradiation. Inhibition of sprouting by gamma radiation alters the metabolism of the tuber compared with that of the non-irradiated tuber. The sprouting tuber uses the stored food reserves to produce growth and becomes completely unsuitable as a food. Research is needed with contemporary analytical equipment to identify the stoichiometric relationship of the radiolytic products in both fruits and tubers.

2.2 Vitamin B$_1$, Thiamine
2.2.1 Function
The thiamine deficiency disease in humans is beriberi in which there are degenerative changes in the intestinal tract, heart, liver, nervous system, skin, and muscle leading to a wide variety of symptoms including lassitude and muscle weakness, anorexia, edema, wasting of tissues, convulsions, and emotional disturbances. The biochemically active form of thiamine, cocarboxylase, has pyrophosphate attached to the hydroxyethyl group of the thiazole nucleus. Lack of the vitamin interferes in the metabolism of glucose, in the formation of pentoses for RNA and DNA synthesis, and in the transmission of neural signals.

2.2.2 Source
The major sources of thiamine are whole grain products, nuts and beans. Thiamine and other vitamins of the B group are located in the bran and are therefore not present in white flours unless they are supplemented; this was discovered when a beriberi-like disease developed in chickens fed

polished rice. Meat is not a major source of thiamine, but pork has a relatively high concentration and supplies about 8% of the vitamin in the average American diet.

2.2.3 Chemistry

Thiamine is composed of a substituted pyrimidine and thiazole nucleus linked by a methylene bridge. It may undergo both oxidation and reduction and is especially vulnerable to cleavage at the carbon-nitrogen bond of the side chain of the pyrimidine nucleus and the thiazole ring.

2.2.4 Radiation Chemistry

Groninger and Tappel[35] found ammonia to be a radiolytic product of thiamine and postulated that it arose from the 6-amino group of pyrimidine. They observed that radiolysis of thiamine produced a ninhydrin-positive and a fluorescent product. Because of decreased ultraviolet absorption they concluded that the pyrimidine ring was destroyed. Dunlap and Robbins[36] found that *Phycomyces blaksleeanus,* which can grow equally well with thiamine or its component parts, did not multiply in irradiated thiamine solutions supplemented with either thiazole or pyrimidine. They concluded that the degradation of thiamine did not proceed by the cleavage of the molecule into its component parts followed by selective degradation of one of those parts. In contrast, Ziporin *et al.*[37] using biological assays with *Saccharomyes cerevisiae* found 96% destruction of thiamine, 68% destruction of thiazole and no loss of pyrimidine. Kishore *et al.*[38] found that the optical absorption of thiamine (due to the ring structures) disappeared upon irradiation. Gregolin *et al.*[39] identified oxythiamine and diphosphooxythiamine, with trace amounts of thiochrome and diphosphothiochrome, as the major oxidation products of X-ray irradiation of diphosphothiamine.

Kishore *et al.*[38,40] studied the radiolysis of thiamine in the presence of N_2O (e_{aq}^- scavenger) and glucose ($OH^{\cdot}$ radical scavenger), and observed extensive degradation with N_2O but none with N_2O and glucose, indicating that thiamine was reacting with the $OH^{\cdot}$ radical. In the presence of oxygen no reaction occurred, indicating that neither O_2^- nor HO_2 reacts with thiamine. Moorthy and Hayon[41] identified a one electron reduction intermediate from thiamine and dihydrothiamine as the end product. The general conclusion is that both oxidative and reductive processes may be involved in the radiolytic destruction of thiamine. Because no radiolysis products of thiamine have been isolated from foods and feeds the mechanism of destruction in them remains unknown.

Maurer and Dittmeyer[42] reported that radiolytic destruction of thiamine increased with decreasing pH, but Wilska-Jeszka and Krakowiak[43] and Luczak[44] found increased stability as the pH decreased. Kishore et al.[40] observed no difference in radiolysis of thiamine at pH 6·8 and 3·0 in aqueous solutions, but in the presence of air and glucose there was less loss at pH 3·0 than there was at pH 6·8. They suggested that this might be due to the conversion by H^+ of e_{aq} to H atoms which then react with O_2 to form HO_2 which does not react with thiamine. Alternatively, the increased radiation stability of thiamine may be due to its increased oxidative stability at lower pH values.

2.2.5 Food Irradiation

The destruction of thiamine by radiation is exponential with dose.[45,46] Since the reaction is principally with the radiolytic products of water, the following rate expression may be written:

$$dT/dt = k[T][P]$$

where T is thiamine and P is the reactive species produced from the irradiation of water. Since the water is in large excess and P is produced at a constant rate, $[P]$ is constant and the equation reduces to the standard form for a first order reaction:

$$dT/dt = k'[T]$$

Since the total dose delivered is a direct function of time, $[T]$ may be expressed as either a function of dose or time:

$$[T]_i = [T]_0\, e^{-k'd_i} = [T]_0 e^{-k''t_i}$$

It is more common to express $[T]$ as a function of dose since dose rates vary from source to source so that time rates are not directly comparable. The total dose term may not by itself be entirely adequate since at low dose rates the possibility exists that the equilibrium may favor the original form. If the concentration of the radical, $[P]$ is equal to or less than $[T]$, the reaction becomes second order and the relative decomposition of the vitamin decreases with an increase in the vitamin concentration.[40]

Oxygen functions as a hydroxyl radical scavenger, and in effect competes with thiamine for the radical. The effect of atmosphere on the destruction of thiamine is in question, however. Diehl[47] found that the exclusion of oxygen during irradiation of rolled oats decreased thiamine loss from 86% to 26%. In contrast, Groninger and Tappel[35] found no effect of N_2 or O_2 atmospheres on thiamine loss in meats. They did find higher concentrations of peroxides but according to Kishore et al.[38] these do not react with thiamine. Neither the presence of air or nitrogen affected the radiolytic destruction of thiamine in meats[45] or milk.[48] In most foods the packaging atmosphere does not seem to affect the radiolytic destruction of thiamine.

Increased radiolytic stability of thiamine was observed at decreased temperatures by Wilson[45] and confirmed by Diehl,[47] Kishore et al.[49] and Fox et al.[46] A sharp increase in vitamin stability to radiolysis occurred over the range at which water freezes in meat.[45-47] Three factors are involved in the phenomenon: the decreased mobility of the products of water radiolysis in the ice matrix, the separation of the tissues into an ice phase, and a highly concentrated solution of the tissue components in water. The mechanism of thiamine destruction may or may not be the same in the unfrozen and frozen states.

There are some factors that should be considered when comparing different destructive processes. When samples are frozen, thawed for irradiation, refrozen and then thawed for analysis, considerable drip loss may occur. Broiling, baking or frying result in large drip losses, and while canning will result in loss of water from the substrate the drip remains in contact with the substrate. Since the B vitamins are water-soluble it is expected that considerable vitamin loss would occur in the drip.

Thomas and Calloway[50] heated turkey to an internal temperature of 79°C and irradiated it to 3×10^6 rep and found on a dry, fat-free basis 106 and 83% retention of thiamine, respectively. Kuprianoff and Lange[51] reported retention values of 35 and 37% for thiamine in foods preserved by heating and irradiation, respectively. Brooke et al.[52] reported that raw haddock irradiated to 25 kGy retained 37% of its initial thiamine content. Steam cooking increased thiamine 6% in the non-irradiated control and 20% in haddock irradiated to 25 kGy. Brooke et al.[53] found no significant change in the thiamine content of clams irradiated to a dose of 45 kGy. Kennedy and Ley[54] found 53% of the thiamine was retained in irradiated, 90% in cooked, and 46% in the irradiated-cooked cod fillets. Diehl[55] found little or no effect of heating (10 min at 100°C) on the thiamine content of dried egg, wheat meal, rolled oats, or ground pork. Wheat retained 80% of its thiamine content after a radiation dose of 0·35 kGy and oats 60% after a dose of 0·25 kGy; neither dried egg nor ground pork lost thiamine because of radiation doses of 0·35 and 1·0 kGy. Williams *et al.*[56] found 128% and 81% thiamine retention in cooked and irradiated beef liver, respectively. Porter and Festing[57] discovered no differences in the thiamine contents of animal diets sterilized by radiation or autoclaving, only that the magnitude of the losses were dependent on the diets. Srinivas *et al.*[58] reported that semi-dried shrimp lost 60% of its initial thiamine; subsequent irradiation to 2·5 kGy resulted in an additional 40% loss. Metlitskii *et al.*[59] reported that bacon lost 50% of its thiamine after broiling and radiation sterilization, 65% after boiling, and 80% after irradiation at 30 kGy. The large scale toxicological study of gamma and electron sterilized chicken meat (135 405 kg) initiated by the US Department of the Army in 1976 provided some direct comparisons of the effects of different pocesses on the vitamin content of chicken meat.[60] All of the chicken was enzyme-inactivated (processed to an internal temperature of 73–80°C) and the study included direct comparisons of frozen, thermally sterilized, gamma (^{60}Co) sterilized, and electron sterilized products. The irradiated products received doses of 45–68 kGy at −25°C ± 15°C. The vitamin analyses for each sample of these products were reported by Black *et al.*[61] and were statistically analyzed for this report. The thiamine contents in both the thermally processed and the gamma-irradiated chicken meat were reduced by over 30% compared with the frozen-controls and the electron-sterilized products. The order of treatment may have influenced the results. Thayer *et al.*[62] found that thiamine was more stable in bacon fried before irradiation than after irradiation. Many irradiated foods will be cooked before consumption which may result in cumulative losses of thiamine. The final result of irradiation may depend on whether the effects of irradiation and cooking are synergistic (non-additive) or cumulative. Kennedy and Ley[54] found the losses to be merely cumulative. There was no evidence of a cooking/irradiation interaction in stored haddock[52] or clams.[53] Diehl[55] observed a synergistic effect between irradiation and cooking in rolled oats, but not in the other foods he studied. Thomas and Calloway[50] observed greater losses of thiamine after cooking irradiated than after cooking non-irradiated raw turkey and further found that the cooking loss of thiamine increased in proportion to the original radiation dose. Increased cooking losses of thiamine in irradiated samples were found by Williams *et al.*[56] in beef liver, by Thayer *et al.*[62] in bacon, by Jenkins *et al.*[63] and Fox *et al.*[46] in pork. Diehl[64] attributed these effects to semi-stable free radicals or secondary oxidative products.

Diehl[55] found that irradiated dried egg lost more thiamine than did non-irradiated dried egg during storage. Although there were large initial losses of thiamine in irradiated wheat meal and rolled oats, the storage losses were about the same for the irradiated and non-irradiated products. Proctor and Goldblith[65] reported increased loss of thiamine in irradiation sterilized as compared with heat sterilized foods. Hozova et al.[66] found a greater storage loss of thiamine in irradiated cooked beef than in non-irradiated cooked beef. On the other hand, Frumkin et al.[67] found that while irradiation (8 kGy) destroyed 55% of the thiamine in culinary meat products, irradiated meat lost less thiamine than did frozen meat during 110 days storage so that the final vitamin contents were the same. Similar results were reported by Sunyakova and Karpova.[68] Jenkins et al.[63] found a small increase in thiamine in stored irradiated and stored non-irradiated cooked pork. El-Bedewy et al.[69] found increasing thiamine concentrations in the drip loss from frozen and irradiated-frozen camel meat during storage.

Brin et al.[70] fed X-ray-treated pork to rats and followed the activity of the enzyme erythrocyte transketolase, which is sensitive to thiamine deficiency. The diets containing the irradiated pork were supplemented with all B vitamins except thiamine. Rats fed the irradiated pork grew less and had lowered transketolase activity in direct response to the radiation dose (0, 27·9 and 55·8 kGy).

2.3 Vitamin B$_2$, Riboflavin

2.3.1 Function
The active forms of riboflavin in living tissue are flavin mononucleotide (FMN) and flavin adenine dinucleotide (FAD). Both forms are involved in electron transfer reactions. Deficiency of the vitamin leads to lowered growth rates; dermatitis; and neurological, circulatory and reproductive disfunctions followed by death.

2.3.2 Source
In order of their relative value sources of riboflavin are yeast, liver, meats, milk, asparagus, broccoli, leafy green vegetables, fruits, beans, nuts, and cereals.

2.3.3 Chemistry
Riboflavin is quite stable to chemical attack, does not react with hydrogen peroxide, and is reversibly reduced to dihydroriboflavin by

sodium dithionite.[71] Riboflavin is light sensitive in alkaline solutions and is converted to lumichrome and lumiflavin.

2.3.4 Radiation Chemistry

Radiolytic destruction of riboflavin is a first order reaction.[72] In contrast, Goldblith and Proctor[73] found that the percent destruction of riboflavin was greater at low doses and interpreted this to mean that the radiolysis was due to secondary reactions ('indirect effect'). An absorption peak at 255 nm was interpreted as evidence of the formation of lumichrome. Maurer[74] postulated a reaction in which the ribose is cleaved, leaving an isoalloxazin radical, which is then cleaved into 1,2-dimethyl-4,5-diaminobenzene and alloxan. Maurer[74] found neither lumichrome nor lumiflavin as radiolytic products. Kishore et al.[40] found that glucose protected riboflavin against radiation destruction indicating that in buffer systems the reaction was an oxidation by $OH^{\cdot}$ radicals. The riboflavin radical formed from the reaction with the hydroxyl radical would subsequently react with H_2.[71] In addition, the one electron reduction product of riboflavin also reacted with H_2O_2.[75] These observations suggest possible pathways for the destruction of riboflavin. However, because the radiolytic intermediates or products have not been identified the mechanism of riboflavin destruction by radiation in foods remains unknown.

2.3.5 Food Irradiation

While riboflavin is not particularly more stable in water or buffer solutions than other vitamins it is the most stable vitamin to irradiation in almost all food substrates. It was the most stable of the B vitamins to irradiation in animal diets, meats, poultry, fruit, and vegetables. It was slightly less stable in fish than niacin, pantothenic acid, or cobalamin. Investigators have found measurable increases in extractable riboflavin in irradiated pork chops,[46] onions and garlic[76] and in a semisynthetic diet.[77] The most probable explanation for increased extractability of riboflavin in irradiated foods would be that irradiation alters the association of riboflavin with proteins. Such protein binding may also protect the vitamin against radiolytic destruction.

Conflicting results have been reported for the stability of riboflavin during storage of irradiated foods. Frumkin et al.[67] and Mameesh et al.[78] reported greater losses of riboflavin during storage of irradiated than of frozen meat products and of cod and dog fish fillets. El-Bedewy et al.[69] observed about equal storage losses of riboflavin in irradiated and frozen camel meat; however, there was increased drip loss of riboflavin from the irradiated meat. No loss of riboflavin occurred during storage of dehydro-irradiated shrimp,[58] of clams,[53] or of haddock.[52]

2.4 Vitamin B$_5$, Niacin

2.4.1 Function

Niacin deficiency results in the development of pellagra characterized by skin and oral lesions, poor growth, diarrhea, and neural degeneration. Niacin, or nicotinic acid, functions in cells as nicotinamide adenine dinucleotide (NAD) and nicotinamide adenine dinucleotide phosphate (NADP), which are coenzymes for dehydrogenases and reductases. The functional form has the nitrogen of the pyridine nucleus reduced, allowing for transfer of both electrons and hydrogen atoms from substrate to substrate.

2.4.2 Source

Peanuts and liver are the best sources of niacin, with red meats, chicken, fish, and rabbit very close behind. Some whole grains and vitamin fortified bakery goods are fairly good sources. Fruits and vegetables are relatively poor sources.

2.4.3 Chemistry

Niacin is stable in air and not destroyed by autoclaving, although it will undergo decarboxylation in alkali at high temperatures. While niacin is stable to most oxidizing agents, when the coenzyme is formed by linking a ribityl residue to the nitrogen of the pyridine ring, the nitrogen is readily oxidizable or reducible.

2.4.4 Radiation Chemistry

Goldblith et al.[79] identified the initial step in the radiolysis of niacin as the cleavage of the carboxyl group followed by the rupture of the pyridine ring. This may be the only study of the radiolysis products of niacin, and these authors did not identify any of the pyridine ring products. Kishore et al.[40] found that niacin in aqueous systems is more unstable during irradiation than thiamine or riboflavin, but the addition of glucose protects it against radiation destruction. Glucose was more effective in protecting niacin than thiamine, pyridoxine, folic acid, and riboflavin. The hydroxyl radical is the primary reactant which destroys niacin and the stability of niacin in foods and feeds is due to the scavenging of the radical by other tissue components. Niacinamide can form both oxidized and reduced radicals which disproportionate upon reacting with hydrogen peroxide. [71,75]

2.4.5 Food Irradiation

Since niacin is no more stable than any of the other B vitamins in aqueous solutions, its stability in irradiated foods may be attributed to the protective effect of other tissue components. This effect is quite specific as the stabilization reactions are not nearly as effective for thiamine and pyridoxine as they are for niacin and riboflavin. Only lowering the temperature to −80°C significantly changed the stability of niacin in foods. Although this temperature stabilized thiamine in ham and pork loin, niacin was destabilized. Since the immobilization of hydroxyl radicals in the ice matrix presumably protects the vitamins from radiolysis, cryogenic temperatures should protect all vitamins equally well against attack by the hydroxyl radical and, since freezing does not provide such protection to niacin this argues that the effect must be more than the immobilization. Since vitamins and other sarcoplasmic components are in the highly concentrated liquid solution outside the ice structure, part of the protection effect will still be that of radical scavenging by other tissue components. Under these conditions the scavenging reactions favor thiamin over niacin.

The apparent niacin contents of both pork and chicken subjected to doses of gamma radiation up to 4 kGy was greater than that of the controls. [46] Luckey et al. [77] found that a steam sterilized rat diet retained 97% of its niacin, but the same diet sterilized by cathode rays retained 115% of its niacin compared with the control suggesting that protein denaturation does not release the vitamins. Diehl [64] reported that while the niacin content of irradiated flour was lower than that of the non-irradiated, the niacin content of the bread baked from the irradiated flour was higher, suggesting that the combined effect of irradiation followed by baking released niacin from some bound form.

2.5 Vitamin B$_{10}$, Biotin

2.5.1 Function

A deficiency of biotin produces skin lesions; lassitude; alopecia; and muscular, circulatory, and neural abnormalities in animals. Biotin is the coenzyme for a number of carboxylases. The enzymes act by forming N-carboxyl biotin from carbonate or organic acids and transferring the carboxyl group to other substrates.

2.5.2 Source

Yeast, liver, soy, rice bran and germ, and egg yolk are the best sources of the vitamin followed by fish, nuts, peas, beans, red meats, milk, fruits, and vegetables.

2.5.3 Chemistry

Biotin is stable in weak acid and alkaline solutions, but it is destroyed by heat in strong acid or basic solutions. Changing the length of the side chain produces strong biotin antagonists as does oxidizing the sulfur to the sulfone.

2.5.4 Radiation Chemistry

Moorthy and Hayon[41] determined that a transient unstable radical intermediate of biotin is formed by reaction with the hydroxy radical, which decays in acid and alkaline solutions to form spectrally distinct species. Watanabe et al.[80] studied a number of aspects of the radiation destruction of biotin and found that the inactivation for a given total dose was independent of the dose rate, and that the reaction was first order. When biotin was irradiated in the presence of nitrous oxide or air the rate of inactivation was increased, indicating that both the hydroxyl radical and the superoxide ion react with the vitamin. This contrasts with the radiolytic inactivation of other B-vitamins where air has little or no effect. Since hydrogen peroxide had no effect on the inactivation of biotin, the radiolytic destruction of biotin must begin with the formation of a radical, subsequently followed by its oxidation possibly to biotin-1-sulfoxide.[81]

2.5.5 Food Irradiation

Biotin in chicken was not significantly affected by freezing, thermal sterilization, gamma sterilization, or electron sterilization.[61] Losses of biotin have not been observed after irradiating dry or semi-dry feeds and dry milk. When water was added to a chick diet,[82] a biotin loss was observed. Kennedy[83] reported that neither doses of 5 nor 50 kGy affected the biotin content of whole egg, but a dose of 2 kGy caused a 10% loss of the vitamin in Manitoba wheat after three months storage at room temperature.

2.6 Folic Acid, Pteroylglutamic Acid

2.6.1 Function

The foliates act as coenzymes for processes in which there is the transfer of a one-carbon unit, as in purine and pyrimidine nucleotide biosynthesis. Folic acid deficiency results in the development of megaloblastic anemia, infertility, and GI disturbances.

2.6.2 Source
Vegetables such as broccoli and spinach are good sources of folate as are organ meats (e.g., liver).[84]

2.6.3 Chemistry
The chemical formula for folic acid is 2-amino-4-hydroxy-6-methyl-eneaminobenzoyl-L-glutamic acid peteridine.[84] The reduced 7,8-dihydro or the coenzyme form 5,6,7,8-tetrahydrofolates occur naturally.

2.6.4 Radiation Chemistry
Kishore et al.[38,49] found that $\approx 95\%$ of folic acid was destroyed by a gamma radiation dose of 20 kGy at room temperature. The same radiation dose delivered at 193 K or in the presence of nitrous oxide and glucose or oxygen and glucose did not inactivate folic acid. The assumption is that the glucose is capable of more efficiently scavenging OH radicals than the vitamin. Moorthy and Hayon[85] determined from pulse radiolysis techniques and spectral data that the electron adds to the pterine ring. This suggests that disproportionation of the free radicals occurs with the formation of the corresponding dihydropyrazine derivatives.

2.6.5 Food Irradiation
Few studies of irradiated foods or feeds have included assays of folic acid. Richardson et al.[86] found no evidence that the folic acid content of a synthetic chick diet was altered by irradiation to 27·9 kGy. Coates et al.[82] found 8% and 22% destruction of folic acid in guinea-pig and cat diets, respectively, after gamma irradiation to 25 kGy. No folic acid loss was noted in enzyme-inactivated chicken meat, sterilized with either gamma or electron radiation in vacuo at −25°C ± 15°C to a dose of 45–68 kGy.[61]

2.7 Vitamin B_6, Pyridoxine
2.7.1 Function
Pyridoxine deficiency produces anemia, weight loss, reduced reproductive performance and neural degeneration leading to hypersensitivity and convulsions. The active forms (pyridoxal-5-phosphate (PLP) and

pyridoxamine-5-phosphate) of the vitamin are coenzymes for trans-aminases, deaminases, carboxylases and sulfhydrases.

2.7.2 Source
Red meats, fowl, fish, some grains, nuts, brown rice, peanuts, white beans, cauliflower, avocado, and raisins are good sources of pyridoxine.

2.7.3 Chemistry
The term pyridoxine refers to the alcohol form of the vitamin, pyridoxol. The vitamin also occurs in two other forms with the alcohol group replaced by an aldehyde (pyridoxal) or methyleneamine (pyridoxamine).

2.7.4 Radiation Chemistry
The radiolytic destruction of pyridoxine appears to start with oxidation by hydroxyl radicals,[40] and the resulting radical reacts with hydrogen peroxide.[71,87] In contrast, radicals produced by reaction with e_{aq} did not react with H_2O_2.[75]

2.7.5 Food Irradiation
Koesters and Kirchgessner[88] studied the interaction of pyridoxine with other vitamins in irradiated dried skim milk and piglet starter feed. The addition of vitamin E and low concentrations of ascorbate had no effect on the pyridoxine loss due to irradiation in milk or feed, but the addition of higher concentrations of ascorbate increased the loss of pyridoxine. Galatzeanu and Antoni[87] mixed powdered pyridoxine and ascorbic acid, and found that ascorbate protected the pyridoxine by formation of a pyridoxine/ascorbic acid complex. If such a pyridoxine/ascorbic acid complex is formed in dried skim milk or feed, it could account for the results observed by Koesters and Kirchgessner.[88]

Richardson et al.[89] heat processed and/or irradiated beef liver, boned chicken, cabbage, green and lima beans, and sweet potatoes to 27·9 or 55·8 kGy. Heat processing caused a greater loss of pyridoxine in liver and chicken than did irradiation, but the reverse was observed in cabbage. During storage the pyridoxine activity decreased in liver, chicken, cabbage, and green beans, but remained the same or increased in lima beans and sweet potatoes. The activity of both heat treated and irradiated foods was 40–60% of the frozen controls at each storage period. Sorman et al.[90] found a greater loss of pyridoxine than of thiamine in irradiated beef. There was a 60% loss of pyridoxine after cooking for 60 min at 121°C with an additional 10% loss following irradiation to 2 kGy.

Pyridoxine loss increased only slightly at doses up to 6 kGy. On storage the cooked samples lost more pyridoxine than did any of the cooked, irradiated samples such that after 42 days the pyridoxine contents of all samples were about equal. In contrast, most other studies showed little or no effect of either processing or storage on the pyridoxine content of foods. [52,53,61,91]

2.7.6 Feeding Studies

Brin et al.[92] fed X-ray treated pork to rats and observed the effects on two pyridoxine dependent plasma transaminases. Neither rat growth nor glutamic-aspartic transaminase activity was affected, but there was a dose-dependent decrease of 40% in the activity of glutamic-alanine transaminase in rats fed pork irradiated to 55·8 kGy. Richardson et al.[86] fed chicks an irradiated (27·9 kGy) solution of vitamins B_1, B_2, B_6, pantothenate, and folic acid supplemented with all but one unirradiated vitamin. The omission of pyridoxine resulted in the third highest mortality rate, second only to thiamine and folic acid. Day et al.[93] found by rat assay that 30 kGy of gamma-radiation destroyed about 25% of the pyridoxine in beef, which is comparable to that found by other methods of assay.

2.8 Pantothenic Acid

2.8.1 Function

Pantothenic acid is an essential component of coenzyme A, and thus functions as an acyltransfer cofactor for enzymatic reactions. Pantothenic acid was identified as a antidermatitis factor for chicks but proved harder to define for man. Adult volunteers fed a pantothenic acid deficient diet developed malaise, abdominal discomfort, and 'burning feet'.[94]

2.8.2 Source

Pantothenic acid is widely distributed in many foods; i.e., in order of decreasing content yeast, liver, egg yolk, kidney, milk, bran, peanuts, and heart.

2.8.3 Chemistry

Pantothenic acid (N-(2,4-dihydroxy-3,3-dimethyl-1-oxobutyl) β-alanine) is a pale yellow oil which is stable at neutral pH but not at acidic or basic pH values.

2.8.4 Radiation Chemistry

Moorthy and Hayon[41] found spectral evidence for hydrogen abstraction by OH radicals at four major sites in the molecule by pulse radiolysis studies of aqueous solutions of pantothenic acid. The proposed sites for H-atom abstraction are at C_2, C_3, C_6 and C_8 positions. Kishore et al.[38] found the reactivity of the vitamin with e_{aq}^- to be low.

2.8.5 Food Irradiation

Pantothenic acid was not affected by gamma irradiation of beef to 27·9 kGy,[89] of clam meats to 45 kGy,[53] of frozen egg yolk to 50 kGy,[83] of haddock fillets to 25 kGy,[52] of powdered eggs and creamed rice to 10 kGy[95] and of enzyme-inactivated chicken to 45–68 kGy by either electron or gamma radiation[61]. Manitoba wheat irradiated to 2 kGy lost 11% of its pantothenic acid content.[83] Coates et al.[82] reported 12% losses of pantothenic acid in chick, guinea-pig, and cat diets subjected to 25 kGy gamma radiation doses at room temperature. In another study chick mash irradiated to 50 kGy did not lose any pantothenic acid.

2.9 Vitamin B_{12}, Cyanocobalamin

2.9.1 Function

Pernicious anemia develops over a period of several years from malabsorption of vitamin B_{12}, the cobalamins, classically as the result of an inherited autosomal dominant trait. Pernicious anemia may also develop in patients who are subjected to gastrectomies without adequate vitamin therapy and in patients suffering from chronic atrophic gastritis.[96,97] The malabsorption occurs because of lack of secretion of the intrinsic factor from the stomach. The anemia, which develops insidiously, results in weakness, tiredness, pale and smooth tongue, and gastrointestinal complaints. In contrast to iron-deficiency anemia, the erythrocytes are few in number, abnormally large, and contain abnormally high amounts of hemoglobin.[97]

2.9.2 Source

Liver, kidney and heart tissues are excellent sources of the cobalamins.[96]

2.9.3 Chemistry

Vitamin B_{12} or cyanocobalamin contains one saturated and three unsaturated pyrroles joined in a large ring containing six conjugated double bonds.[96] The cyano group is attached to cobalt, which is linked coordinately to nitrogen in 5,6-dimethylbenzidazole. Ribose is linked to nitrogen 1 of the 5,6-dimethylbenzidazole by an α-glycosidic linkage. The D-1-amino-2-propanol moiety is esterified to the nucleotide and joined by an amide linkage to the porphyrin-like nucleus. Ellenbogen[96] bases his description on several reports by Hodgkin.

Simic[4] describes vitamin B_{12} as having three oxidation states for cobalt. The normal stable state is Co(III). Co(II) and Co(I) react with oxygen. All three forms are highly colored and have distinctive visible and ultraviolet absorption spectra. Vitamin B_{12} is relatively stable. Aqueous solutions of the vitamin are neutral though vitamin B_{12} is most stable at a slightly acidic pH (4·5–5). Aqueous solutions deteriorate in the presence of reducing agents such as ascorbic acid and sulfite.[4]

2.9.4 Radiation Chemistry

The reactions of the hydrated electron (e_{aq}) with the cobalamins were studied by the use of pulse radiolysis.[98,99] Blackburn et al.[98] reported that e_{aq} reacts with the cobalt atom of cyanocobalamins reducing it to Co(II). The rate at which this reaction takes place is close to the diffusion controlled rate. The Co(II) also reacts with e_{aq} and is reduced to Co(I) with an efficiency of about 80%.[99] This is attributed to the reaction of e_{aq} with amide groups on the molecule. Carboxyl radicals (CO_2) and several organic radicals did not react efficiently with B_{12}(III) to produce B_{12}(II), but these radicals did reduce the B_{12}(II) to B_{12}(I).[98] Blackburn et al.[100] demonstrated the formation of cob(I)alamin from cob(II)alamin during pulse radiolysis of N_2O saturated solutions containing sodium formate. Cyanocobalamin did not react with cob(I)alamin.

Simic[4] reported that oxidizing free radicals such as the $\cdot CO_3^-$ radical and the tertiary-butanol radical destroy vitamin B_{12}. He indicated that these radicals form a Co-C bond similar to that of the coenzyme and considers it unlikely that the OH radical will react with B_{12} in meats because of the very small concentration of B_{12}. If such a reaction did take place the large number of reactive sites probably would lead to destruction of the vitamin.

2.9.5 Food Irradiation

In an abstract, Alexander and Salmon[101] indicated that 40% of vitamin B_{12} in raw beef irradiated with gamma radiation to 27·9 kGy was lost. A subsequent publication, apparently describing the complete study, did not mention vitamin B_{12}.[93] Brooke et al.[53] did not observe a loss of vitamin B_{12} in raw or steamed clams exposed to 3·5 or 4·5 kGy of gamma radiation nor was a loss of vitamin B_{12} found after various periods of storage in raw or steamed haddock fillets exposed to 1·5 or 2·5 kGy of gamma radiation.[52] The presence or absence of air did not affect the results. de Groot et al.[102] did not observe any loss of vitamin B_{12} in fresh chicken meat pasteurized by exposure to 3·0 or 6·0 kGy of gamma-radiation. Fox et al.[46] reported no measurable losses in the content of vitamin B_{12} in pork chops gamma irradiated to doses up to 6·65 kGy at temperatures between −20°C and +20°C. No treatment related effect on the concentration of vitamin B_{12} was observed in direct comparisons of frozen, thermally sterilized, gamma (^{60}Co) sterilized, and electron sterilized enzyme inactivated chicken.[60]

Coates et al.[82] found no losses of vitamin B_{12} in either air- or vacuum-packed chick mash exposed to 50 kGy of ^{60}Co gamma-rays. Porter and Festing[57] found no loss of vitamin B_{12} in a breeding diet for mice sterilized with gamma radiation (25 kGy) but did observe an 11% loss in the irradiated normal diet for mice; however, only one sample of each was analyzed. Both diets underwent significant losses (22% and 33%, respectively) of vitamin B_{12}, when sterilized by autoclaving.

3 FAT SOLUBLE VITAMINS

3.1 Vitamin A
3.1.1 Function
Vitamin A is a constituent of human blood, and is necessary for normal growth, spermatogenesis, and for maintenance of differentiated epithelia. A deficiency or absence results in night blindness and ultimately death.

3.1.2 Source
Vitamin A is present in foods of animal origin including milk, butter, egg yolk, kidney, liver (especially of halibut and shark), and fish. Oxidative cleavage of plant carotenes in the mammalian intestine produces one or more molecules of vitamin A. Good sources include carrots, spinach, oranges, and papayas. Maximum provitamin A activity is derived from the ingestion of the beta isomer of carotene, because splitting of the centrally located double bond by intestinal enzymes results in the formation of two molecules of retinol. The other isomers of carotene: α-carotene, γ-carotene, and β-cryptoxanthin are half as potent as β-carotene because only a single retinol molecule is liberated.

3.1.3 Chemistry

The International Union of Pure and Applied Chemistry has adopted the name retinol ($C_{20}H_{29}OH$) to signify vitamin A and its derivatives. Vitamin A or retinol, 9,13-dimethyl-7-(1,1,5,-trimethyl-6-cyclo-hexene-5-yl)-7,9,11,13-nonatetraene-15-ol, occurs only in animal tissue, and not in plants. Vitamin A is present in two major forms, A_1 and A_2. (A_2 contains an additional double bond between carbon atoms 3 and 4.) The compound which predominates in most higher animals is retinol A_1. Rat feeding studies indicate that A_1 is twice as potent as A_2. Vitamin A_2 is present in highest concentrations in fish liver oils. The chemical structure of vitamin A_1 indicates that it is an alcohol with a 6-membered alicyclic ring and a side chain containing 2 isoprene units.

The carotenes have approximately twice the molecular weight of vitamin A. The most reactive site on the carotene molecule is the C_7 position, and since β-carotene can be both an electron donor and acceptor, either a direct or indirect attack could result in the saturation of the conjugated polyene chain.

3.1.4 Radiation chemistry

Goldblith and Proctor[73] reported a 40% loss of pure dry β-carotene at a dose of 24 kGy. Subsequent studies to determine the effects of 0·7–20 kGy of gamma radiation on crystalline β-carotene at room temperature under nitrogen or air indicated no destruction to β-carotene nor the formation of other radiolytic products.[103,104] Lukton and MacKinney[104] attributed the stability of crystalline β-carotene to the absence of a lipid fraction from which side reactions could be initiated.

The stability of β-carotene decreases in solution; the extent of the instability is dependent on the solvent. Chalmers et al.[105] separately dissolved crystalline β-carotene and an ester of vitamin A in n-hexane under nitrogen and found that they were completely destroyed when exposed to 150 mCi radon for 48 h. They conducted a series of studies using X-rays and concluded that for each ion pair produced, an equal number of molecules were destroyed. Goldblith and Proctor[73] reported that irradiating a pure solution of β-carotene in petroleum ether with cathode rays (electrons) to doses of 6, 12, 18 and 25 kGy resulted in β-carotene losses of 50%, 80%, 94%, and 99%, respectively. Only 40% destruction of carotene occurred at 25 kGy in absence of a solvent. Slightly greater losses of carotene occurred in more dilute solutions. Goldblith and Proctor[73] concluded from this and a similar study with riboflavin that the greater destruction occurring in more dilute solutions was indicative of indirect action of the radiation. Knapp and Tappel[106] found that vitamin A acetate was twice as stable as the α- or β-carotenes when dissolved in iso-octane and irradiated at 21°C in either air or nitrogen. In solution the carotenoids were stabilized to a small extent by the presence of oxygen during irradiation. Lukton and MacKinney[104] suggested that these compounds are more readily attacked by solvent free radicals, than by hydroperoxides. The presence of vitamin E during irradiation was more protective to carotene than to vitamin A acetate. The stabilities of the fat soluble vitamins in iso-octane to gamma radiation were as follows: vitamin K_3 > vitamin A acetate > vitamin D_3 > carotene > vitamin E. Under aerobic conditions the relative stabilities are the same except that vitamin A acetate is less stable than vitamin D_3.[106]

Snauwaert et al.[103] conducted a comprehensive study to determine the mechanism for the degradation of β-carotene under anaerobic conditions as a function of solvent, concentration of solute (5–100 μM), and dose (0·03–2·1 kGy), using gamma rays. G values were determined at 12 different dose levels. They found that the G value for the destruction of β-carotene in each solvent was inversely proportional to the dose (0·03–2·1 kGy in benzene and hexane). The G values in each solvent were: benzene, 0·55–0·23; n-hexane, 0·86–0·23; cyclohexane, 1·13–0·70 (0·03–0·5 kGy). The authors observed that at a given dose the percent loss increased with increasing concentration of carotene, and that the solvent affected the result. It was postulated that in cyclohexane radiolysis of the carotene results from direct action, whereas in benzene indirect effects prevail.

β-Carotene was irradiated to 2·5–10 kGy with a ^{60}Co source at room temperature in carrot, corn, and salmon oils.[104] A 10-kGy radiation dose decreased β-carotene 10% in carrot oil, 23% in corn oil, and 66% in salmon oil. Polister and Mead[107] studied the effect of vitamins (A, C, D, E) on the radiation-induced autoxidation of methyl linoleate and found that vitamin E was effective at concentrations so low that the extent of its destruction could not be measured.

3.1.5 Food Irradiation

Lukton and MacKinney[104] irradiated raw tomato purees in air and under nitrogen to 59, 120 and 200 kGy. In air the loss in β-carotene was only 1% up to 120 kGy, and 11% at 200 kGy. Irradiating tomato puree under nitrogen produced higher losses. At 120 and 200 kGy there were 5% and 18% losses, respectively. Raw carrot purees irradiated in air to 10 kGy had a 5% increase in the concentration of β-carotene; in purees irradiated to 40 kGy there was a 20% reduction of β-carotene. Cooked carrot purees irradiated to 10 or 20 kGy contained 5% less β-carotene compared to unirradiated cooked controls.

Franceschini et al.[108] studied the effect of gamma radiation on carotenoid retention and on the color of carrots, sweet potatoes, green beans, and broccoli. The vegetables were irradiated under the following conditions: (a) dose: 18 kGy; (b) temperature: ambient, frozen; (c) atmosphere: vacuum, air, nitrogen; (d) packaging: dry, water, brine; (e) container: tin can (plain and enameled); (f) storage temperature: 0°F, 70°F, 100°F; (g) duration of storage, 0·5, 1, 3, 6 and 12 months. The general findings from this study indicated that greater losses in β-carotene and carotenoids occurred when samples were irradiated in cans containing air in the headspace. The effects of any of these processing conditions on the carotenoid contents of carrots, sweet potatoes, and broccoli were not significant. Green beans stored at 100°F showed high retention of carotene when irradiated at room temperature, but exceedingly low retention when irradiated frozen. Pigments decreased during storage for 1 year, initially, followed by an increase and then gradual decrease.

The effect of irradiation on the carotene and ascorbic acid in mangoes and papayas has also been studied. Gamma-irradiating mangoes (preclimacteric) at 0·25 kGy did not affect the formation of carotenoids. Different lengths of storage (13–43 days) and different ripening conditions (7–20°C) were included in the study. The irradiated mangoes showed slightly higher levels of carotene in comparison to the unirradiated controls.[109] Irradiating cultivars of mangoes and papayas in the green mature state and strawberries in the edible ripe state with 2 kGy of gamma radiation one to two days after harvesting resulted in slight increases in the carotene content. Changes in nutritional values as a result of irradiation were not significant.[29] Beyers et al.[29] found that papayas lost 5, 30 and 73% of their carotene after irradiating (0·8–2·0 kGy), freezing, and canning, respectively.

The β-carotene and vitamin A in whey butter-fat were completely destroyed by exposure to a 150-mCi radon source for 48 h at 45°C.[105] Since oxygen and almost all of the moisture were excluded from the butter fat, it was concluded that the destruction of the vitamin A and provitamin A occurred by direct action. Gamma sterilizing cheese, butter, milk, and margarine at −2°C with a 80 000-roentgen/h ^{60}Co source, caused severe dose-related losses of vitamin A.[72] Research by Diehl[110] on irradiated dairy and meat products indicated that loss of vitamin A was greatly affected by dose, irradiation and storage temperature, duration of storage, and atmosphere (air, nitrogen, or vacuum). Irradiation and storage at 0°C instead of at ambient temperature reduced vitamin A losses. Exclusion of air (vacuum, nitrogen) or irradiation at cryogenic temperatures were even more effective in preventing vitamin destruction. Cream cheese irradiated to 50 kGy and stored at ambient temperature in air lost 60% of its vitamin A content. Storage under nitrogen reduced the loss to 20% and vacuum packaging or irradiation at dry ice temperatures to 5%. Studies by de Groot et al.[102] found that vitamin A was not affected in chickens irradiated to either 3 or 6 kGy, stored for 4–7 days at 5°C, cooked, and finally stored at −20°C. It was noted, however, that peroxide levels increased after irradiation and on storage at both 5° and −20°C.

Coates et al.[82] investigated the effect of radiation dose, packaging, antioxidants, and moisture on chick, guinea pig and cat diets supplemented with vitamin A. Irradiation of the cereal based chick and guinea pig diets to doses of 20–30 kGy resulted in vitamin A losses of 6–12% and β-carotene losses of 5–25%. Vacuum packaging the feeds decreased the destruction of the vitamins. Adding 0·0125% of the antioxidant Santoquin prevented any loss of vitamin A up to 50 kGy. Irradiation of the cat diet, which was not cereal based, to 25 kGy destroyed almost all vitamin A and 64% of the carotene. Porter and Festing[57] compared the effects of irradiation (25 kGy) and vacuum autoclaving on the vitamin A in a mouse breeding diet and an FFG diet. An assay of one sample from each diet indicated that in the breeding diet the vitamin A level was unaffected by irradiation and that the β-carotene level exceeded that of the control, whereas in the FFG diet losses of 15% and 45% occurred in vitamin A and carotene, respectively. Autoclaving lab chows destroyed 40% of their vitamin A. Rats fed irradiated (25 or 45 kGy) diets did not have different serum levels of vitamin A, but vitamin A deposits in the livers were reduced (27% at 25 kGy, 37% at 45 kGy).[111] The levels of vitamin A stored in the liver of rat pups weaned from parents that were fed the irradiated chow were 35–40% lower than in control animals.

3.2 Vitamin D

3.2.1 Function

Vitamin D is required for intestinal absorption and renal reabsorption of calcium and phosphate derived from the diet. A vitamin D deficiency can result in rickets in children or osteomalacia in adults.

3.2.2 Source

The best sources of vitamin D in foods are fish liver oils and salt water fish, such as sardines and salmon. Vitamin D fortified milk and margarine are additional sources. Smaller quantities of this vitamin are present in egg yolk, butter, and beef. Plants are a poor dietary source for these vitamins.

3.2.3 Chemistry

The D vitamins are derivatives of the steroid nucleus and, thus, differ structurally from the other lipid soluble vitamins. Because the B ring in the D vitamins is broken at C_9-C_{10}, these compounds are designated as secosteroids. The two most important vitamins in this group are vitamin D_2 (9,10-secoergosta-5,7,10(19),22-tetraene-3β-ol) or calciferol and vitamin D3 (9,10-secocholesta-5,7,10(19)-triene-3β-ol) or cholecalciferol.

3.2.4 Radiation Chemistry

Information on the radiation chemistry of vitamin D is sparse. Knudson and Moore[112] discovered that ultraviolet light irradiation of ergosterol produced a more potent antirachitic compound than did cathode ray irradiation. Hoffman and Daniels[113] determined that cathode rays had a strong destructive effect on vitamin D after its initial formation from ergosterol. Steroids have been demonstrated to be subject to attack by hydroxyl radicals with the formation of 7-keto compounds.[114-116] Ergosterol and vitamin D_2 are stable to irradiation in the absence of air. In air, peroxide formation was found to be linear up to a dose of 100 kGy.[117] Rexroad and Gordy[118] found that the free radicals from vitamin D_2 (powder) generated by ionizing radiation *in vacuo* decayed rapidly in air.

Knapp and Tappel[106] irradiated vitamin D_3 in iso-octane to a maximum dose of 14 kGy and found it to be much more stable when irradiated in air rather than under nitrogen. Vitamin D_3 was more stable than vitamin A acetate, carotene, or vitamin E, aerobically or anaerobically. Tappel[119] determined that the specific inactivation doses (63% reduction/initial concentration) for vitamin D_3 in nitrogen or oxygen are $2 \cdot 6 \times 10^9$ and $6 \cdot 3 \times 10^9$, respectively. Vitamin D_3 was unaffected by gamma radiation when it was irradiated in salmon oil possibly because of the presence of tocopherol or other antioxidants in the oil.[106] Teichert and Horubala[120] irradiated vitamin D_3, ergosterol, and 7-dehydrocholesterol in 90% ethanol, benzene, cod liver oil, and butter in air with gamma rays to 0·5–50 kGy. The breakdown of D_3 due to irradiation was similar in each of the four solvents tested. Vitamin D was most stable in benzene > cod liver oil > butter > 90% ethanol, leading to the conclusion that degradation of D_3 occurred by direct action in benzene and cod liver oil. Water increased the degradation of each of the compounds studied. Basson[121] cited results from a study conducted by Snauwaert et al.[103] in which radiolytic products of vitamins D_2 and D_3 in petroleum ether were isolated and characterized. Four reaction products were formed, each with unaltered side chains. Cleavage presumably occurred within the triene system, with loss of the hydroxyl group indicated by the inability to silylate these products.

3.2.5 Food Irradiation
Teichert[122] investigated the possible production of toxic sterols by ionizing irradiation of vitamin D or other provitamins, such as are produced during excessive ultraviolet irradiation of milk. Gamma irradiation to 0·5–50 kGy of vitamin D_3, 7-dehydrocholesterol, and ergosterol in 90% ethanol produced 6, 12, and 10 radiolytic products, respectively. Several of the radiolytic compounds were common to all of the initial starting materials. Since none of the radiolytic products was hydroxylated, it was concluded that exposure to low level gamma radiation did not result in the formation of toxic sterols.

3.3 Vitamin E
3.3.1 Function
The exact biochemical role of the group of compounds classified as vitamin E is uncertain. These vitamins are powerful antioxidants, which can terminate free radical initiated autoxidation reactions thus protecting the integrity of the lipophilic cellular membranes. It is believed that vitamin E may also have an unknown role in metabolism. Vitamin E deficiencies produce reproductive disorders and muscular dystrophy in animals. Vitamin E deficiencies in human beings, as indicated by low blood tocopherol levels, have been linked to premature birth and to development of cystic fibrosis. Vitamin E may be involved with the structural integrity of muscle and the vascular system.

3.3.2 Source
Vitamin E (α-tocopherol) is synthesized only by plants. At least eight tocopherols are present in nature and include: α- (most potent), β-, γ-, δ-tocopherol and α-, β-, γ-, δ-tocotrienol. The richest sources of the

tocopherols are from vegetable oils, seeds, nuts, vegetables, and fruits. In animal products levels are highest in butter, salmon, and liver.

3.3.3 Chemistry

The tocopherols and tocotrienols all have a 6-chromanol ring structure. The tocopherols differ from one another only in the number and position of the methyl groups in the aromatic ring. They all have a 16 carbon isoprenoid chain attached at position C-2. The four tocotrienols are derivatives of the tocopherols; these compounds have side chains containing three double bonds. Tocopherols are stable to heat in the absence of oxygen, but oxidize slowly in the presence of air. Oxidation is accelerated by light, heat, alkali, and some metal salts.

3.3.4 Radiation Chemistry

Knapp and Tappel[106] found vitamin E to be the most sensitive of all the fat soluble vitamins to gamma radiation. The specific inactivation doses for α-tocopherol, irradiated in iso-octane, were $4·5 \times 10^8$ and $5·2 \times 10^8$ in air or nitrogen, respectively.[119] Less then 5% of the vitamin E remained after it was irradiated as a 1% solution in iso-octane to 400 kGy. Infrared spectral analysis indicated that the main product was 5-exomethylene tocopher-6-one derived by abstraction of two hydrogens. When vitamin E was irradiated in tributyrin, it was found to be only slightly less stable than when irradiated in iso-octane. Vitamin E was twice as stable to radiation in lard as in tributyrin,[123] because of many competing side reactions. Rose et al.[124] and Chipault and Mizuno[125] found that more radiolytic destruction of tocopherol occurred in saturated than in unsaturated solvents. Rose et al.[124] postulated that the free radicals reacted with the unsaturated solvents before they could interact with tocopherol. Diehl,[126] in contrast, found greater radiolytic destruction of tocopherol in solvents with a greater degree of unsaturation. Diehl reasoned that irradiation would promote autoxidation from the unsaturated fatty acids along with the production of peroxide and other reaction products. Significant tocopherol decomposition would therefore be expected to occur in triolein and trilinolein. At a dose of 10 kGy in air the following losses of tocopherol were observed in various solvents: in tributyrin, 51%; in tristearin, 55%; in triolein, 71%; in trilinolein, 100%. Significantly less α-tocopherol destruction occurred when it was irradiated under nitrogen. The data and conclusions reached by Diehl are in general agreement with the observation that susceptibility and rate of lipid oxidation increase with the degree of unsaturation.[127]

3.3.5 Food Irradiation

The effects of irradiation on the vitamin E level in foods, particularly after processing, are complex because multiple reactions due to irradiation and processing occur simultaneously. Irradiating α-tocopherol dissolved in tributyrin in the presence of air to 10, 50, or 100 kGy produced a 51%, 78%, or 95% loss of tocopherol, respectively.[126] The vitamin E content of wheat was decreased by irradiation at ambient temperatures in the presence of air.[128] Diehl[129] found that the loss of tocopherol in oatmeal steadily increased from 7% at −180°C to 46% at 50°C as the irradiation temperature was increased. Irradiating food in the presence of air or oxygen usually results in greater destruction of α-tocopherol. The peroxides and hydroperoxides that are formed during irradiation of foods can destroy α-tocopherol.[6]

Kraybill[130] reported that there was no loss of α-tocopherol in beef irradiated under an atmosphere of nitrogen; under an oxygen atmosphere 60% and 80% losses of tocopherol were found when the beef was irradiated to 30 kGy or 300 kGy. Oats that were packaged, irradiated to 1 kGy, and stored for 8 months under nitrogen lost only 5% of their tocopherol content compared with a 56% loss in oats irradiated and stored in air.[126] Vacuum packaging was as effective as nitrogen in protecting tocopherol for the first 3 months of storage but was less effective after 3 months. Packaging under carbon dioxide rather than air provided no protection to α-tocopherol.

Diehl[64] found that irradiation of hazel nuts to 1 kGy produced an 18% loss of α-tocopherol, while baking produced a 13% loss. When the nuts were irradiated and then cooked, the total loss of tocopherol was 67%.[64] Diehl's[64] proposed explanation for the apparent synergistic loss was that free radicals and secondary reaction products may attack vitamin E during processing or subsequent storage. Freshly killed chickens irradiated to 3–6 kGy and stored did not lose vitamin E.[102]

A 3-year study was conducted with beagles fed diets irradiated to 28 or 58 kGy.[131] The foods tested were gamma-irradiated pork, tuna fish, and chicken stew and electron-irradiated, frozen beef (packaged aerobically in cellophane bags). One half of the dogs' caloric intake was supplied by the meat, the other half was obtained from dry dog food containing 5·6 units α-tocopherol per pound of meal. Vitamins A, E and K and cod liver oil were also administered monthly. The dogs that were fed diets of pork, chicken stew or tuna fish were normal in all the parameters that were tested: growth, hematology, and reproduction. Beagles fed beef had somewhat smaller litters and slightly higher mortality. McCay and

Rumsey[131] speculated that peroxides resulting from irradiation destroyed vitamin E in the beef leading to reproductive impairment. There was no correlation between the two dose levels that were tested and these effects.

Barna and Kramer[111] fed irradiated (25 or 45 kGy) commercial animal chow to rats and measured the levels of vitamin E in their serum and livers. Although the vitamin E content in the food was reduced 4·6% and 11·1% at 25 and 45 kGy, respectively, the vitamin E levels in the serum and in the livers were the same as in the control animals. Rat pups, which were weaned from parents on the irradiated diets, also had normal serum and liver vitamin E levels.

Although considerable work has been conducted on the effects of ionizing radiation on vitamin E in model systems and in foods following irradiation and after prolonged storage, scant information is available about the radiation products.

3.4 Vitamin K

3.4.1 Function

Vitamin K is required for the formation of prothrombin. Vitamin K deficiency may lead to improper coagulation of the blood and hemorrhaging. The exact biochemical role of vitamin K in the synthesis of prothrombin isn't clear. Adult human beings require about $0·5-1·0$ $\mu g/kg$ of vitamin K per day, which is attainable in the normal diet. Vitamin K is readily metabolized and is not stored to any appreciable extent in man.

3.4.2 Source

The best sources of vitamin K are green leafy vegetables, broccoli, spinach, lettuce, and brussels sprouts containing over 100 $\mu g/100$ g. Beef liver contains $50-100$ $\mu g/100$ g while skeletal muscle supplies less than 10 $\mu g/100$ g. Cheese, butter, and eggs average between $10-50$ $\mu g/100$ g.

3.4.3 Chemistry

All vitamin Ks are derivatives of naphthoquinone. The two naturally occurring forms of these vitamins are: vitamin K_1, 2-methyl-3-phytyl-1,4-naphthoquinone, phylloquinone; vitamin K_2, 2-methyl-3-all-*trans*-polyprenyl-1,4-naphthoquinone, menaquinone. Vitamin K_3, 2-methyl-1,4-napthoquinone, menadione is synthetic and is frequently used in irradiation studies. Additional forms of vitamin K exist, including K_5-K_7, and K-S(II). All of these vitamins have antihemorrhagic properties and are substituted 1,4-napthoquinones.

The term Coenzyme Q (ubiquinone) includes benzoquinones with isoprenoid side chains of various lengths. Because of their structural similarity to vitamin K they are frequently discussed together. Ubiquinones are abundant in heart, liver, and kidney cells. Their functions involve electron transport and oxidative phosphorylation. The chemical reactions of these quinones are similar to those of vitamin K.

3.4.4 Radiation Chemistry

Napthoquinones and benzoquinones react readily with both hydrated electrons and reducing electrons. The formation of the semiquinones can be readily followed, since they absorb strongly around 400 nm. Semiquinone radicals disproportionate to produce hydroquinones. Free radicals may also attach to quinones resulting in their destruction.[4]

The natural vitamin Ks and vitamin K_3 are unstable to ultraviolet light and sunlight in aqueous solution.[132,133] Knapp and Tappel[106] determined that vitamin K_3 was the most stable lipid soluble vitamin, and Tappel[119] reported that K_3 in iso-octane was more stable when irradiated with gamma rays in air than in nitrogen. The specific inactivation doses were: in oxygen, 3×10^{10} and in nitrogen, 2×10^9 at a concentration of 9×10^{-5} M. Knapp and Tappel[106] suggested that the destruction of vitamin K might be due to a reductive attack by alkyl radicals or hydrogen atoms, resulting in reduction or etherification of the benzoquinone oxygen. Bancher et al.[134] gamma irradiated spinach up to 500 kGy and found no destruction of vitamin K_1, ubiquinone-50, plastoquinone A, or α-tocopherol. These compounds may be stabilized in spinach by formation of bonds with non-polar groups within the plant cell biomembranes. When Bancher et al.[134] irradiated ubiquinone-45 in tributyrin to doses between 1–500 kGy, there was a 7% loss at 1 kGy and 100% destruction at 100 kGy. Based on IR data, the main radiolysis product had a hydroxyl group on the isoprenoid side chain, and the benzoquinoide structure was unaffected by irradiation.

3.4.5 Food Irradiation

A study to assess the effects of ionizing radiation on the wholesomeness of beef resulted in the death of 70% of the male rats (14), and one female rat.[135] None of the 40 rats in the control study died due to hemorrhage. The diet consisted of raw beef that was gamma irradiated to 55·8 kGy and then stored for 6 months at 24°C. The experimental beef (35%) was incorporated into a balanced diet consisting of 35% wheat flour and 30% other nutrients some rich in vitamins A, B, and E. Oral supplements of

1 μg/day of vitamin K_3 prevented rat deaths due to hemorrhaging, indicating that irradiating beef destroyed or made vitamin K unavailable. A follow up study[136] determined that irradiation reduced the vitamin K content in the food from ≈ 0.12 to $0.03-0.06$ μg/g of dry matter, an inadequate level resulting in hypothrombinemia and hemorrhagic deaths. Furthermore, no antagonist to vitamin K was produced or could be detected as a result of gamma irradiation as had been speculated. Matschiner and Doisy[137] determined that fresh beef contained 0.4 μg/g of vitamin K in fat and 0.07 μg/g in fresh tissue. Vitamin K was not detected in the irradiated beef. The blood prothrombin levels collected from these rats reflected their diets. Rats fed irradiated pork diets had hemorrhage disorders similar to the animals in the irradiated beef study.[138]

Richardson and Woodworth[139] used a chick bioassay to determine the sensitivity of the different forms of vitamin K to irradiation and found that the natural vitamins K_1 and K_5 were more stable to ionizing radiation than vitamin K_3. Plough et al.[140] conducted a short term feeding study, with human beings as the test subjects to determine if vitamin K deficiencies would result from the ingestion of irradiated meat. The subjects ate, as 32% of their total caloric intake, diets containing irradiated (30 kGy) ground pork for 15 days. The diets did not include a vitamin K supplement. The results of the clinical findings, including prothrombin time, were all normal. This finding would be as expected, since the vegetables included in the diet should have amply supplied the minimum daily vitamin requirements.

ACKNOWLEDGMENTS

The authors are grateful for the helpful comments and suggestions of Dr. John M. Woychik, Dr. William Obermeyer, and Mrs. Suzanne C. Thayer.

REFERENCES

1. Sugiura, K. and Benedict, S.R., The action of radium emanation on the vitamines of yeast, J. Biol. Chem., 1919, **39**, 421–433.
2. Sugiura, K., The effect of x-rays on the vitamin B content of wheat seedlings, Radiology, 1933, **21**, 438–448.
3. Goldblith, S.A., Preservation of foods by ionizing radiations, J. Am. Diet. Assoc., 1955, **31**, 243–249.

4. Simic, M.G., Radiation chemistry of water-soluble food components. In: *Preservation of Food by Ionizing Radiation*, D.E.S. Josephson and D.M.S. Peterson (eds.), Boca Raton, Florida, 1983, **2**, pp. 1–70.

5. Swallow, J., Fundamental radiation chemistry of food components. In: *Recent Advances in the Chemistry of Meat*, A.J. Bailey (ed.), London, 1983, pp. 165–177.

6. von Sonntag, C., *The Chemical Basis of Radiation Biology*, Taylor and Francis, London, 1987.

7. Johnson, F.C., The antioxidant vitamins, *CRC Crit. Rev. Food Sci. Nutr.*, 1979, **11**, 217–309.

8. Jaffe, G.M., Vitamin C. In: *Handbook of Vitamins*, L.J. Machlin (ed.), New York, NY, 1984, pp. 199–244.

9. Rosenberg, H.R., *Chemistry and Physiology of the Vitamins*, Interscience Publishers, Inc., New York, NY, 1942.

10. Todhunter, E.N., McMillan, T. and Ehmke, D.A., Utilization of dehydro-ascorbic acid by human subjects, *J. Nutr.*, 1950, **42**, 297–308.

11. Barr, N.F. and King, C.G. The γ-ray induced oxidation of ascorbic acid and ferrous ion, *J. Am. Chem. Soc.*, 1956, **78**, 303–305.

12. Proctor, B.E. and O'Meara, J.P., Effect of high-voltage cathode rays on ascorbic acid, *Ind. Eng. Chem.*, 1951, **43**, 718–721.

13. Ogura, H., Murata, M. and Kondo, M., Radiolysis of ascorbic acid in aqueous solution, *Radioisotopes*, 1970, **19**, 89–91.

14. Bielski, B.H., Comstock, D.A. and Bowen, R.A., Ascorbic acid free radicals. I. Pulse Radiolysis study of optical absorption and kinetic properties, *J. Am. Chem. Soc.*, 1971, **93**, 5624–5629.

15. Laroff, G.P., Fessenden, R.W. and Schuler, R.H., The electron spin resonance spectra of radical intermediates in the oxidation of ascorbic acid and related substances, *J. Am. Chem. Soc.*, 1972, **94**, 9062–9073.

16. Fessenden, R.W. and Verma, N., A time-resolved electron spin resonance study of the oxidation of ascorbic acid by hydroxyl radical, *Biophys. J.*, 1978, **24**, 93–101.

17. Sadat-Shafai, T., Ferradini, C., Julien, R. and Pucheault, J., Radiolytic study of the action of perhydroxyl radicals with ascorbic acid, *Radiat. Res.*, 1979, **77**, 432–439.

18. Bielski, B.H., Allen, A.O. and Schwarz, H.A., Mechanism of disproportionation of ascorbate radicals, *J. Am. Chem. Soc.*, 1981, **103**, 3516–3518.

19. Matsuyama, A. and Umeda, K., Sprout inhibition in tubers and bulbs. In: *Preservation of Food by Ionizing Radiation*, E.S. Josephson and M.S. Peterson (eds.), Boca Raton, Florida, USA, 1983, **3**, pp. 159–213.

20. Salkova, E.G., Effect of the radiation from Co^{60} on the vitamin C content of potatoes, *Doklady Akad. Nauk SSSR*, 1957, **114**, 474–476.

21. Schreiber, J.S. and Highlands, M.E., A study of the biochemistry of irradiated potatoes stored under commercial conditions, *Food Res.*, 1958, **23**, 464–472.

22. Panalaks, T. and Pelletier, O., The effect of storage on ascorbic acid content of gamma radiated potatoes, *Food Res.*, 1960, **25**, 33–36.

23. Schwimmer, S., Weston, W.J. and Makower, R.U., Biochemical effects of gamma radiation on potato tubers, *Arch. Biochem. Biophys.*, 1958, **75**, 425–434.

24. Cotter, D.J. and Sawyer, R.L., The effect of gamma irradiation on the incidence of black spot, and ascorbic acid, glutathione and tyrosinase content of potato tubers, *Am. Potato J.,* 1961, **38**, 58–65.
25. Romani, R.J., Van Kooy, J., Lim, L. and Bowers, B., Radiation physiology of fruit-ascorbic acid, sulfhydryl and soluble nitrogen content of irradiated citrus, *Radiat. Bot.,* 1963, **3**, 363–369.
26. Maxie, E.C., Eaks, I.I. and Sommer, N.F., Some physiological effects of gamma irradiation on lemon fruit, *Radiat. Bot.,* 1964, **4**, 405–411.
27. Lee, C.Y. and Salunkhe, D.K., Effects of gamma radiation on freeze-dehydrated apples (*Pyrus malus*), *Nature,* 1966, **210**, 971–972.
28. Maxie, E.C. and Sommer, N.F., Changes in some chemical constituents in irradiated fruits and vegetables. In: *Preservation of Fruit and Vegetables by Radiation,* Vienna, 1968, pp. 39–56.
29. Beyers, M., Thomas, A.C. and Van Tonder, A.J., γ-Irradiation of subtropical fruits. 1. Compositional tables of mango, papaya, strawberry, and litchi fruit at the edible-ripe stage, *J. Agric. Food Chem.,* 1979, **27**, 37–42.
30. Beyers, M. and Thomas, A.C., γ-Irradiation of subtropical fruits. 4. Changes in certain nutrients present in mangoes, papayas, and litchis during canning, freezing and γ-irradiation, *J. Agric. Food Chem.,* 1979, **27**, 48–51.
31. Thomas, A.C. and Beyers, M., γ-Irradiation of subtropical fruits. 3. A comparison of the chemical changes occurring during normal ripening of mangoes and papayas with changes produced by γ-irradiation, *J. Agric. Food Chem.,* 1979, **27**, 157–163.
32. Moshonas, M.G. and Shaw, P.E., Irradiation and fumigation effects on flavor, aroma and composition of grapefruit products, *J. Food Sci.,* 1982, **47**, 958–960, 964.
33. Moshonas, M.G. and Shaw, P.E., Effects of low-dose γ-irradiation on grapefruit products, *J. Agric. Food Chem.,* 1984, **32**, 1098–1101.
34. Nagai, N.Y. and Moy, J.H., Quality of gamma irradiated California Valencia oranges, *J. Food Sci.,* 1985, **50**, 215–219.
35. Groninger, H.S. and Tappel, A.L., The destruction of thiamine in meats and in aqueous solution by gamma radiation, *Food Res.,* 1957, **22**, 519–523.
36. Dunlap, C.E. and Robbins, F.C., The effect of roentgen rays, radon and radioactive phosphorus on thiamin chloride, *Am. J. Roentgenol.,* 1943, **50**, 641–647.
37. Ziporin, Z.Z., Kraybill, H.F. and Thach, H.J., Vitamin content of foods exposed to ionizing radiations, *J. Nutr.,* 1957, **63**, 201–209.
38. Kishore, K.M., Moorthy, P.N. and Rao, K.N., Radiation protection of vitamins in aqueous systems, *Radiat. Eff.,* 1976, **27**, 167–171.
39. Gregolin, C., Rossi, C.S. and Siliprandi, N., Effect of x-irradiation on diphosphothiamine, *Radiat. Res.,* 1960, **13**, 185–191.
40. Kishore, K., Moorthy, P.N. and Rao, K.N., Radiation protection of vitamins in aqueous systems. Part III Effects of different variables, *Radiat. Eff.,* 1978, **38**, 97–105.
41. Moorthy, P.N. and Hayon, E., One-electron redox reactions of water-soluble vitamins. 4. Thiamine (vitamin B1), biotin and pantothenic acid, *J. Org. Chem.,* 1977, **42**, 879–885.

42. Maurer, H.J. and Dittmeyer, R., Beitrage zur Wirkung ionisierender Strahlen auf Vitamine I. (vorlaufige) Mitteilung: Thiamin und Riboflavin, *Strahlentherapie*, 1957, **102**, 531–534.

43. Wilska-Jeszka, J. and Krakowiak, W., Some effects of ionizing radiation on aqueous solutions of thiamine, *Nukleonika*, 1975, **20**, 511–516.

44. Luczak, M., Effect of low and high gamma (cobalt-60) ray doses on some biologically active components of whole milk powder (B1, B2, B12 and biotin), *Roczniki Instytutu Przemyslu Mleczarskiego.*, 1970, **12**, 71–86.

45. Wilson, G.M., The treatment of meats with ionising radiations. II — Observations on the destruction of thiamine, *J. Sci. Food Agric.*, 1959, **10**, 295–300.

46. Fox, Jr., J.B., Thayer, D.W., Jenkins, R.K., Phillips, J.G., Ackerman, S.A., Beecher, G.R., Holden, J.M., Morrow, F.D. and Quirbach, D.M., Effect of gamma irradiation on the B vitamins of pork chops and chicken breasts, *Int. J. Radiat. Biol.*, 1989, **55**, 689–703.

47. Diehl, J.F., Verminderung von strahleninduzierten Vitamin-E und B$_1$ verlusten durch Bestrahlung von Lebensmitteln bei tiefen Temperaturen und durch Ausschluss von Luftsauerstoff (Reduction of radiation-induced vitamin losses by irradiation of foodstuffs at low temperatures and by exclusion of atmospheric oxygen], *Z. Lebensm. Unters. Forsch.*, 1979, **169**, 276–280.

48. Ford, J.E., Gregory, M.E. and Thompson, S.Y., The effect of gamma irradiation on the vitamins and proteins of liquid milk. In: Int. Dairy Congr. (Proc.) 16th, Copenhagen, Sect. A., 1962, pp. 917–923.

49. Kishore, K., Moorthy, P.N. and Rao, K.N., Radiation protection of vitamins in aqueous systems. Part II. A comparative study in fluid and frozen aqueous systems, *Radiat. Eff.*, 1976, **29**, 165–170.

50. Thomas, M.H. and Calloway, D.H., Nutritive value of irradiated turkey II. Vitamin losses after irradiation and cooking, *J. Am. Diet. Assoc.*, 1957, **33**, 1030–1033.

51. Kuprianoff, J. and Lang, K., Strahlenkonservierung und Kontamination von Lebensmitteln, *Beit. z. Ernahrungs.*, 1960, 3.

52. Brooke, R.O., Ravesi, E.M., Gadbois, D.M. and Steinberg, M.A., Preservation of fresh unfrozen fishery products by low-level radiation. 5. The effects of radiation pasteurization on amino acids and vitamins in haddock fillets, *Food Technol.*, 1966, **20**, 1479–1482.

53. Brooke, R.O., Ravesi, E.M., Gadbois, D.F. and Steinberg, M.A., Preservation of fresh unfrozen fishery products by low-level radiation. III. The effects of radiation pasteurization on amino acids and vitamins in clams, *Food Technol.*, 1964, **18**, 1060–1064.

54. Kennedy, T.S. and Ley, F.J., Studies on the combined effect of gamma radiation and cooking on the nutritional value of fish, *J. Sci. Food Agric.*, 1971, **22**, 146–148.

55. Diehl, J.F., Thiamine in bestrahlten lebensmitteln. I. Einfluss verschiedener bestrahlungsbedingungen und des zeitablaufs nach der bestrahlung, *Z. Lebensm. Unters. Forsch.*, 1975, **157**, 317–321.

56. Williams, C., Yen, J. and Fenton, F., Effects of radiation and of cooking on the quality of baby beef liver, *Food Res.*, 1958, **23**, 473–491.

57. Porter, G. and Festing, M., A comparison between irradiated and autoclaved diets for breeding mice, with observations on palatability, *Lab. Anim.*, 1970, **4**, 203–213.
58. Srinivas, H., Vakil, U.K. and Sreenivasan, A., Nutritional and compositional changes in dehydro-irradiated shrimp, *J. Food Sci.*, 1974, **39**, 807–811.
59. Metlitskii, L.V., Rogachev, V.M. and Krushchev, V.G., *Radiation Processing of Food Products*. Izdatel'stvo 'Ekonomika', Moscow, 1978.
60. Thayer, D.W., Christopher, J.P., Campbell, L.A., Ronning, D.C., Dahlgren, R.R., Thomson, G.M. and Wierbicki, E., Toxicology studies of irradiation-sterilized chicken, *J. Food Protection*, 1987, **50**, 278–288.
61. Black, C.M., Christopher, J.P., Cuca, G.C., Dahlgren, R.R., Israelson, E.L., Miranti, R.A., Monti, K.L., Reutzel, L.F., Ronning, D.C. and Troup, C.M., *Final report: A chronic toxicity, oncogenicity, and multigeneration reproductive study using CD-1 mice to evaluate frozen, thermally sterilized, cobalt-60 irradiated, and 10 MeV electron irradiated chicken meat*. Vols. 1–14. Raltech Scientific Services, St. Louis, MO. Available: National Technical Information Service, Springfield, VA. PB84-187012, 1983.
62. Thayer, D.W., Shieh, J.J., Jenkins, R.K., Phillips, J.G., Wierbicki, E. and Ackerman, S.A., Effect of gamma ray irradiation and frying on the thiamine content of bacon, *J. Food Qual.*, 1989, **12**, 115–134.
63. Jenkins, R.K., Thayer, D.W. and Hansen, T., Effect of low-dose irradiation and post-irradiation cooking and storage on the thiamine content of fresh pork, *J. Food Sci.*, 1989. in press.
64. Diehl, J.F., Effects of combination processes on the nutritive value of food. In *Combination Processes in Food Irradiation*, Proceedings of a Symposium, Colombo, 24–28 November 1980. IAEA-SN-250, Vienna, 1981, pp. 349–366.
65. Proctor, B.E. and Goldblith, S.A., Effects of ionizing radiations on food nutrients. In: *Nutritional Evaluation of Food Processing*, R.S. Harris and H. von Loesecke, New York, 1960, pp. 133–144.
66. Hozova, B., Sorman, L., Salkova, Z. and Richter, P., (Combined effects of heat and ionizing radiation on contents of thiamin and riboflavin in representative types of canned products (beef in natural juice and raw beef with cauliflower).) *Sbornik UVTIZ, Potravinarske Vedy*, 1986, **4**, 197–204.
67. Frumkin, M.L., Koval'skaya, L.P. and Gel'fand, S.Y., Technological principles in the radiation treatment of food products, *Pishchevaya Promyshlennost*, 1973, 506–512.
68. Sunyakova, Z.M. and Karpova, I.N., A comparative study of the action produced by gamma-rays and thermal sterilization on the content of thiamine, riboflavin, nicotinic acid and tocopherol in the beef, *Vopr. Pitaniya*, 1969, **25**, 52–55.
69. El-Bedewy, L.A., Abdallah, N.M., Mohsin, S.M. and Morsi, M.K.S., Changes in the vitamin content, physical characteristics and microbial count during cold storage and freezing of irradiated camel meat, *Egypt. J. Food Sci.*, 1978, **6**, 13–25.

70. Brin, M., Ostashever, A.S., Tai, M. and Kalinsky, H., Effects of feeding X-irradiated pork to rats, on their thiamine nutrition as reflected in the activity of erythrocycte transketolase, *J. Nutr.*, 1961, **75**, 29–34.

71. Kishore, K., Moorthy, P.N. and Rao, K.N., Role of hydrogen peroxide in the radiolysis of B-group vitamins in neutral aqueous solutions. Part I. Reactions with vitamin radicals produced by hydroxyl radical reaction, *Radiat. Eff. Lett.*, 1982, **67**, 153–159.

72. Kung, H.-C., Gaden, Jr., E.L. and King, G.C., Vitamins and enzymes in milk. Effect of gamma-radiation on activity, *J. Agric. Food Chem.*, 1953, **1**, 142–144.

73. Goldblith, S.A. and Proctor, B.E., Effect of high-voltage X-rays and cathode rays on vitamins (riboflavin and carotene), *Nucleonics,* 1949, **5**, 50–58.

74. Maurer, H.-J., Beitrage zur wirkung ionisierender strahlung auf vitamine 2. Mitteilung: Riboflavin. Einfluss der dosisleistung (zeitfaktor) und strahlungen verschiedener intensitat — reactionsmechanismus, *Strahlentherapie,* 1958, **106**, 294–299.

75. Moorthy, P.N., Kishore, K. and Rao, K.N., Role of hydrogen peroxide in the radiolysis of B-group vitamins in neutral aqueous solutions. Part II. Reactions with reduced vitamin radicals, *Radiat. Eff. Lett.*, 1982, **67**, 161–166.

76. Le Clerc, A.M., Effets des radiation ionisantes sur la teneur en vitamines de quelques produits alimentaires, *Ann. Nutr. Aliment,* 1963, **17**, B449–B461.

77. Luckey, T.D., Wagner, M., Reyniers, J.A. and Foster, Jr., J.A., Nutritional adequacy of a semi-synthetic diet sterilized by steam or by cathode rays, *Food Res.*, 1955, **20**, 180–185.

78. Mameesh, M.S., Boge, G., Myklestad, H. and Braekkan, O.R., Studies on the radiation preservation of fish. I. The effect on certain vitamins in fresh fillets of cod and dogfish and in smoked fillets of cod and herring. Thiamine, Riboflavin, Niacin, Cobalamin. In: *Reports on Technological Research concerning Norwegian Fish Industry,* 1964, pp. 1–10.

79. Goldblith, S.A., Proctor, B.E., Hogness, J.R. and Langham, W.H., The effects of cathode rays produced at 3000 kilovolts on niacin, *J. Biol. Chem.*, 1949, **179**, 1163–1167.

80. Watanabe, H., Aoki, S. and Sato, T., Gamma-ray inactivation of biotin in dilute aqueous solution, *Agric. Biol. Chem.*, 1976, **40**, 9–15.

81. Sjoestedt, M. and Ericson, L.E., The effect of gamma radiation on a number of water soluble vitamins, *Acta Chemica Scandinavica,* 1962, **16**, 1989–1998.

82. Coates, M.E., Ford, J.E., Gregory, M.E. and Thompson, S.Y., Effects of gamma-irradiation on the vitamin content of diets for laboratory animals, *Lab. Anim.,* 1969, **3**, 39–49.

83. Kennedy, T.S., Studies on the nutritional value of foods treated with γ-radiation. I.-Effects on some B-complex vitamins in egg and wheat, *J. Sci. Food Agric.,* 1965, **16**, 81–84.

84. Brody, T., Shane, B. and Stokstad, E.L.R., Folic acid. In: *Handbook of Vitamins Nutritional, Biochemical, and Clinical Aspects,* L.J. Machlin (ed.), New York, 1984, pp. 459–496.

85. Moorthy, P.N. and Hayon, E., One-electron redox reactions of water-soluble vitamins. II. Pterin and folic acid, *J. Org. Chem.*, 1976, **41**, 1607–1613.

86. Richardson, L.R., Martin, J.L. and Hart, S., The activity of certain water-soluble vitamins after exposure to gamma radiations in dry mixtures and in solutions, *J. Nutr.*, 1958, **65**, 409–418.

87. Galatzeanu, I. and Antoni, F., Les effets des radiation gamma sur la vitamine B6, *Int. J. Appl. Radiat. Isot.*, 1966, **17**, 369–390.

88. Koesters, W.W. and Kirchgessner, M., Effect of UV and gamma irradiation on vitamin B6 content and protein constituents of feeds, *Landwirtsch Forsch*, 1976, **29**, 194–203.

89. Richardson, L.R., Wilkes, S. and Ritchey, S.J., Comparative vitamin B6 activity of frozen, irradiated and heat-processed foods, *J. Nutr.*, 1961, **73**, 363–368.

90. Sorman, L., Hozova, B. and Rajniakova, A., (Thiamin and vitamin B6 retention during food preservation by a combination of heat and ionizing radiation), *Sbornik UVTIZ Potravinarske Vedy*, 1986, **4**, 191–196.

91. Calloway, D.H. and Thomas, M.H., Nutrient content and processing characteristics of irradiated food used in long term animal feeding studies. *QMFCIAF Report N.R. 17–61*, Quartermaster Food and Container Institute for the Armed Forces. Quartermaster Research and Engineering Command, Chicago, 1961.

92. Brin, M., Ostashever, A.S., Tai, M. and Kalinsky, H., Effects of feeding X-irradiated pork to rats on their pyridoxine nutrition as reflected in the activity of plasma transaminases, *J. Nutr.*, 1961, **75**, 35–38.

93. Day, E.J., Alexander, H.D., Sauberlich, H.E. and Salmon, W.D., Effects of γ-radiation on certain water-soluble vitamins in raw ground beef, *J. Nutr.*, 1957, **62**, 27–38.

94. Fox, H.M., Pantothenic acid. In: *Handbook of Vitamins Nutritional, Biochemical and Clinical Aspects*, L.J. Machlin (ed.), New York, 1984, pp. 437–457.

95. Gounelle, H., Gulat-Marnay, C. and Fauchet, M., Effets des radiations ionisantes sur la teneur de divers aliments en vitamines du groupe B et C, *Ann. Nutr. Aliment.*, 1970, **24**, 41–49.

96. Ellenbogen, L., Vitamin B_{12}. In: *Handbook of Vitamins Nutritional, Biochemical and Clinical Aspects*, L.J. Machlin (ed.), New York, 1984, pp. 497–547.

97. Anonymous, Anemia due to vitamin B_{12} deficiency (Pernicious Anemia). In: *The Merck Manual of Diagnosis and Therapy*, R. Berkow and A.J. Fletcher (eds.), 1987, pp. 1109–1111.

98. Blackburn, R. and Erkol, A.Y.A.P.G.O., One-electron reactions in some cobalamins, *J. Chem. Soc. Faraday Trans. 1*, 1974, **70**, 1693–1701.

99. Blackburn, R., Kyaw, M., Phillips, G.O. and Swallow, A.J., Free radical reactions in the coenzyme B_{12} system, *J. Chem. Soc. Farady Trans. 1*, 1975, **71**, 2277–2287.

100. Blackburn, R., Kyaw, M. and Swallow, A.J., Reaction of Cob(I)alamin with nitrous oxide and Cob(III)alamin, *J. Chem. Soc. Farady Trans. 1*, 1977, **73**, 250–255.

101. Alexander, H.D. and Salmon, W.D., Effects of gamma radiation and heat on certain nutrients in ground beef, *Fed. Proc.*, 1958, **17**, 468.

102. de Groot, A.P., Van der Mijll Dekker, L.P., Slump, P., Vos, H.J. and Willems, J.J.L., Composition and nutritive value of radiation pasteurized chicken. *Central Institute for Nutrition and Food Research*, Netherlands Organization for Applied Scientific Research, Report No. R3787; 1972.

103. Snauwaert, F., Tobback, P., Anthonissen, A. and Maes, E., Influence of gamma irradiation on the provitamin A (β-carotene) in solution. *Radiation Preservation of Food*, Proceedings of a Symposium. Bombay, 1972. I.A.E.A., Vienna 1973. STI/PUB/317, 29–46.

104. Lukton, A. and MacKinney, G., Effect of ionizing radiations on carotenoid stability, *Food Technol.*, 1956, **10**, 630–632.

105. Chalmers, T.A., Goodwin, T.W. and Morton, R.A., Action of ionizing radiations on carotene and vitamin A, *Nature*, 1945, **155**, 513.

106. Knapp, F.W. and Tappel, A.L., Comparison of the radiosensitivities of the fat-soluble vitamins by gamma irradiation, *Agric. Food Chem.*, 1961, **9**, 430–433.

107. Polister, B.H. and Mead, J.F., Effect of certain vitamins and antioxidants on irradiation-induced autoxidation of methyl linoleate, *J. Agr. Food Chem.*, 1954, **2**, 199–202.

108. Franceschini, R., Francis, F.J., Livingston, G.E. and Fagerson, I.S., Effects of gamma ray irradiation on carotenoid retention and color of carrots, sweet potatoes, green beans and broccoli, *Food Technol.*, 1959, **13**, 358–365.

109. Thomas, P. and Janave, M.T., Effects of gamma irradiation and storage temperature on carotenoids and ascorbic acid content of mangoes on ripening, *J. Sci. Food Agric.*, 1975, **26**, 1503–1512.

110. Diehl, J.F., Vitamin A in bestrahlten Lebensmitteln, *Z. Lebensm. Unters. Forsch.*, 1979, **168**, 29–31.

111. Barna, J. and Kramer, M., Vitamin A and E levels in the liver and serum of rats kept on irradiated food, *Acta Physiol. Acad. Sci. Hung.*, 1972, **41**, 383.

112. Knudson, A. and Moore, C.N., Comparison of the antirachitic potency of ergosterol irradiated by ultra-violet light and by exposure to cathode rays, *J. Biol. Chem.*, 1929, **81**, 49–64.

113. Hoffman, R.M. and Daniels, F., The formation of vitamin D by cathode rays, *J. Biol. Chem.*, 1936, **115**, 119–130.

114. Weiss, J. and Keller, M., Chemical action of ionizing radiations on steroid compounds, *Experientia*, 1950, **6**, 379.

115. Keller, M. and Weiss, J., Chemical actions of ionizing radiations in solutions. Part VI. Radiation chemistry of sterols. The action of X-rays on cholesterol and 3β-hydroxypregn-5-en-20-one, *J. Chem. Soc.*, 1950, 2709–2714.

116. Coleby, B., Keller, M. and Weiss, J., Chemical actions of ionizing radiations in solution. Part XII. The action of X-rays on some steroids in organic solvents, *J. Chem. Soc.*, 1954, 66–71.

117. Fazakerley, H., Gamma irradiation of some unsaturated steroids, *Int. J. Appl. Radiat. Isot.*, 1960, **9**, 130–132.

118. Rexroad, H.N. and Gordy, W., Electron-spin resonance studies of radiation damage to certain lipids, hormones, and vitamins, *Proc. Natl. Acad. Sci. U.S.A.*, 1959, **45**, 256–269.

119. Tappel, A.L., Relationship of irradiation induced fat oxidation and flavor, color and vitamin changes in meat, fat soluble vitamins, other vitamins. Final Report *S*-520. Quartermaster food and container institute for the Armed Forces. Research and Engineering Command, Quartermaster Corps, U.S. Army. Chicago, IL, 1957.

120. Teichert, M. and Horubala, A., The effects of gamma ^{60}Co on vitamin D-3 and certain sterols in foods, *Acta Aliment. Polon.*, 1917, **3**, 361–368.

121. Basson, R.A. *Recent Advances in Radiation Chemistry of Vitamins*, Elsevier Biomedical Press, Amsterdam, 1983.

122. Teichert, M., Radiolysis of vitamin D3, 7-dehydrocholesterol and ergosterol in alcohol-water solutions, *Acta Aliment. Polon.*, 1978, **4**, 263–272.

123. Knapp, F.W. and Tappel, A.L., Some effects of γ-radiation or linoleate peroxidation on α-tocopherol, *J. Am. Oil Chem. Soc.*, 1961, **38**, 151–156.

124. Rose, D., Lips, H.J. and Cyr, R., Destruction of α-tocopherol by γ-irradiation, *J. Food Sci.*, 1961, **26**, 153–155.

125. Chipault, J.R. and Mizuno, G.R., Effect of ionizing radiations on antioxidants in fats, *J. Agric. Food Chem.*, 1966, **14**, 221–224.

126. Diehl, J.F., Einfluss verschiedener bestrahlungsbedingungen und der lagerungen und der lagerung auf strahleninduzierte vitamin E verluste in Lebensmitteln [Influence of irradiation conditions and of storage on radiation-induced vitamin E losses in foods], *Chem. Mikrobiol. Technol. Lebensm.*, 1979, **6**, 65–70.

127. Gray, J.I., Measurement of lipid oxidation: A review, *J. Am. Oil Chem. Soc.*, 1978, **55**, 539–546.

128. Vakil, U.K., Aravindakshan, M., Srinivas, H., Chauhan, P.S. and Sreenivasan, A., Nutritional and wholesomeness studies with irradiated foods: India's program. In *Radiation Preservation of Food*, IAEA-SM-166/12, Vienna, 1973, pp. 673–702.

129. Diehl, J.F., Food Irradiation. In: *Proceedings of the International Congress of Refrigeration (12th Madrid)*, 1967, pp. 3–14.

130. Kraybill, H.F., The effect of ionizing radiation on vitamins and other physiologically active compounds. In: *Report on the meeting on the wholesomeness of irradiated foods*, Rome, 1962, pp. 44–60.

131. McCay, C.M. and Rumsey, G.L., Effect of irradiated meat upon growth and reproduction of dogs, *Fed. Proc.*, 1960, **19**, 1027–1030.

132. Almquist, H.J., Purification of the antihemorrhagic vitamin, *J. Biol. Chem.*, 1936, **114**, 241–245.

133. Davis, R.H., Mathis, A.L., Howton, D.R., Schneiderman, H. and Mead, J.F., Studies on the photochemistry of 2-methyl-1,4-naphthoquinone, *J. Biol. Chem.*, 1949, **179**, 383–388.

134. Bancher, E., Washuttl, J. and Schiffauer, R., Die wirkung von γ-bestrahlung auf ubichinon 45 in tributyrinlosung, *Monatschefte für chemie*, 1974, **105**, 71–73.

135. Metta, V.C., Mameesh, M.S. and Johnson, B.C., Vitamin K deficiency in rats induced by the feeding of irradiated beef, *J. Nutr.,* 1959, **69**, 18–22.
136. Mameesh, M.S., Metta, V.C., Rama Rao, P.B. and Johnson, B.C., On the cause of vitamin K deficiency in male rats fed irradiated beef and the production of vitamin K deficiency using an amino acid synthetic diet, *J. Nutr.,* 1962, **77**, 165–170.
137. Matschiner, J.T. and Doisy, E.A.J., Vitamin K content of ground beef, *J. Nutr.,* 1966, **90**, 331–334.
138. Read, M.S., Current aspects of the wholesomeness of irradiated food, *J. Agr. Food Chem.,* 1960, **8**, 342–349.
139. Richardson, L.R., Woodworth, P. and Coleman, S., Effect of ionizing radiations on vitamin K, *Fed. Proc.,* 1956, **15**, 924–926.
140. Plough, I.C., Bierman, E.L., Levy, L.M. and Witt, N.F., Human feeding studies with irradiated foods, *Fed. Proc.,* 1960, **19**, 1052–1054.

INDEX

327